HEALTHCARE TECHNOLOGIES SERIES 43

Evolving Predictive Analytics in Healthcare

IET Book Series on e-Health Technologies

Book Series Editor: Professor Joel J.P.C. Rodrigues, College of Computer Science and Technology, China University of Petroleum (East China), Qingdao, China; Senac Faculty of Ceará, Fortaleza-CE, Brazil and Instituto de Telecomunicações, Portugal

Book Series Advisor: Professor Pranjal Chandra, School of Biochemical Engineering, Indian Institute of Technology (BHU), Varanasi, India

While the demographic shifts in populations display significant socio-economic challenges, they trigger opportunities for innovations in e-Health, m-Health, precision and personalized medicine, robotics, sensing, the Internet of things, cloud computing, big data, software defined networks, and network function virtualization. Their integration is however associated with many technological, ethical, legal, social, and security issues. This book series aims to disseminate recent advances for e-health technologies to improve healthcare and people's wellbeing.

Could you be our next author?

Topics considered include intelligent e-Health systems, electronic health records, ICT-enabled personal health systems, mobile and cloud computing for e-Health, health monitoring, precision and personalized health, robotics for e-Health, security and privacy in e-Health, ambient assisted living, telemedicine, big data and IoT for e-Health, and more.

Proposals for coherently integrated international multi-authored edited or co-authored handbooks and research monographs will be considered for this book series. Each proposal will be reviewed by the book Series Editor with additional external reviews from independent reviewers.

To download our proposal form or find out more information about publishing with us, please visit https://www.theiet.org/publishing/publishing-with-iet-books/.

Please email your completed book proposal for the IET Book Series on e-Health Technologies to: Amber Thomas at athomas@theiet.org or author_support@theiet.org.

The Institution of Engineering and Technology

Evolving Predictive Analytics in Healthcare

New AI techniques for real-time interventions

Edited by
Abhishek Kumar, Ashutosh Kumar Dubey,
Surbhi Bhatia, Swarn Avinash Kumar and
Dac-Nhuong Le

The Institution of Engineering and Technology

Published by The Institution of Engineering and Technology, London, United Kingdom

The Institution of Engineering and Technology is registered as a Charity in England & Wales (no. 211014) and Scotland (no. SC038698).

First published 2022

The Institution of Engineering and Technology
Futures Place
Kings Way, Stevenage
Hertfordshire, SG1 2UA, United Kingdom

www.theiet.org

British Library Cataloguing in Publication Data
A catalogue record for this product is available from the British Library

ISBN 978-1-83953-511-6 (hardback)
ISBN 978-1-83953-512-3 (PDF)

Typeset in India by MPS Limited

Contents

About the editors

Abhishek Kumar is an assistant professor in the Department of Computer Science and Engineering at the Institute of Engineering and Technology, Chitkara University, India. He has a Doctorate in Computer Science from the University of Madras and is a Post-Doctoral Fellow of the Ingenium Research Group, Universidad De Castilla-La Mancha, Spain. He has published over 80 journal papers and edited 21 books. His research areas include artificial intelligence, image processing, computer vision, data mining, and machine learning.

Ashutosh Kumar Dubey is an associate professor in the Department of Computer Science and Engineering at the Institute of Engineering and Technology, Chitkara University, India. He is a Senior Member of IEEE and ACM. His research interests include machine learning, health informatics, and cloud computing.

Surbhi Bhatia is an associate professor at the College of Computer Sciences and Information Technology, King Faisal University, Saudi Arabia. She is PMP certified from PMI, USA. She has published 75 papers in reputed journals and conferences and edited 9 books. She has 9 patents from the USA, Australia, and India and has completed 5 research projects from the Ministry of Education and DSR, Saudi Arabia. Her research interests include sentiment analysis, data analytics, and machine learning.

Swarn Avinash Kumar is an applied research scientist at Meta Platforms, Inc. (previously Facebook), USA. He has previously worked in software engineering roles at Lyft Inc., Google and Amazon. He has filed multiple patents in the autonomous industry and the machine learning sphere. His research interests include robotics, deep learning, natural language processing, and computer vision.

Dac-Nhuong Le is an associate professor in the Department of Computer Science and head of the Faculty of Information Technology at Haiphong University, Vietnam. He has more than 20 years of academic teaching experience and has published more than 80 papers, presentations, and book chapters. His research interests include evolutionary multi-objective optimization, network communication and security, cloud computing, and virtual/argument reality.

Chapter 1

COVID-19 detection in X-ray images using customized CNN model

Deepshikha Jain[1], Venkatesh Gauri Shankar[1], Bali Devi[2] and Surbhi Bhatia[3]

Around the middle of December 2019, the outbreak of an epidemic triggered by a virus gripped the city of China, Wuhan. In the first week of January 2020, the World Health Organization (WHO) identified and named the virus SARS-CoV-2, a new type of coronavirus popularly later known as COVID-19. The genesis of COVID-19 traces the fact that it was discovered as a coronavirus in 2019. COVID-19 soon spread worldwide through millions of passengers traveling from Wuhan to other parts of Europe and the USA. COVID-19 assumed the proportions of a pandemic and soon spread across the world. The virus has a typical characteristic of changing or mutating as it infects people. The latest variant, Omicron, is now affecting people and spreading in the starting quarter of 2022. This is the most concerning aspect of its characteristic. Previously, researchers built models using machine learning, such as k-nearest neighbor, support vectors machine, multilayer perceptrons, decision trees, and random forests, and many deep learning models, such as CNN-LSTM, CNN VGG-16, VDSNet, RES NET50, UNET, CSEN, CheXNet, Inception V3, and DenseNet. This paper will include the work done for detecting COVID-19 patients by building customized deep learning models, such as artificial neural network (ANN) and convolutional neural network (CNN). We applied the ANN model after extracting the gray-level co-occurrence matrix features from images, whereas the CNN model was directly applied to the X-ray image dataset. The CNN models are used to analyze the images in which mathematical convolution is performed to extract the features. Various parameters, such as accuracy, sensitivity, recall, precision, and F1-score, were calculated to predict the model's efficiency.

[1]Department of Information Technology, Manipal University Jaipur, Rajasthan, India
[2]Department of Computer Science and Engineering, Manipal University Jaipur, Rajasthan
[3]King Faisal University, Saudi Arabia

1.1 Introduction

COVID-19 is a respiratory tract infection. The virus can spread in either the upper or lower respiratory tracts. The upper tracts of the respiratory system are the sinuses, nose, and throat. The windpipe and lungs are the lower tracts of the respiratory system, which is why they are important [1,2]. In any way, the virus attacks the human respiratory tract. This virus can also cause diseases such as Middle East respiratory syndrome and sudden acute respiratory syndrome, which can be fatal if they are not taken care of quickly [3–5]. The virus has a typical characteristic of changing or mutating as it infects people. This is the most concerning aspect of its characteristic.

Furthermore, the COVID-19 virus is inorganic. The moment it encounters human metabolism, the COVID-19 virus becomes active [6,7]. Then, this virus takes over the cellular metabolism and becomes a "Molecular Hijacker" inside the human body. It hijacks the cellular metabolism of various human organs, specifically the respiratory tract and the cardiovascular system. After that, the coronavirus replicates into thousands of other coronavirus particles [8–10].

Preventives: COVID-19 can be prevented through the following easy-to-follow and straightforward measures [8,11–14]:

- Direct physical human contact is the principal carrier of the COVID-19 virus. Therefore, maintaining a distance of 6 feet or 2 m with another person can prevent infection.
- Frequent washing of the hands, eyes, and exposed arms with soap or sanitizer with 90% alcohol can kill the virus. Similarly, the frequently used surfaces, doorknobs, and handles should be sanitized from time to time.
- Masks are the most potent tool to avoid coronavirus infection. The simple cotton cloth mask prohibits the entry of viral droplets between two facing persons. The coronavirus can stay as an aerosol in the air for even hours after discharge from a person's nose, mouth, or spit.
- Whenever needed, the elbow or handkerchief should be used to obstruct the release of viral droplets in the air.
- Dining out is prohibited.
- Sharing food products is not allowed.
- The personal eyes, nose, and mouth should not be touched daily.
- Stay home and avoid crowded places.

Coronavirus-affected persons are kept isolated in homes and separated from other family members [10,14,15]. Home quarantine is encouraged because the medical infrastructure throughout Europe and the USA has wholly crashed amidst COVID-19. Likewise, less affluent but high-density populated countries, such as India, have weak health infrastructure and overwhelming pressure on patients. Hence, home quarantine is manageable and preferred. In the case of advanced stages of COVID-19 complications, patients are kept under observation in isolation wards in hospitals [5,6].

Researchers are working overtime to produce a booster vaccine against COVID-19 worldwide. The latest variant Omicron is now affecting people and spreading since the starting quarter of 2022 [8]. The Serum Institute of India and the University of Oxford are working to find the mutation and booster vaccine against COVID-19 [16]. As expected, a reliable vaccine booster will probably be available in the world by 2022 or, at best, by the end of 2022 to stop the mutation of COVID-19 or any new variant. The clinical trials need to be intensive and extensive before any new variant will come.

1.2 Related work

Islam *et al.* [17] came up with a network that automatically accepts early diagnoses of COVID-19 patients very quickly. This network is called a combined deep convolutional neural network (CNN) and long short-term memory network. The researchers also used X-rays of the chest and looked at almost 4,575 images in a dataset. They did much research and came up with different ways to measure how well the proposed system worked.

Bharati *et al.* [18] proposed a new hybrid method determined as a visual geometry group (VGG) data-based spatial transformer network (STN) with CNN (VDSNet). The researchers used tools like TensorFlow and Keras. In the Kaggle repository, they found a chest X-ray image dataset from the National Institute of Health. They used the new hybrid model to look at the data. According to the researchers, the hybrid model claims to cut medical costs and be helpful for the efficient detection of COVID-19. The proposed VDSNet framework for detecting the presence of coronavirus in the lungs from X-ray images has shown validation accuracy of better than 73%. Researchers have also claimed that distinct sample datasets have shown an accuracy value of just 70.8%. Therefore, according to the researchers, their hybrid system has improved the accuracy, whereas the training time for the laboratory technicians has been significantly reduced.

Pereira *et al.* [11] used hierarchical classification and ENT (ear, nose, and throat) multiclass classification methods to provide more efficient detection and differentiation between real and false pneumonia. Their paper uses the resampling algorithm in the schema to rebalance the distribution of classes. There are known texture descriptors and a model that has already been trained to use in this framework to get features from chest X-ray computed tomography (CT) scan. To make sure that different texture descriptors and base classifiers work together, they should use techniques called "early and late fusion."

Harsh Panwar *et al.* [12], in their paper, studied chest X-ray CT scans, and the deep learning neural network-based method named nCOVnet was proposed for designing a fast screening method that can be used to detect COVID-19 among affected patients. Researchers have attempted to train images with a deep learning technique to help medical doctors diagnose faster. The CNN has been highly efficient in analyzing the CT scan chest X-rays of COVID-19 patients. The sample size in this research paper is minimal. The researchers used the data of only 19 patients. It is further interesting to observe in their paper a solution based on the application of the model only on 19 patients. Regardless of the sample size, the researchers claim high accuracy.

Dansana *et al.* [19] used VGG-16, Inception V2, and decision tree methods to compare the models. They correctly did this by looking at X-ray images of people who had different pneumonia and using them as a dataset. Apostolopoulos [20] and his team did this. In this study, the authors used samples of common bacterial pneumonia that had been confirmed to have COVID-19 and normal viral fever to detect COVID-19 automatically. The authors used the transfer learning method. In total, they looked at 1,427 X-ray images, 224 of which had COVID-19 infections, 700 of which had confirmed common bacterial pneumonia, and 504 of which had no infections. COVID-19 patients will be better diagnosed with the help of artificial intelligence. Combining machine learning methods that use automatic feature extraction and classification from images for correct object detection or medical imaging classification, deep learning makes it easier for computers to learn how to do these things.

Cohen *et al.* [13] developed a predictive model to study the severity score in COVID-19 pneumonia patients. The model and the diagnostic tool developed by them are supposed to gauge the escalation and de-escalation of the COVID-19 infection and the efficacy of the treatment, especially for patients who have been seriously infected and required treatment in intensive care units of hospitals. The authors have admitted a lack of sufficient databases from the hospitals in this model, making it difficult to conduct a large-scale evaluation of their newly designed predictive model. Furthermore, sales pressure on the hospitals has been enormous, so the hospitals could not provide a sufficient number of samples. Accordingly, the researchers observe that their model has been trained on a smaller database that exposes the evaluation of their model to the risk of sample bias.

Wu *et al.* [21] used a multiview fusion model for screening COVID-19-related pneumonia. The researchers claim that it is a multicenter study. The methodology collected CT scan images of only 495 patients from the three hospitals in China where the coronavirus outbreak occurred in the middle of December 2019. The authors divided 395 samples into various categories, wherein 80% of the samples, 294 cases, were drawn from COVID-19, and the remaining 101 patients were drawn from pneumonia patients. Furthermore, they drew 50 more cases of COVID-19 and another type of pneumonia. They used a deep learning network to train a multiview fusion model that could be used to screen patients with the most lung areas in axial, coronal, and sagittal views. A way to determine how good the model would be is by looking at both the validation and testing groups.

The early detection of COVID-19 from chest X-ray images has been of interest to researchers worldwide. Ahishali *et al.* [22] also attempted to find an efficient approach to diagnosing diagnostic treatment unit (DTU) to make a predictive assessment of the early detection of COVID-19. They used complete classifiers and deep learning approaches in their study. Early COVID-19 detection will help prevent the spread of the disease and isolate infected patients from healthy people.

The authors made a new benchmark dataset called QaTa-COV-19 that was based on early COVID-19 with some extra images. People who were in the early stages of COVID-19 were included in the study. Their goal was to make sure that COVID-19 was found as soon as possible through X-ray images alone because CT scan facilities and reverse transcription-polymerase chain reaction have been the diagnostic methods with low sensitivity and took a long time to find COVID-19 patients. Their dataset consisted of 175 early-stage COVID-19 patients whose constitution was 40 images with no infection, 135 images with limited infections, and 1,579 samples from the normal class, meaning the patients had common flu conditions.

Selvan *et al.* [9] conducted a mid-day study through lung segmentation from a chest X-ray. Generally, in all patients suffering from COVID-19, pulmonary opacification is caused by respiratory ailments and subsequent inflammation in the lungs. Currently, it has been challenging for medical doctors to distinguish different regions of the lung that become invisible or opaque because of regional opacification. When the researchers did this work, they used an automated risk score for COVID-19 that came from chest X-rays. The researchers looked at the areas that were very dark and did not have any data. They developed a new image segmentation network that used a deep generative model to look at the data.

Mostafa *et al.* [23] used a deep learning technique for detecting COVID-19 by studying chest X-rays. They used deep learning to classify and detect COVID-19 through the CNN method using publicly available COVID-19 chest X-rays. The hospitals were very cooperative in sharing the X-ray and CT scans of the COVID-19 patients. The researchers found that chest X-ray and CT scans did not have clear contrast pictures of the images. It was felt that if a technique for image enhancement and segmentation can be used to process the images, it shall improve the detection process. The researchers found that the X-ray attenuation across the lungs was lower than other contents, such as bones; the lung image became darker than the other chest parts observed under radiography. Most of the research used linear or S-form transformation to change the brightness and contrast. This was not effective. The authors developed an algorithm that used the deep learning technique to improve the quality of the image, and this led to the discovery of an algorithm called the contrast limited adaptive histogram equalization algorithm, which works well for chest X-rays. It was thought that this method would make it easier for patients to obtain a quick diagnosis and treatment.

1.2.1 Key contributions and proposed work

This paper will include the work done for detecting COVID-19 patients by building customized deep learning models using CNNs. We applied the artificial neural network (ANN) model after extracting the gray-level co-occurrence matrix (GLCM) features from images, whereas the CNN model was directly applied to the X-ray image dataset. The CNN models are used to analyze the images in which mathematical convolution is performed to extract the features.

- Our proposed model used the CNN concept, modified it accordingly, and named it a customized CNN.
- This chapter methodology has taken chest X-ray images and extracted GLCM features from the same images.
- We compared our model with the existing CNN and ANN models.

- We have tested and validated our model with evaluation and assessment metrics.
- We also evaluated the model with an AU_ROC score.

1.3 Materials and methods

(a) **Materials:** The dataset used in this work is X-ray images, which can be accessed through Kaggle [24] and GitHub [25]. Sample images are shown in Figures 1.1 and 1.2.

(b) **Methods:** Deep learning is a technique that works like the human brain. We use the methodology of two basic deep learning models: ANN and CNN.

A simple ANN is shown in Figure 1.3. Now, the layer that interacts with features of the dataset is the first layer of the neural network, known as the input

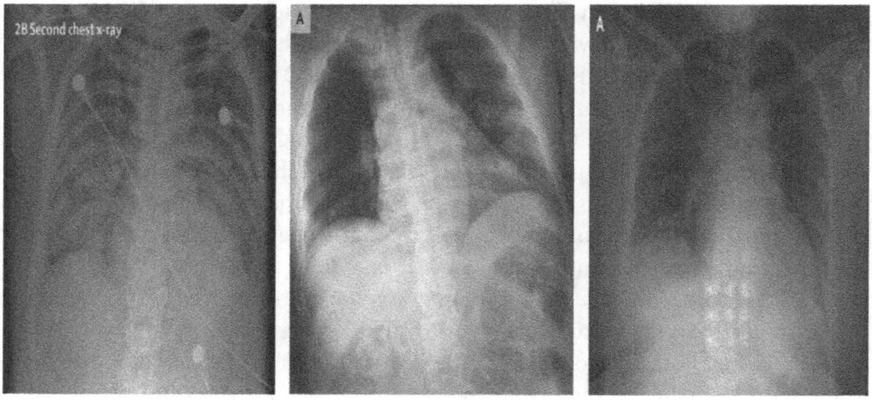

Figure 1.1 X-ray images of COVID-19 patients

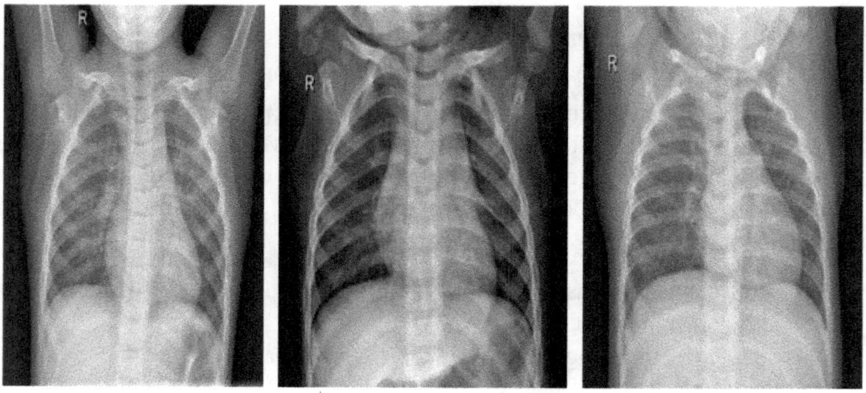

Figure 1.2 X-ray images of normal patients

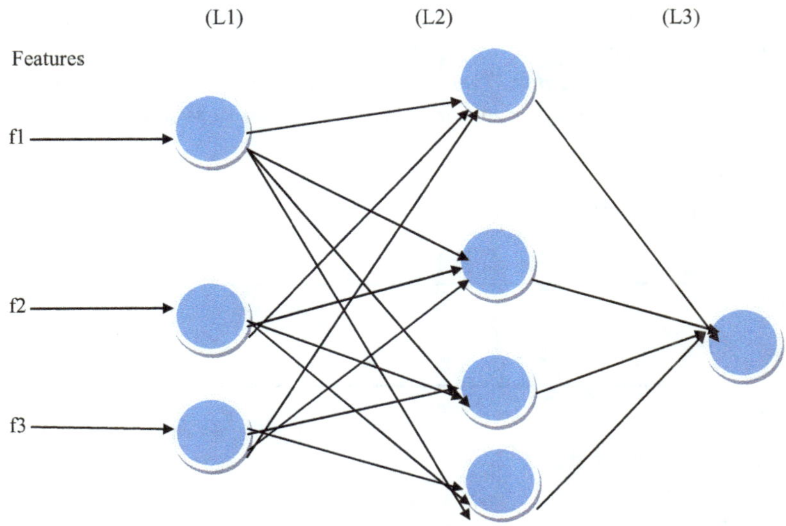

Figure 1.3 A simple ANN

layer. Let f1, f2, and f3 be three features given to the input layer. Each layer consists of several neurons. In Figure 1.3, we take three features given to the input layer consisting of three neurons. The next layer is the hidden layer, consisting of the number of neurons. There can be more than one hidden layer, but in Figure 1.3, we take only one hidden layer consisting of four neurons. Processing takes place in hidden layers. The features shown in the lower image, i.e., f1, f2, and f3 from layer L1, are passed to all hidden layer neurons. Finally, once processing takes place inside hidden layers, the output of layer L2 is passed to L3, consisting of neurons required for predicting the output or classifying the output. In Figure 1.3, there is only one neuron taken in the output layer. If a multiclass classification is at the output, we can take that many output nodes. For the binary type of classification, we can only take one node in the output layer to classify whether the object is of a particular class.

Now, after ANN, we discuss the convolutional neural network. The CNN technique is as follows:

- We use CNNs to analyze the visual images.
- Mathematical convolution is the main operation performed in CNN, which measures the overlapping of two functions.
- It uses feature matrices or filters to extract or detect features or attributes in an image, such as curves and vertical and horizontal edges.
- The convolution operation is performed once the feature matrix or filter slides over the input.
- Matrix multiplication is performed at each pixel location of an image (where a matrix of pixels for a particular image is given as input). The result is aggregated on a filter map or feature map.

Convolution layer: Feature extraction and creating feature maps are the primary tasks of this layer. It is the most important layer of CNN. "*Mathematical operation that generates a third function using two functions is called convolution operation.*" This is done by computing the dot product between filter entries and input volume.

Example:

This is an input matrix of size 3*4:

A	B	C	D
E	F	G	H
I	J	K	L

This is the filter of size 2*2:

M	N
O	P

The output will be a 2*3 matrix computed as follows:

AM+BN+EO+FP	BM+CN+FO+GP	CM+DN+GO+HP
EM+FN+IO+JP	FM+GN+JO+KP	GM+HN+KO+LP

Filters and kernels are the basic units of this architecture. Moving ahead one by one, the input volume (containing the width and height) is convolved with each filter. By doing this, filter properties are acquired by the CNN network, which helps them detect features at a specific position in the input.

The following parameters are adjusted for the filter:

- Size of the filter.
- Stride is the rate of kernel movement by 2 increments, and a stride of 1 means movement of the kernel by 1 increment.
- Padding: It is applied if we want to keep the output size the same as the input size after the convolution operation. Zero padding ensures that the kernel is passed ideally on the edges.
- Output layer: Number of filters or kernels that are applied.

  Pooling layer: It is inserted between successive convolution layers. To reduce the size of the image or control the overfitting problem. Pooling can be of different types:

  - MAX POOLING: Features such as sharp edges and curves can be depicted by taking the largest element value of the rectified feature map.

Example:

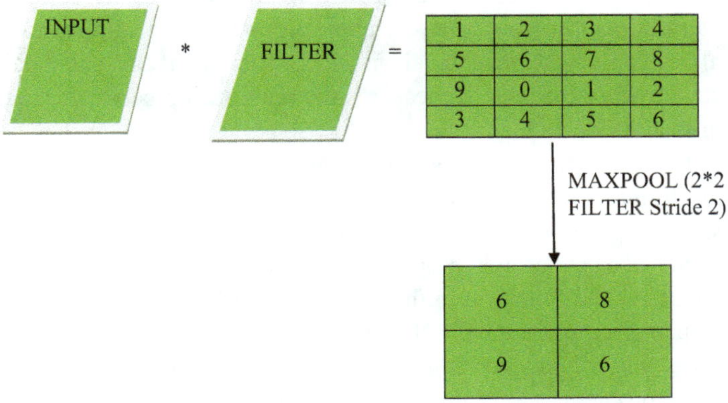

AVERAGE POOLING: To identify the features smoothly, we use them. Straight lines and small curves can be detected by it, but it is challenging to detect sharp edges.

SUM POOLING: Used when we sum all the elements in the feature map.

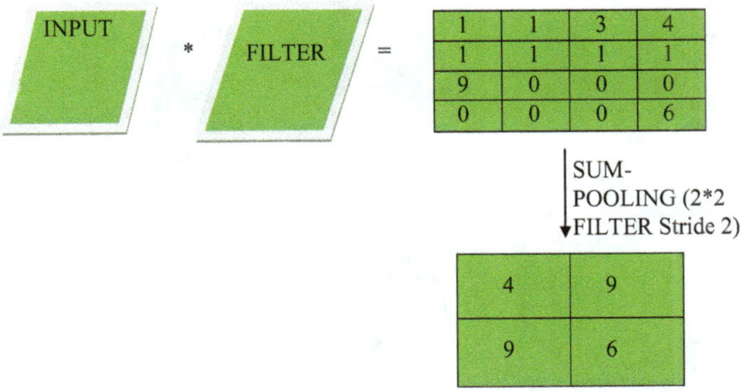

Flattening layer: Once all the features are extracted from convolution layers and the dimensions are reduced by pooling layers, we are now ready to reduce our multi-dimensional data to single-dimensional data that can be fed into the fully connected layer.

Fully connected layer: The traditional multilayer perceptron layer may use the softmax function at the output layer.

Therefore, a typical CNN network consists of the above four layers.

1.3.1 Feature extraction and selection

After ANN and CNN model, we discuss the feature extraction method: GLCM features for texture analysis of images have been used in this paper.

Textures are the arrangement of color or intensities occupying an image or a region of an image, i.e., if we encounter the arrangement of pixel values of the image dramatically or there is a large variation in pixel values. We can say that the texture is rough. Textures appear homogenous to our eyes [26–28].

Texture also involves the distribution of gray levels in an image. Now, we can use textures to extract different features of an image, such as contrast, homogeneity, dissimilarity, energy, and correlation. One of the application areas of texture analysis is medical image analysis. GLCM is one of the statistical methods used to extract texture features. It measures the relationship between groups of two pixels in an image [29–31].

The ANN and CNN model block diagrams implemented in this research work are shown in Figure 1.4. In the ANN model, each hidden layer is connected with bias, and each neuron in it is applied to the RELU activation function. Additionally, the output layer consists of a sigmoid activation function. Each hidden layer consists of 10 neurons and 1 neuron in the output layer [32,33].

The proposed customized CNN model is presented in Figure 1.5, and the overall process flow of our model is shown in Figure 1.10.

Thus, the problem was formulated by building an ANN model by first extracting the GLCM features and then developing a customized CNN model that performed better than the ANN model using X-ray images.

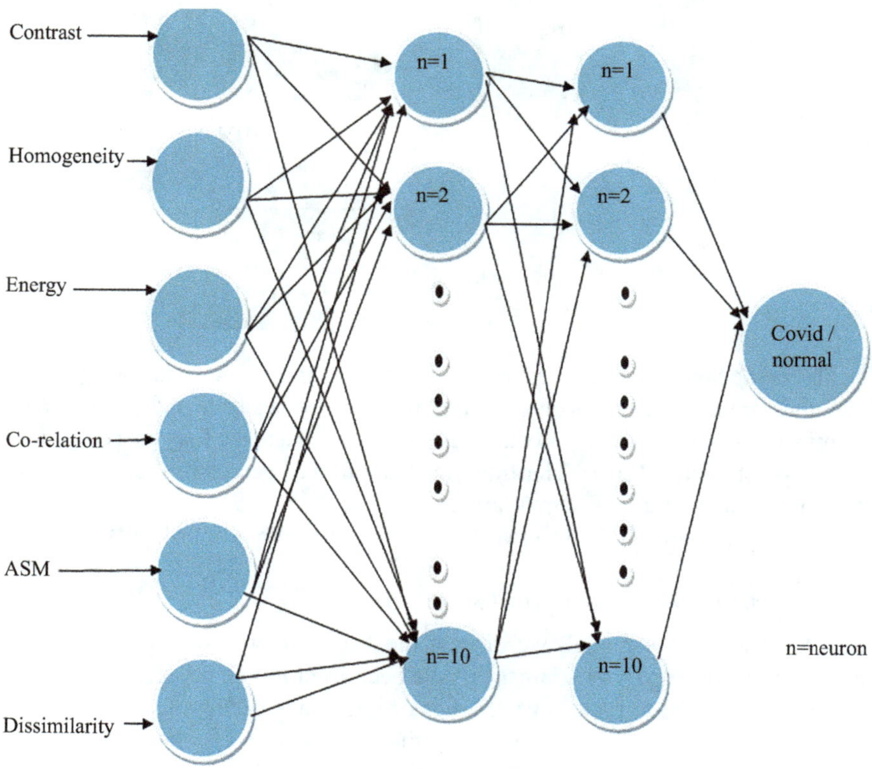

Figure 1.4 CNN model with GLCM features

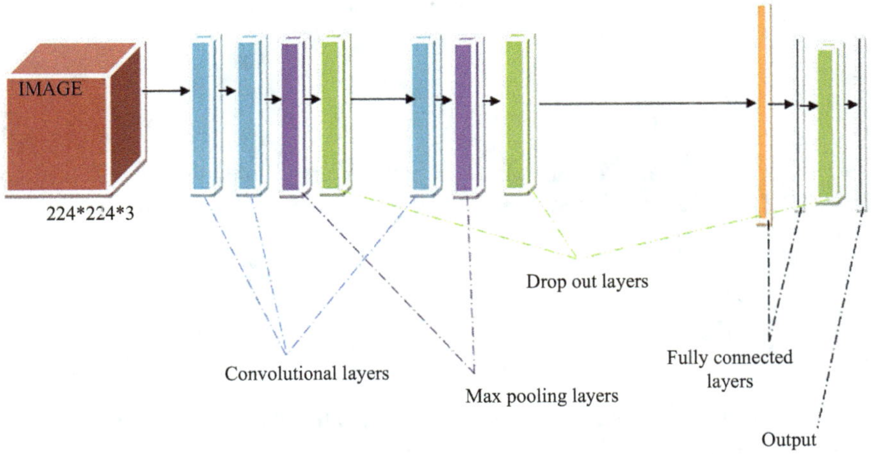

Figure 1.5 Proposed customized CNN model

1.4 Results and discussion

The result was computed based on the following parameters:

- Accuracy: This is the ratio of correctly classified records to the total number of classified records. The formula is as follows:

$$\frac{T.P. + T.N.}{T.P. + T.N. + F.P. + F.N.}$$

- Precision: It is considered when we want to know how many of those who tested positive for coronavirus are COVID-19 patients. It is the total number of correctly classified positive records to the total number of predicted positive records. The formula is given by:

$$\frac{T.P.}{T.P. + F.P.}$$

- Recall: It is considered when we want to know of all the people who are suffering from COVID-19 and how many of those were correctly predicted? The formula is as follows:

$$\frac{T.P.}{T.P. + F.N.}$$

- F1-score: This is harmonized mean of recall and precision. It is considered when we take both precision and recall into consideration. The formula is as follows:

$$2 * \frac{precision * recall}{precision + recall}$$

- Specificity: It is also known as the "true negative rate" and is given by the formula:

$$\frac{T.N.}{T.N. + F.P.}$$

- False positive rate: It tells us what percentage of people not affected by COVID-19 were predicted to be COVID-19 patients. The formula is given by:

  1-Specificity or

$$\frac{F.P.}{T.N. + F.P.}$$

where

True positives (T.P.): the data record for which the classifier predicted "Covid" and the cell was actually Covid.

True negatives (T.N.): the data record for which the classifier predicted "Normal" and the cell was actually Normal.

False positives (F.P.): the data record for which the classifier predicted "Covid" but the cell was actually Normal.

False negatives (F.N.): the data record for which the classifier predicted "Normal" but the cell was actually Covid.

The confusion matrix (Table 1.1) obtained for the dataset of COVID and normal X-ray images for the ANN model is as follows:

From the above matrix, we obtain the following:

- T.P.: 26
- T.N.: 29
- F.P.: 1
- F.N.: 1

The evaluation result of ANN shows how the model is successfully trained and gives the following results:

- Accuracy: 0.96
- Precision: 0.97
- Recall/sensitivity/true positive rate: 0.97
- F1-score: 0.97
- Specificity: 0.97
- False positive rate: 0.03
- Area under the curve (AUC)=0.98

Table 1.1 ANN confusion matrix

	Actual values	
Predicted values	26	1
	1	29

The confusion matrix obtained for the customized CNN model is given in Table 1.2.

From the above matrix, we obtain the following:

- T.P.: 28
- T.N.: 30
- F.P.: 0
- F.N.: 2

The evaluation result of the CNN shows how the model is successfully trained and gives the following results:

- Accuracy: 0.97
- Precision: 1.00
- Recall/sensitivity/true positive rate: 0.933
- F1-score: 0.96
- Specificity: 1.00
- False positive rate: 0
- AUC=0.9989

For comparative analysis, we constructed a bar chart (Figures 1.6 and 1.7) of the results for both models.

Table 1.2 Customized CNN confusion matrix

	Actual values	
Predicted values	28 2	0 30

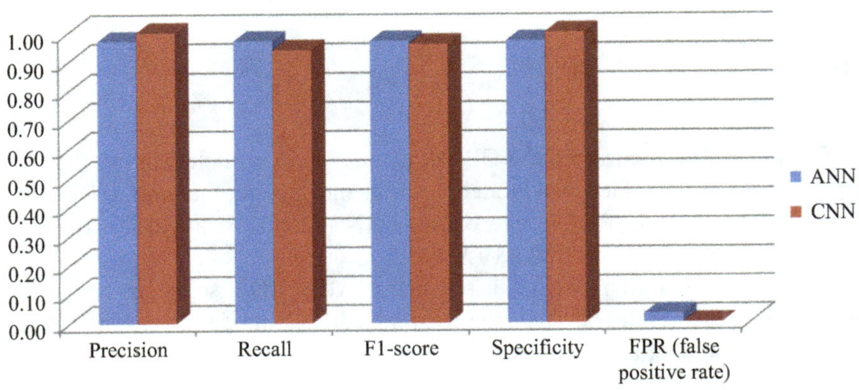

Figure 1.6 Comparison chart of ANN and customized CNN

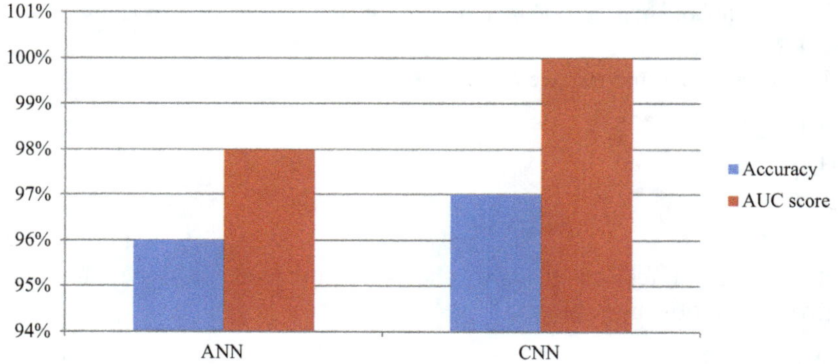

Figure 1.7 Bar chart of accuracy and AUC score of ANN and customized CNN

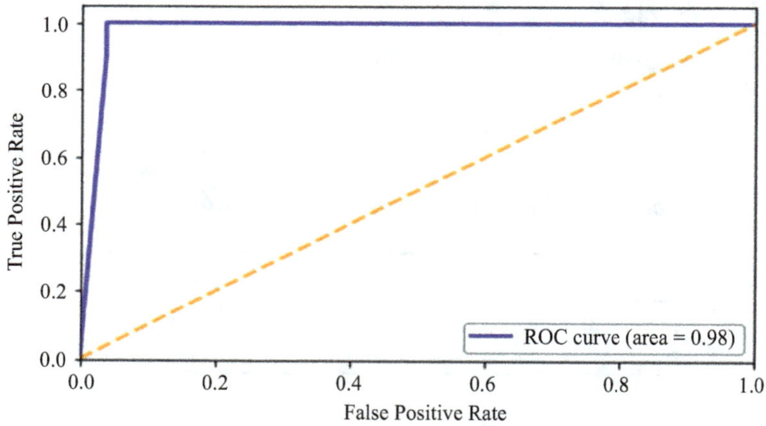

Figure 1.8 AU_ROC curve for the ANN model

Even though we extracted the GLCM features before implementing the ANN model, we can see that the CNN model performs much better than the ANN model. Additionally, the AUC_ROC curve of the CNN is better than that of the ANN model, as shown in Figures 1.8 and 1.9.

The customized CNN model has a better AUC_ROC score than the ANN model with GLCM features. Thus, the customized CNN model gives better results than the ANN model, as shown in Table 1.3.

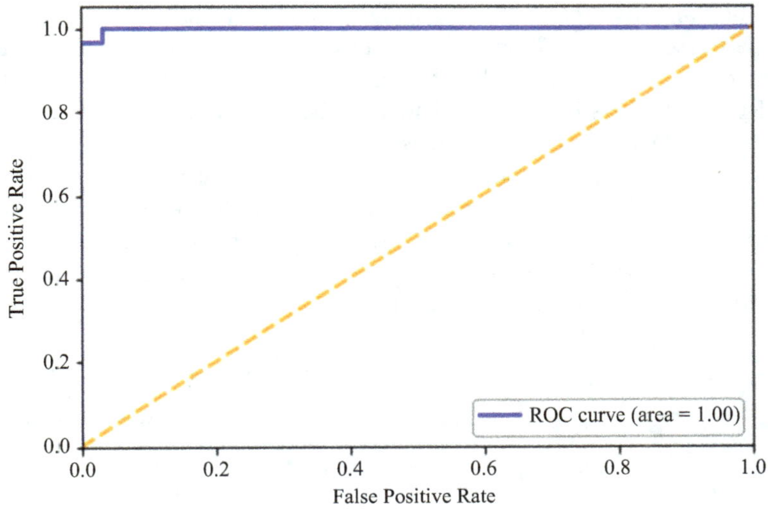

Figure 1.9 AU_ROC curve for the CNN model

Table 1.3 Comparative results of ANN and customized CNN

Evaluation metrics	ANN	Customized CNN
Precision	0.97	1.00
Recall	0.97	0.94
F1-score	0.97	0.96
Specificity	0.97	1.00
False positive rate	0.03	0

1.5 Conclusion and future scope

In recent years, COVID-19 has emerged as one of the most contagious illnesses in the world. Although many countries have developed their vaccination now, still people are waiting to get one. Indeed, this may help eradicate the disease. As many people are affected, it becomes difficult for technicians and doctors to detect so many patients in less time. In this dissertation work, I have used deep learning techniques for detecting COVID-19-affected patients from their X-ray images. Even for patients going for an X-ray test is much cheaper than CT scans. Thus, this work proves to be fruitful for both doctors and patients to detect the presence of disease. First, I developed an ANN model using GLCM features such as contrast, energy, dissimilarity, homogeneity, angular second moment, and correlation. The model performed pretty well, giving an accuracy score of 96%. Then, a customized CNN model is developed, which performs even better than the ANN model. The AUC_ROC curve was between 0.5 and 1 for both models, clearly distinguishing between COVID-affected X-ray images and X-ray images for regular patients. In the future, discriminative models can be implemented using different features, improving the model's accuracy, recall, and precision. Researchers can implement the models on the different datasets to check how the models perform with a different kind of dataset. Future research may also include implementing a model on a much larger dataset using the same features.

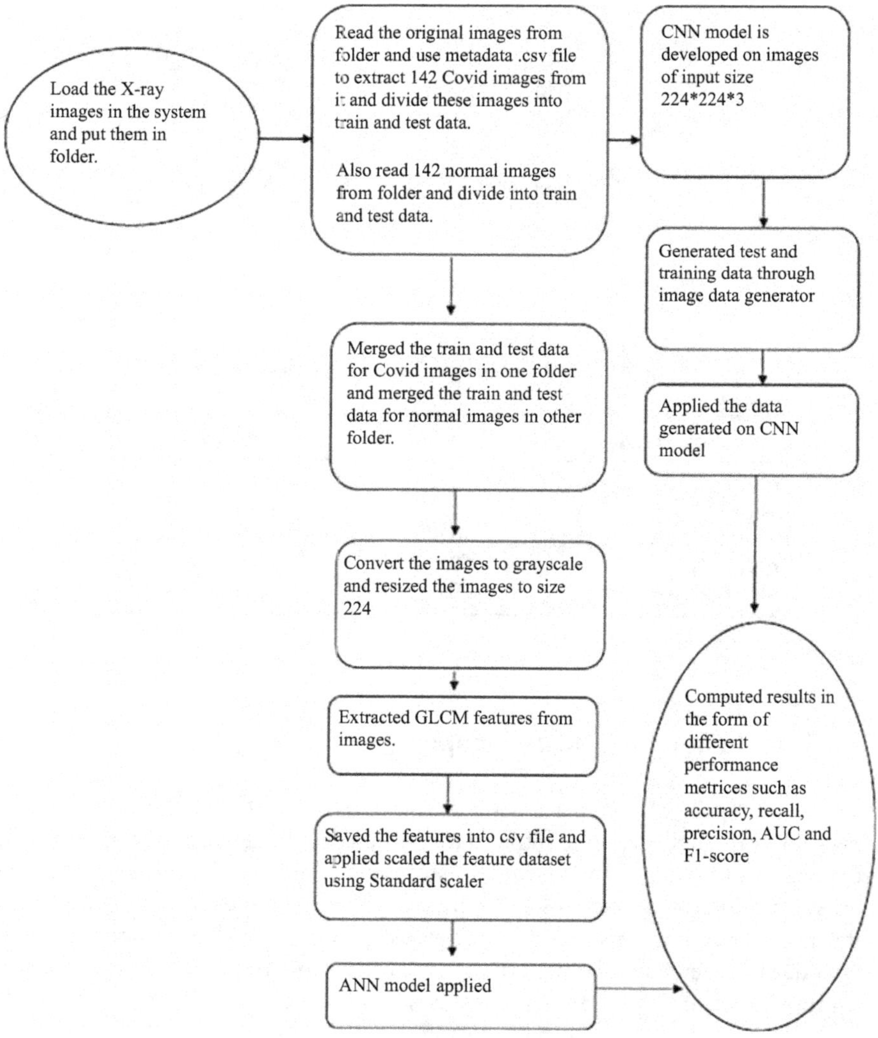

Figure 1.10 Flow diagram of the proposed neural network-based model

References

[1] D. Arias-Garzón, J. A. Alzate-Grisales, S. Orozco-Arias, *et al.*, "COVID-19 detection in X-ray images using convolutional neural networks," *Mach. Learn. Appl.*, 6, 100138, 2021, DOI: 10.1016/j.mlwa.2021.100138.

[2] G. Sasubilli and A. Kumar, "Machine learning and big data implementation on health care data," *4th Int. Conf. Intell. Comput. Control Syst. (ICICCS)*, 2020, pp. 859–864, DOI: 10.1109/ICICCS48265.2020.9120906.

[3] A. K. Das, S. Ghosh, S. Thunder, *et al.*, "Automatic COVID-19 detection from X-ray images using ensemble learning with convolutional neural network," *Patt. Anal. Appl.* 24, 1111–1124, 2021. Available at: https://doi.org/10.1007/s10044-021-00970-4.

[4] M. E. H. Chowdhury, T. Rahman, A. Khandakar, *et al.*, "Can A.I. help in screening viral and COVID-19 pneumonia?" vol. 8, pp. 132665–132676, 2020.

[5] S. Akter, F. M. J. M. Shamrat, S. Chakraborty, A. Karim, and S. Azam, "COVID-19 detection using deep learning algorithm on chest X-ray images," *Biology*, 10(11), 1174,2021. Available at: https://doi.org/10.3390/biology10111174.

[6] N. A. Alam, M. Ahsan, M. A. Based, J. Haider, and M. Kowalski, "COVID-19 detection from chest X-ray images using feature fusion and deep learning," *Sensors*, 21(4), 1480, 2021. Available at: https://doi.org/10.3390/s21041480.

[7] S. Singh, P. Sapra, A. Garg, and D. K. Vishwakarma, "CNN based Covid-aid: Covid 19 detection using chest X-ray," 2021 5th Int. Conf. Comput. Method. Commun. (ICCMC), 2021, pp. 1791–1797, DOI: 10.1109/ICCMC51019.2021.9418407.

[8] WHO, https://www.who.int/health-topics/coronavirus#tab=tab_1, Last accessed: January 05, 2022.

[9] R. Selvan, E. B. Dam, N. S. Detlefsen, and S. Rischel, "Lung segmentation from chest X-rays using variational data imputation," May 2020, [Online]. Available at: http://arxiv.org/abs/2005.10052.

[10] D. Jain, V. Gauri Shankar, and B. Devi, "A robust approach of COVID-19 Indian data analysis using support vector machine," 2022, DOI: https://doi.org/10.1007/978-981-16-8546-0_29.

[11] R. M. Pereira, D. Bertolini, L. O. Teixeira, C. N. Silla, and Y. M. G. Costa, "COVID-19 identification in chest X-ray images on flat and hierarchical classification scenarios," *Comput. Methods Programs Biomed.*, 194, 2020, pp. 105532–105562, DOI: 10.1016/j.cmpb.2020.105532.

[12] H. Panwar, P. K. Gupta, M. K. Siddiqui, R. Morales-Menendez, and V. Singh, "Application of deep learning for fast detection of COVID-19 in X-rays using nCOVnet," *Chaos Solitons Fract.*, 138, 109944, 2020, DOI: 10.1016/j.chaos.2020.109944.

[13] J. P. Cohen, L. Dao, K. Roth, "Predicting COVID-19 pneumonia severity on chest X-ray with deep learning," *Cureus* June 2020, DOI: 10.7759/cureus.9448.

[14] S. Rajaraman, J. Siegelman, P. O. Alderson, L. S. Folio, L. R. Folio, and S. K. Antani, "Iteratively pruned deep learning ensembles for COVID-19 detection in chest X-rays," *IEEE Access*, 8, 115041–115050, 2020, DOI: 10.1109/ACCESS.2020.3003810.

[15] P. M. Shah, F. Ullah, D. Shah, "Deep GRU-CNN model for COVID-19 detection from chest X-rays data," *IEEE Access*, DOI: 10.1109/ACCESS.2021.3077592.

[16] Serum Institute, https://www.seruminstitute.com/, Last accessed: January 02, 2022.

[17] M. Z. Islam, M. M. Islam, and A. Asraf, "A combined deep CNN-LSTM network for the detection of novel coronavirus (COVID-19) using X-ray images," *Inform. Med. Unlocked*, 20, 100412, 2020, DOI: 10.1016/j.imu.2020.100412.

[18] S. Bharati, P. Podder, and M. R. H. Mondal, "Hybrid deep learning for detecting lung diseases from X-ray images," *Informatics Med. Unlocked*, 20, 100391, 2020, DOI: 10.1016/j.imu.2020.100391.

[19] D. Dansana, R. Kumar, A. Bhattacharjee, *et al.* "Early diagnosis of COVID-19-affected patients based on X-ray and computed tomography images using deep learning algorithm," *Soft Comput.*, 0123456789, 2020, DOI: 10.1007/s00500-020-05275-y.

[20] D. Apostolopoulos and T. A. Mpesiana, "Covid-19: automatic detection from X-ray images utilizing transfer learning with convolutional neural networks," *Phys. Eng. Sci. Med.*, 43 (2), 635–640, 2020, DOI: 10.1007/s13246-020-00865-4.

[21] X. Wu, H. Hui, M. Niu, *et al.*, "Deep learning-based multiview fusion model for screening 2019 novel coronavirus pneumonia: a multicenter study," *Eur. J. Radiol.*, 128 (March), 1–9, 2020, DOI: 10.1016/j.ejrad.2020.109041.

[22] M. Ahishali, A. Degerli, M. yamaç, *et al.*, "Advance warning methodologies for COVID-19 using chest X-ray images," June 2020, pp. 41052– 41065, [Online]. Available at: http://arxiv.org/abs/2006.05332.

[23] A. Mostafa, A. Elbehery, O. Ashraf. Mohamed, and A. Mahmoud, "Detecting COVID-19 via chest X-rays using deep learning," 0–6, May 2020, DOI: 10.13140/RG.2.2.16248.88324.

[24] Kaggle Data Set, https://www.kaggle.com/paultimothymooney/chest-xray-pneumonia, Last accessed: December 22, 2021.

[25] GitHub, https://github.com/ieee8023/covid-chestxray-dataset, Last accessed: November 17, 2021.

[26] S. R. Swarna, A. Kumar, P. Dixit, and T. V. M. Sairam, "Parkinson's disease prediction using adaptive quantum computing," *2021 Third International Conference on Intelligent Communication Technologies and Virtual Mobile Networks (ICICV)*, 2021, pp. 1396–1401, DOI: 10.1109/ICICV50876.2021.9388628.

[27] V. G. Shankar, D. S. Sisodia, and P. Chandrakar, "DataAutism: an early detection framework of autism in infants using data science." In: Sharma N., Chakrabarti A., Balas V. (eds), *Data Management, Analytics, and Innovation.*

Advances in Intelligent Systems and Computing, vol. 1016, 2020; Springer, Singapore. Available at: https://doi.org/10.1007/978-981-13-9364-8_13.

[28] V. G. Shankar, B. Devi, U. Sachdeva, and H. Harsola, "Real-time human body tracking system for posture and movement using skeleton-based segmentation". In: *Micro-Electronics and Telecommunication Engineering. Lecture Notes in Networks and Systems*, vol. 179, 2021, pp. 499–510, Springer, Singapore. Available at: https://doi.org/10.1007/978-981-33-4687-1_48.

[29] V. Goel, V. Jangir, and V. G. Shankar "DataCan: robust approach for genome cancer data analysis". In: N. Sharma, A. Chakrabarti, and V. Balas (eds), *Data Management, Analytics, and Innovation. Advances in Intelligent Systems and Computing*, vol. 1016, 2020, pp. 153–166, Springer, Singapore. Available at: https://doi.org/10.1007/978-981-13-9364-8_12.

[30] B. Devi, V. G. Shankar, S. Srivastava, K. Nigam, and L. Narang, "Racist tweets-based sentiment analysis using individual and ensemble classifiers." In: *Micro-Electronics and Telecommunication Engineering. Lecture Notes in Networks and Systems*, vol. 179, 2021; Springer, Singapore. Available at: https://doi.org/10.1007/978-981-33-4687-1_52.

[31] B. Devi, V. G. Shankar, S. Srivastava, and D. K. Srivastava, "AnaBus: a proposed sampling retrieval model for business and historical data analytics," *Adv. Intell. Syst. Comput.*, 1016, 2020, pp. 179–187. Available at: https://doi.org/10.1007/978-981-13-9364-8_14.

[32] P. K. Kotturu and A. Kumar, "Comparative study on machine learning models for early diagnose of Alzheimer's disease: multi correlation method," *2020 5th Int. Conf. Commun. Electron. Syst. (ICCES)*, 2020, pp. 778–783, DOI: 10.1109/ICCES48766.2020.9137872.

[33] A. K. Dubey, S. Narang, A. Kumar, S. M. Sasubilli, and V. García-Díaz, "Performance estimation of machine learning algorithms in the factor analysis of covid-19 dataset," *Comput. Mater. Continua*, 66(2), 1921–1936, 2021.

Chapter 2

Introducing deep learning in medical diagnosis

N. Padmapriya[1], N. Kumaratharan[2], M. Pavithra[1], R. Rajmohan[1], P. Kanimozhi[1] and D.J. Samuel[3]

Medical imaging refers to various techniques that include interior visual representations of the human body, which help clinicians and physicians to identify, diagnose, and manage diseases more quickly and effectively. Deep learning (DL), in contrast to conventional machine learning (ML) algorithms, automatically detects insightful representations without the need for domain experts' expertise. DL has attracted increasing attention in recent years using neuroimages such as structural magnetic resonance imaging, functional MRI, and positron emission tomography, and traditional ML has been superior in the diagnosis of brain disorders using the computer. DL is currently propelling medical image processing to new heights. Classification, detection, registration, and segmentation are the four main categories of medical image processing. The aim of the process is to categorize medical images into multiple groups. The detection task entails locating and identifying landmarks or lesions in a complete medical picture. Efficient treatments for neurodegenerative disorders are desperately needed around the world. The intricacy of the biological mechanisms generating neuronal degeneration, and also the patient population's diversity, make developing early diagnostic methods and successful therapies for these diseases extremely difficult. DL, a cutting-edge ML technique, detects complicated characteristics in complex high-dimensional data, particularly in computer vision. As significant advancements in neuroimaging techniques have resulted in massive multimodal neuroimaging data, DL's application to the detection and automatic categorization of Alzheimer's disease is becoming more feasible and has recently got a lot of attention. This chapter proposes an efficient method for classifying two widely diagnosed neurodegenerative disorders: Parkinson's and Alzheimer's disease.

While DL architectures are widely used for the classification of various brain tumors and other types of cancers, they are less commonly used for the classification of neurodegenerative diseases. Since the number of patients suffering from

[1]Department of Computer Science and Engineering, IFET College of Engineering, India
[2]Department of Electronics and Communication Engineering, Sri Venkateswara College of Engineering, India
[3]MD Anderson Cancer Center, The University of Texas, USA

neurodegenerative diseases is growing, it is critical to provide an automated system that can perform the classification of these diseases. This method can also aid radiologists in improving the quality and accuracy of their diagnoses of neurodegenerative diseases. By presenting different forms of deep neural networks and recent developments, a detailed description of familiar network structures and DL methods are provided. After that, we go over DL approaches for the computer-assisted study of neurodegenerative brain conditions. We also address the shortcomings of previous research and suggest potential future directions.

2.1 Introduction

Medical imaging tools such as magnetic resonance imaging (MRI) scans, mammography, computed tomography scans, ultrasound, positron emission tomography, and X-rays have been used to detect, diagnose, and manage disease in the last few decades [1]. At the clinic, human experts such as radiographers and clinicians performed primarily medical imaging interpretations. Moreover, given the wide range of pathologies and the possible exhaustion of human expertise, Computer-assisted therapies have begun to benefit physicians and researchers [43]. Despite advancement in computational medical image processing has fallen behind that of medical imaging technologies, the economy is changing as machine learning (ML) algorithms are implemented.

When employing ML, many tasks in medical picture segmentation require discovering or training relevant features that accurately characterize the regularities or structures hidden in data. Classifying data of task-related attributes was mostly focused on experts' understanding of the target fields, making it much more difficult for non-experts to adopt ML to their own research. Meanwhile, presumably from training instances, efforts were made to establish a segmentation approach based on defined dictionaries [2,3].

Many disciplines of science are impacted by the theory of parsimony, which states that the simplest interpretation for a data set should indeed be favoured over more complex ones [42]. The effectiveness of this strategy for extraction of features and classification techniques in medical image processing has been demonstrated using sequencing penalization and translation learning [4,5]. Owing to their limited representational power, the extraction of features or dictionary learning approaches published in the literature can find meaningful patterns or consistencies in data with a deep architecture.

As a result, extraction of features has shifted away from humans and toward computers, allowing for more accurate results. The following elements are still largely responsible for the amazing success of deep learning (DL): (a) improvements in rising central processing units and graphical processing units, (b) the accessibility of massive amounts of information, and (c) advances in training algorithms [6,7]. DL outperforms ordinary artificial neural networks (ANNs) in terms of allowing the creation of networks with additional layers (greater than two).

Deep neural networks (DNNs) [8] have the ability to learn hierarchical image representation, which allows them to retrieve higher-level information from lower-level information. Because deep learning approaches allow deep hierarchical representations to be learned simply from data, it has achieved record-breaking performance in various artificial intelligence technologies [9,10] and grand obstacles. Image fusion [11], image segmentation [12], image annotation [13], image registration [14], computer-aided diagnosis [15], microscopic image analysis, and lesion/landmark recognition [16] all have employed DL in medical imaging [17]. When there is a large amount of data accessible during the training stage, DL algorithms are particularly successful.

DL is a sub-branch of ML that involves combining multiple layers of data processing elements to produce deep architectures that retrieve many levels of data abstraction. Although DL was introduced in the 1980s, it has recently obtained prominence for two purposes [18]. Recent advances in ML approaches include the following: Deep architecture training was incredibly difficult and produced bad outcomes before 2006. The challenge stems from the fact that gradient-based optimization, which begins with random weight matrix initialization, typically produces disappointing results [19].

Hinton *et al.* proposed greedy layer-wise learning [20] as a useful strategy for training deep networks. This paper is widely recognized as a turning point in the fascinating subject of ML deep architectures. The architectures of the deep belief network (DBN), convolutional neural network (CNN), deep Boltzmann machine, stacked autoencoder, and recurrent neural network (RNN) have been effectively utilized in various areas, which include medical image analysis, computer vision, machine health monitoring, driverless cars, and natural language processing. The rest of the paper is structured as follows: Section 2.2 briefly presents a literature survey. The overview of DL models is explained briefly in Section 2.3. Section 2.4 explains the proposed work. Preprocessing of the dataset is explained in Section 2.5. Implementation and results are presented in Section 2.6. Section 2.7 concludes the paper.

2.2 Literature survey

A quadratic Bayesian classifier was proposed by Banaie *et al.* The total accuracy of the results was 86.95% correct categorization [21]. Priya *et al.* classified neurodegenerative disorders into three stages: gait problems, cognitive impairment, and retro genesis. Their research attempted to determine the disease severity, and they used a quadratic Bayes normal classifier to do so [22]. Zeng *et al.* used the support vector machine (SV10M) to achieve an accuracy of 86.85% in illness categorization [23,24].

Using an Elman's RNN, an overall classification rate of 87.1% is achieved [25]. Zhao suggested a deterministic learning theory-based technique. The results demonstrated a high level of categorization accuracy of 93.75% [26].

Zeng *et al.* used the empirical mode decomposition method to convert gait rhythm time-series data into intrinsic mode. The area under the curve values of

0.949, 0.900, and 0.934 for PD, HD, and ALS, respectively, reveal satisfactory accuracy in binary classification [27]. Bilgin decomposed the compound force signal for feature determination using discrete wavelet transform “bior2.6” and applied a naive Bayesian classifier to obtain 90.93% accuracy in binary classification.

There have been a lot of studies on neuroimaging, medical, perceptual, and biomarker datasets for aging and Parkinson’s disease in the medical field for years [28]. An aging population is a global phenomenon that affects both developing and industrialized countries. About 1% of individuals over the age of 60 years suffer from neurodegenerative disorders. The significance of detecting this sort of diagnostic medical information is that it may lead to the creation of a forecasting model to classify or identify PD patients and healthy controls, or to assess the stage of the illness.

Kim *et al.* [29] proposed that one of the main obstacles in image categorization problems is achieving good classification. This is mostly owing to the images’ descriptive and discriminative capabilities which generate feature vectors. The classification model is trained based on using the vector features. Investigators have examined ANN to address challenging picture categorization problems for many years. Convolutional neural network [FAST-regional convolution neural network (FAST-RCNN)], a DL approach, has recently demonstrated good results for interpreting picture content in image categorization.

Mamun [30] has proved FAST-RCNN’s better performance in addressing crucial picture classification issues in applications like identification of the traffic sign and outperforming human competence in benchmarking tests. Achieving high classification accuracy is difficult because of the great visual variety between similar classes and very high resemblance between different classes. To prevent the over-fitting problem, the convolutional layer method was employed to limit the number of variables in the FAST-RCNN model. Saha [31] discusses the difficulties of image reduction classification and the feature selection process. The research shows how difficult it is to choose the most discriminant information for constructing a classification model.

2.3 Overview of DL algorithms

Models based on DL: When researching various DL models, it is easy to get caught up in the “buzzwords” that are commonly used and lose sight of the learning procedure’s true goal. The following two categories are the most useful for categorizing model kinds [32,33]:

Discriminant architectures: Patterns are described using these models, which are based on posterior distributions of classes. Techniques like classification and regression can be applied to this. Producing an output for input is the paradigm of discriminant models. Bottom-up networks can be considered discriminant models. Inputs are given and outputs are generated by propagating them up the network. This is the primary distinction between it and its generative equivalent, which has no outputs. Supervised DL can be used to describe CNN, and long short-term memory (LSTM), DNN, and RNNs are examples of these models.

Generative architectures: In these models, unlike classification/regression techniques, there are no values to forecast for the input information. The goal is to learn high-order features by extracting meaningful correlations between features. From training data, generative models will be able to learn a distribution that helps to generate samples. The term "generative" refers to the fact that all these models begin at the upper layer and work their way through the network to provide inputs. Unsupervised feature learning is hence the primary domain of these systems.

2.3.1 Convolutional neural network

It is commonly used to process two-dimensional (2D) data. As an outcome, it is a great choice for data like photos and recordings. A CNN (also referred to as a ConvNet) in image processing can retrieve an image's high-order characteristics as well as provide exceptional content knowledge. When it comes to learning representations of data, convolutional networks have shown to be very effective [34].

To keep things simple, we start with a 1D dataset and then show how the model communicates its efficiency on 2D input. To categorize a dataset, complete these steps: As shown in Figure 2.1, we utilize a basic neural network that connects all of the inputs to a fully connected region, in which every input sampling corresponds to each neuron in the hidden layer.

2.3.2 Recurrent neural network

The most typical application of an RNN is to process data in the form of a sequence. As a result, it is well suited to speech analysis, language processing, and translation. ConvNets have the drawback of accepting with a given number of computing steps the number of levels in the structure and the number of units in each level taking a fixed-size vector as input and returning a fixed-size vector as output to complete this mapping. RNNs differ from other types of neural networks in that they work with sequences of vectors in both the input and output.

Memoryless neural networks (including ConvNets) are used in traditional neural networks. It is unclear how a conventional neural network will categorize the weather by relying on predicted readings. They work with a fixed-size input and output and execute computations with a predetermined number of hidden layers and units [35].

Figure 2.2 depicts how RNNs deal with this problem by adding memory in the form of a loop. An RNN can be thought of as a collection of multiple neural

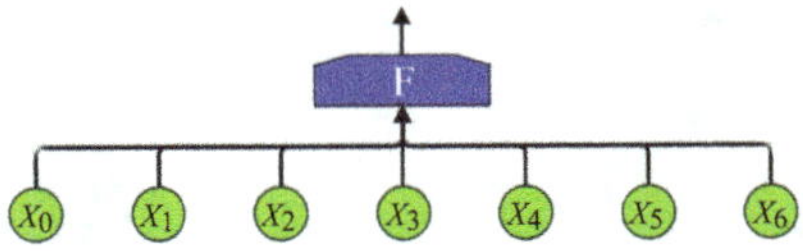

Figure 2.1 CNN's sample input model

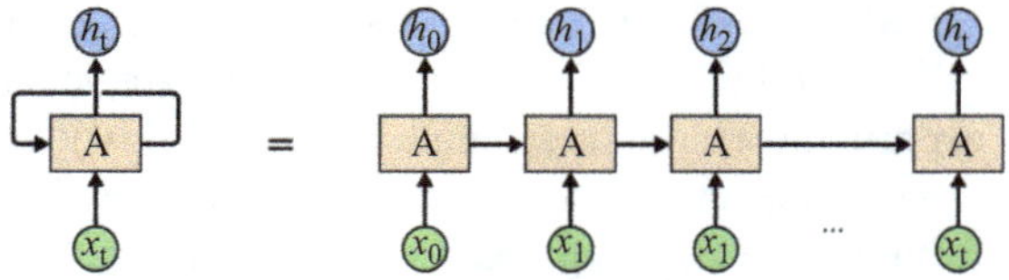

Figure 2.2 Recurrent neural network

networks each with its own set of parameters that are taken from the previous network and function as a memory.

2.3.3 Long short-term memory

LSTM is an advancement over RNNs in addressing the issue of long-term reliance. LSTM models, rather than basic RNNs, are used in most real-world applications [36]. Like RNNs, LSTMs have a chain-like layout (when unrolled). Rather than a single neural network layer in the recurring module, LSTMs feature four neural network layers functioning in perfect harmony, as illustrated in Figure 2.3.

From one node to the next, each line holds a full vector. The neural networks in the yellow boxes have the given activation function. Point-wise operations such as vector addition are shown by the pink circles. Concatenation is indicated by lines merging. Content is cloned to separate locations when lines fork. The horizontal line that runs through the top of the component is the core concept of LSTMs. The line represents a cell state that transmits information from one phase to another. Addition and multiplying gates govern the data saved in the cell state vector. Each of the four layers of the neural network is in charge of a different function in the cell. A cell's operation is divided into three stages: First, the neural network layer from the left determines which data from the state vector will be discarded. The sigmoid layer examines h_{t-1} and x_t and then outputs a value between the range of 0 and 1 for each value in the cell state vector; this information is then transferred to the top line. The value 1 indicates the determination to "totally keep this," whereas 0 indicates the decision to "completely eliminate this." The output of the first layer

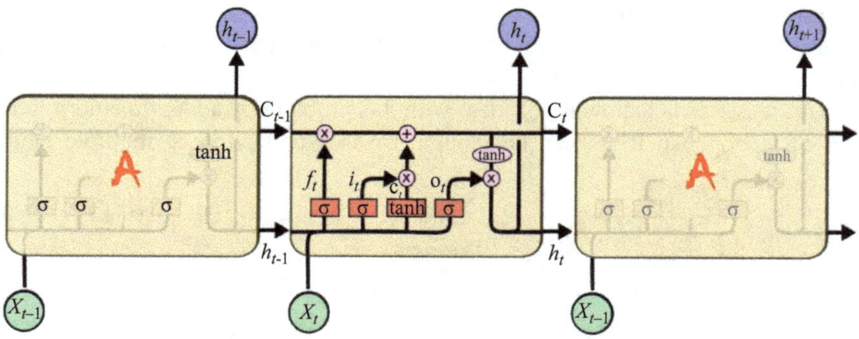

Figure 2.3 Long short-term memory

is shown below as f_t:

$$f_t = \sigma\left(W_f.[h_{t-1}, x_t] + b_f\right)$$

Second, the next two levels determine which data is going to be stored in the cell state vector, and which values will be updated is decided by the sigmoid layer or input gate layer:

$$i_t = \sigma(W_i.[h_{t-1}, x_t] + b_i)$$

On the other hand, the tanh layer produces vectors of new candidate values which could be included in the state:

$$\widetilde{C}_t = \tanh\left(W_C.[h_{t-1}, x_t] + b_C\right)$$

The new cell state vector C_t is computed as follows:

$$C_t = f_t * C_{t-1} + t_i * C_t$$

The final layer of the cell determines output h_t. Both the prior sigmoid layer and the freshly computed cell state vector have an impact on the output value.

$$o_t = \sigma(W_o)[h_{t-1}x] + b_0$$

$$h_t = o_t * \tanh\left(C_t\right)$$

2.3.4 Restricted Boltzmann machine

The restricted Boltzmann machine (RBM) is the first generative architecture that is not the same as the Boltzmann machine. Figure 2.4 displays the distinction and the following section will go over the specifics. An RBM is frequently used in unsupervised learning tasks, including dimensionality reduction, feature learning, and collaborative filtering [37].

An RBM has an input layer and a hidden layer. These two layers are connected haphazardly. An RBM is limited in that no two devices in the same layer can be linked. A Boltzmann machine and an RBM are distinguished by this feature. The former has been around for a long time, but it was not until this recent incident that it became well known.

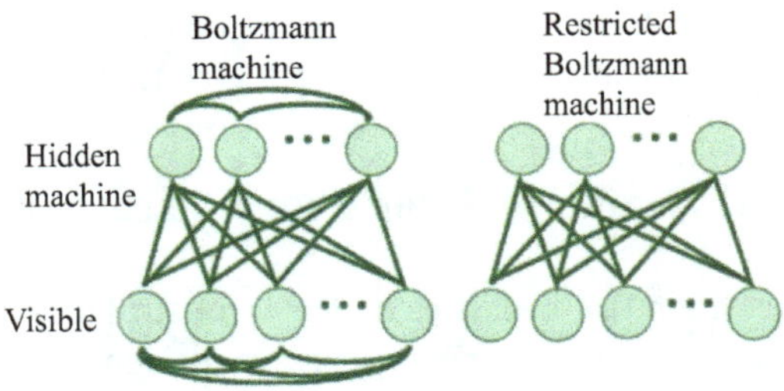

Figure 2.4 Restricted Boltzmann machine

A Boltzmann machine is utterly untrainable without the restriction of having any intra-layer connections, and it essentially folds into chaos. As a result, an RBM is a two-layer neural network with numerous inter-layer connections but no intra-layer connections. Each interconnection is given a weight to sustain during the learning process. An RBM can adapt its variables (hidden layer nodes) to represent the allocation of the training data by adjusting these weights. After training these hidden units, specimens that match the allocation of the training examples can be generated. This method has been used to reimburse for a scarcity of accessible data in various fields.

The training technique for an RBM differs from the methods used for discriminant models in a few ways. While the last step of the procedure involves executing stochastic gradient descent to reduce the error rate, the method used to compute the error differs. Contrastive divergence is a method used in an RBM. Each iteration can be broken down into three steps in the simplest sense. The hidden layer is constructed from the input nodes using probability values that reduce the free energy of the model. Based on that minimization function, a hidden layer with specific activations will be created.

2.3.5 Deep belief networks

For learning a representation of certain input data, DBNs are used. Their aim is quite similar to that of the RBM; however, researchers no longer use RBMs in practice. The DBN can be viewed as the RBM's logical next stage of development. It is the model's next iteration and innovation, and it has acquired wide acceptance as a replacement for the RBM.

Few claimed that DBNs are pointless because RBMs have a representational capacity for every function approximation. Further research has only led to the conclusion that adding an extra layer increases the information acquisition over a shallower one. This means that there is no damage in adding another layer, and from what we can see, it is for detecting higher-level abstractions in the data [38].

A stack of feedforward RBMs makes up a DBN. The theory behind this architecture is that combining a number of simpler models that are learned sequentially provides an effective approach to learning a complicated model.

Learning DBN investigations has dominated various research articles and cannot be sufficiently described within the scope of this discussion. However, in general, learning is given in a greedy-layer-wise fashion. To conclude, all learning occurs at the local level. The network can be trained recursively, and the complexity can be kept under control by employing a greedy-layer-wise method.

2.4 Proposed DL framework for neuro disease diagnosis

DL is a complex algorithm that evaluates higher-level attributes and retrieves them directly from data using a complex neural network design. DL is a technique for lowering the dimensionality of data or reducing the number of features [21]. It is typically used to solve problems using analog input data, such as text documents, pixel images, and video or audio recordings. DL aims to learn feature hierarchies in which higher-level characteristics are produced by the composition of lower-level features.

2.4.1 FAST-RCNN

The FAST-RCNN applies to a huge range of tasks, which include picture classification, visual recognition, and handwritten digit recognition. FAST-RCCN is one kind of multilayer neural network which helps in the minimization of preprocessing of the data for the extraction of patterns from the image pixels [25]. To simplify the learning model, FAST-RCNN models consolidate weights into smaller kernel filters. The proposed architecture of FAST-RCNN for neurodegenerative disease is illustrated in Figure 2.5.

The FAST-RCNN is made up of a CNN that has had its last convolution layers replaced by an "ROI pooling" level and its last FC level modified by two branches: a (K + 1) category soft-max layer branch and a classification bounding box regression branch. The backbone CNN receives the complete image and extracts the features from the final convolution layer. The resulting extracted features are substantially less than the source file size, based on the underlying CNN used. The duration of the underlying CNN, which is typically 16 in the case of a visual geometry group framework, determines this. Since you do not have to input 2,000 area predictions to the CNN each time, "FAST-RCNN" is better than RCNN. Rather, a feature map is formed from the convolutional, which is performed once for each image. Normalization layers, convolution layers, fully linked layers, max-pooling layers, and dense layers are used in the FAST-RCNN [39].

2.4.1.1 Convolution layer

FAST-RCNN is a neural network that treats tiny quantities of the source images as the sublayer of a hierarchical system, similar to a neural network. This layer consists of nerve cells with trainable weights and biases that can be thought of as a collection of convolutional filters. The fully connected layer is the first layer in the hierarchy of layers. The CONV layer is the most important building block in this system because it is responsible for taking the image as input and generating image features from it. Essentially, filtration is a structure of values, referred to as

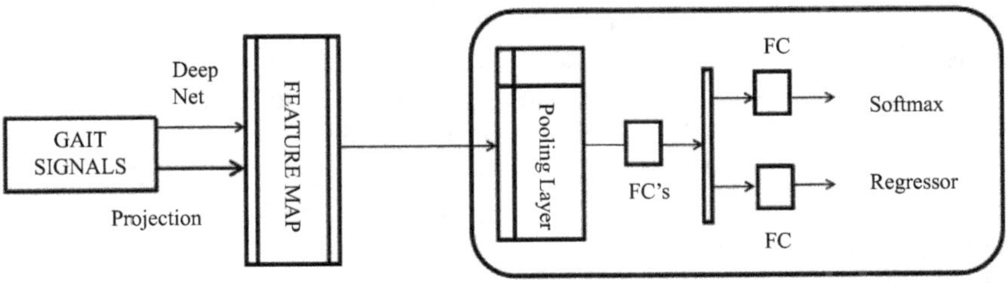

Figure 2.5 The architecture of a FAST-RCNN model used in medical data

weights, which has been provided with training to identify particular aspects. Each filter has spatially tiny size, but it encompasses the entire quantity of the input. When a data point between both the weight training of the nerve cells and a local region or update of an input volume to which they are attached is computed, this layer is called an inner product layer. A 2D initialization plot of the sensor is then generated by convolving each filter all across the thickness and height of the object quantity with interlacing regions [22]. A neuron that makes it look into a tiny window for the information and needs to share variables with neurons in the same activation layer can be construed as an outcome of a neuron. When all fitters all along depth have their activation maps layered, the total output volume is formed from the total output volume. Because even the smallest spots of the source images quantity contain millions of pixels, the amount of correlation weights that must be trained is extremely large. Once this is done, the weight values are decided to share from across input data space, with the one that belongs to a single layer sharing the same weights as the others. This pooling layer contributes to a reduction in the total amount of learnable parameters, resulting in a much more effective model in the end. There are several types of inputs available, including original input volumes and details from other neurons that are fed into the consecutive CONV layers.

2.4.1.2 Max pooling

Between each CONV layer, a pooling layer is frequently placed. The pooling layer's job is to minimize the spatial size of sample representation. This is done to control overfitting and limit the massive amount of network hyperparameters. By using max pooling operation, the pooling layer works independently on each spatial dimension of the inputs and reshapes it spatially. The max pooling function divides the input data from the CONV layer's predecessor into overlapping frames and produces the maximum benefit for each part of the region. With a convolution kernel of 2*2, the nearby elements in the convolution layer output matrices are mixed in the pooling layer [23]. The limit of the input sequence is then modified with this set of multiple adjacent elements, which results in the formation of one aspect in the output data matrix. When inaccuracy backpropagation occurs, it is predicted that the steepest descent signal will only be guided back to specific nerve cells that have contributed to the accumulating consequence during the inaccuracy backpropagation phase.

2.4.1.3 Normalization layer

The normalization layer, also known as the activation layer, is the final layer in a combination of max pooling layers and CONV layers as shown in Figure 2.6. This layer is utilized in batch normalizing at hidden layers and at the input for characteristic scaling. This layer uses an aspect of non-linear activation, including such max $(0,x)$ thresholding at zero, for each individual element. The input is multiplied by all of the neurons' individual weights, and the sum is then simulated using the encoder to see if it is more than or less than the threshold [26]. If it exceeds the threshold, it returns a 1; otherwise, it returns 0. This is true for the activation function of the rectified linear unit (ReLu). Upon the comparison with the threshold, the sigmoid activation method contains a range from 0 to 1 (a fraction). This layer has no effect on the image's size.

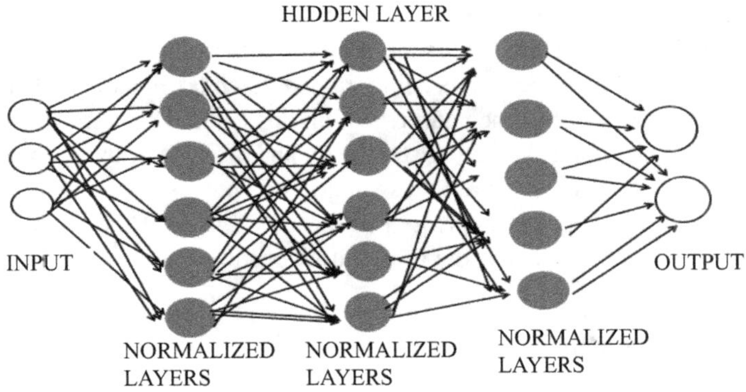

Figure 2.6 Normalization layer in DL

Activation function – ReLu: The ReLu is a fairly basic method that produced better experimental findings. The non-linear function is depicted in the graph below. ReLu specifically described as $f(\max) = \max(0, x)$. For negative values, the value is 0; for positive values, it develops linearly.

Activation function – Sigmoid: The following equation defines the sigmoid function:

$$(x) = \frac{1}{1 + e\hat{}x}$$

- **Sigmoid** has minimal output fluctuations in (0,1) when the input fluctuates in (-,), as shown in the graph below. The function is linear in mathematics. The graph below depicts an example of a sigmoid function. When there are numerous classes in the stochastic process, the SoftMax activation function is typically used.
- **Flatten:** The flatten layer comes right after a succession of pooling layers and convolution layers. Convolution layers provide 3D activation maps; therefore, all that is required is to determine whether an image belonging to a particular classification is the output. By this layer, 3D maps are converted to 1D vectors [27].

2.4.2 Ten fully connected layer

The fully connected layer is the outermost layer of the FAST-RCNN network, which is a basic neural network in which every contribution is connected to every outcome by a trainable weight. The fully connected layer is the outermost layer of a FAST-RCNN network. It is possible to build a fully connected network from the information from all locations and all extract information from the layers beneath by using FAST-computation RCNNs [24]. The nerve cells in this layer will be connected to all the statistics from the volume that came before it. At this layer, we used the sigmoid activation function to create the desired effect.

Drop-out algorithm: In each layer of the FAST-RCNN network, neurons are inhibited at random. This is known as the drop-out algorithm, and it is used to progress the model's performance as it is being trained. Each layer starts with a drop-out map and is of the same size as the neurons within this layer. This is carried out at the beginning of every learning step to register the on and off state of the relevant neuron [28]. During analysis, every neuron is switched on; however, during the training phase, the activation signal is suppressed to the percentage of the average turn-on rate. To increase model accuracy, the drop-out algorithm is used.

2.5 Preprocessing of dataset

Data preprocessing is the method in which information is encrypted in order to transform it into a mathematical state that can be learned by a computer system with ease. Data preprocessing methodologies are a subset of data mining algorithms that are used to generate end products from raw data that has been normalized, has no missing value, and other characteristics [29]. ML and DL tasks, ranging from computation to computer vision, require data preprocessing in order to be completed successfully.

Data preprocessing consists of

- dataset collection
- predefined libraries
- dataset
- handling null values
- categorical data encoding
- dataset splitting
- feature extraction

Dataset collection

The creation of the input data is the initial process in the DL data preprocessing process. Before beginning the building and developing of DL networks, researchers must first gather the necessary data. This particular neurodegenerative dataset was obtained from various sources, including BIOGPS, JPND, ALZFORUM, and NIAGADS AND PHYSIONET. It was used to evaluate the conceptual approach. In order to create a dataset, data will be collected from numerous and different neurodegenerative disease resources, which will be consolidated in the appropriate format to form a single dataset. The formats of datasets vary depending on their intended use.

The Python frameworks that have been predefined can be used to execute specific preprocessing stage tasks. In DL, importing all of the necessary library resources is the second step in the data preprocessing process [31]. Use three essential software packages for this data preprocessing stage in order to determine neurodegenerative disorders using DL algorithms in our case. These three libraries are as follows:

- **NumPy**: NumPy is a Python package that is essential for performing science-based calculations. As a result, it can be used to incorporate each and every type of arithmetic operation into a computer program. It also enables users to include huge complex and multifaceted arrays and matrices in the user's framework, which is particularly useful.

- **Matplotlib**: Matplotlib is a 2D plotting repository written in Python which can be used to visualize any type of chart in the language [30]. It is capable of delivering publishing figures in a variety of physical copy configurations as well as interactive environments across a number of platforms.

Handling null values

It is important to maintain that null values are identified and managed appropriately, which is essential in data preprocessing; otherwise, it creates the risk of marking inaccurate and mistaken conclusions from the data and the information given by data. With regard to dealing with missing data, the following approaches need to be considered.

Deleting a specific row

The data must be deleted and no additional bias must be introduced after the data has been deleted. This technique is beneficial for features that contain quantitative information, such as age, salary, year of birth, and other character traits of a similar nature. It is possible to use this section to compare the average, median, as well as mode of a particular trait, column, or row in order to reconstruct the calculated value for the attribute data in that feature, column, or row. There is the potential for this method to enhance the variability of a dataset while simultaneously preventing any data loss from occurring. Therefore, when compared to the first method, it produces superior results. Using the deviation of values that are close to each other as an approximation is another method of approximation. For linear data, on the other hand, this method is the best method.

Categorical data encoding

When researchers talk about categorical data, researchers are referring to data that is divided into different classifications inside a dataset. There is only one categorical data in our dataset, and that is the country. A huge number of ML models are built around mathematical formulas. You could indeed intuitively know that maintaining the categorical variables in the equation will cause some problems because the formulas will only require numbers, and thus the categorical variable will cause some problems.

Dataset splitting

Data preprocessing in DL continues with the division of the dataset as the next step. To prepare a dataset for an ML model, it is necessary to divide it into two separate sets: a training set and a testing set. In ML, a training set refers to a subset of a dataset that is used to train an ML model. The ML model makes predictions based on the data from the test set. Typically, the dataset is divided into two parts: a 70:30 ratio and an 80:20 ratio. One would then use 70% or 80% of the data for training purposes while discarding the remaining 30% or 20%. When it comes to splitting a dataset, the method used depends on the shape and size of the dataset in question. The dataset splitting ratio achieved for the implementation of the proposed model is 80:20, which means that 80% of the dataset is used for training and 20% is used for testing.

Feature extraction

In DL, feature selection methods signal the completion of the data preprocessing stage. Standardization is a technique used to standardize the significant factors of a data source within a given range of values. In other words, the feature set narrows the range of factors available for comparison, allowing us to compare them on a level-playing field [40].

2.6 Implementation and results

The studies were conducted out in the Anaconda prompt tool with the help of the TensorFlow packages and a Jupyter notebook. Tests are performed on a computer with an Intel Core i5 processor, 8 GB of RAM, and a 64-bit processor [41]. According to the gait speed signals, we conducted the experiments with patients who had diseases classified into three types: Parkinson's disease, Huntington's disease, and Alzheimer's disease. A representation of the dataset considered for disease detection is shown in Figure 2.7, and the data are tabulated in Table 2.1.

Here is an illustration of how the precision estimation for various architectures can be achieved by differing the hidden layer. The precision of hidden layer 3 is indicated in Table 2.2, and the correlating plot is visualized in Figure 2.8.

Figure 2.9 illustrates the precision acquired for hidden layers with variant architectures of CNN, and Table 2.3 lists the accuracy obtained for each architecture.

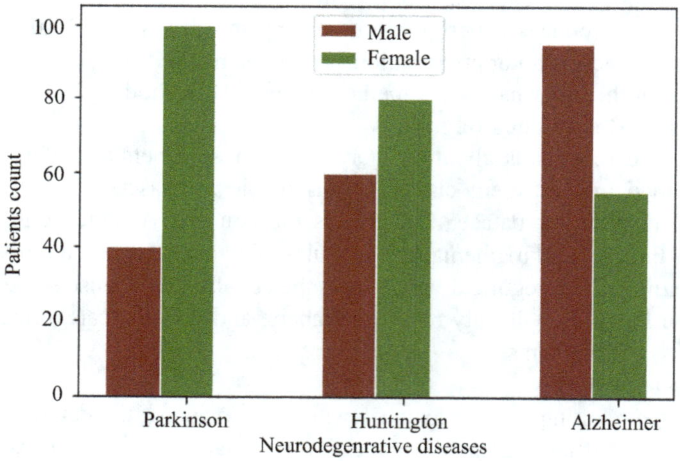

Figure 2.7 Illustration of patient data

Table 2.1 Patient data

Disease	Male	Female	Age
Healthy	35	35	>40
Parkinson	40	100	>50
Huntington	60	80	>60
Alzheimer	95	55	>40

Table 2.2 Determination of accuracy by varying hidden layer 3

Method	**CNN**	**RCNN**	**FRCNN**
Parkinson	78.7	80.35	**85.87**
Huntington	70.5	78.9	**82.23**
Alzheimer	77.41	85.75	**88.67**

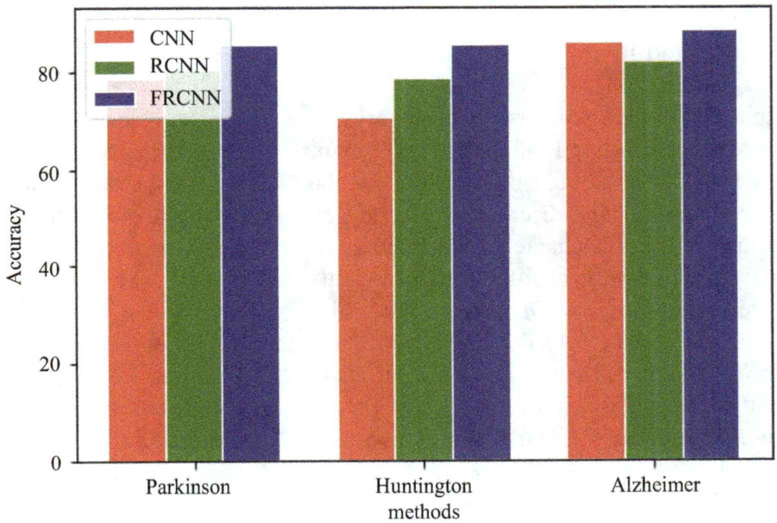

Figure 2.8 Architecture comparison for hidden layer 3

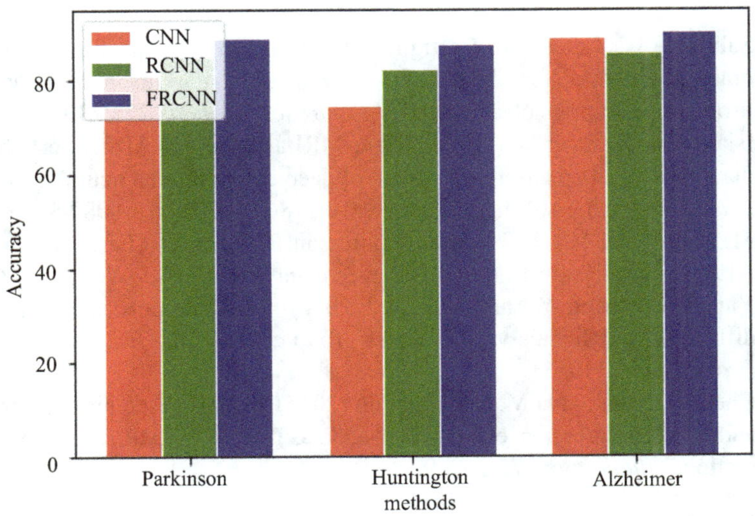

Figure 2.9 Architecture comparison for hidden layer 4

Table 2.3 Determination of accuracy by varying hidden layer 4

Method	CNN	RCNN	FRCNN
Parkinson	80.7	85	**88.78**
Huntington	74.5	82.9	**85.45**
Alzheimer	79.23	87.57	**90.14**

2.7 Conclusion

In various fields, DL, which is a subcategory of artificial intelligence, has made an impact. Medical imaging is among the most significant application fields for DL, in which DL algorithms could be used for the classification of the dataset and prediction of datasets. The discriminative and generative structures of DL are discussed in detail in this chapter. Also discussed is the significance of CNN and its variants in the treatment of neurodegenerative disorders. The use of GPU-automated machines for DL allows medical professionals to analyze images. CNN and its variants, FAST-RCNN, are used to track the progression of disease symptoms in humans. In other words, various CNN algorithms are used to model and predict the possibility of new illness progression due to the patient's current clinical illness state. FAST-RCNN is also used to classify new brain disorders based on the patient's gait features, and various CNN techniques are also used to detect the development of diseases with high accuracy.

References

[1] Salvatore C., Cerasa A., Castiglioni I., *et al.* (2014) "Machine learning on brain MRI data for differential diagnosis of Parkinson's disease and Progressive Supranuclear Palsy," *J. Neurosci. Methods* 222:230–7.

[2] Islam J. and Zhang Y. (2018) "Brain MRI analysis for Alzheimer's disease diagnosis using an ensemble system of deep convolutional neural networks," *Brain Inform.* 5:1–14. https://doi.org/10.1186/s40708-018-0080-3.

[3] Mathew N.A., Vivek R.S., and Anurenjan P.R. (2018) "Early diagnosis of Alzheimer's disease from MRI images using PNN," *Proc. IC4*, pp. 161–6.

[4] Yap P.H., Zhang Y., and Shen D. (2016) "Multi-tissue decomposition of diffusion MRI signals via L0 sparse group estimation," *IEEE Trans. Image Process.* 25:4340–53.

[5] Chen Y., Juttukonda M., Su Y., *et al.* (2015) "Probabilistic air segmentation and sparse regression estimated pseudo CT for PET/MR attenuation correction," *Radiology* 275:562–9.

[6] Vincent P., Larochelle H., Lajoie I., Bengio Y., and Manzagol P.A. (2010) "Stacked denoising autoencoders: learning useful representations in a deep network with a local denoising criterion," *J. Mach. Learn. Res.*11:3371–408.

[7] Ioffe S. and Szegedy C. (2015) "Batch normalization: accelerating deep network training by reducing internal covariate shift." In: *Proc. 32nd Int. Conf. Mach. Learn.*, New York: ACM, pp. 448–56.

[8] Swarna S.R., Kumar A., Dixit P., and Sairam T.V.M. (2021) "Parkinson's disease prediction using adaptive quantum computing," *2021 Third Int. Conf. Intell. Commun. Technol. Virtual Mobile Netw. (ICICV)*, pp. 1396–1401, doi: 10.1109/ICICV50876.2021.9388628.

[9] Sasubilli G.and Kumar A. (2020) "Machine learning and big data implementation on health care data," *Fourth Int. Conf. Intell. Comput. Control Syst. (ICICCS)*, pp. 859–64, doi: 10.1109/ICICCS48265.2020.9120906.

[10] Sasubilli S.M., Kumar A., and Dutt V. (2020) "Machine learning implementation on medical domain to identify disease insights using TMS," *2020 Int. Conf. Adv. Comput. Commun. Eng. (ICACCE)*, pp. 1–4, doi: 10.1109/ICACCE49060.2020.9154960.

[11] Sasubilli S.M., Kumar A., and Dutt V. (2020) "Improving health care by help of internet of things and bigdata analytics and cloud computing," *2020 Int. Conf. Adv. Comput. Commun. Eng. (ICACCE)*, pp. 1–4, doi: 10.1109/ICACCE49060.2020.9155042.

[12] Kotturu P.K. and Kumar A. (2020) "Comparative study on machine learning models for early diagnose of Alzheimer's disease: multi correlation method," *2020 5th Int. Conf. Commun. Electron. Syst. (ICCES)*, pp. 778–83, doi: 10.1109/ICCES48766.2020.9137872.

[13] Dubey A.K., Narang S., Kumar A., Sasubilli S.M., and García-Díaz V. (2021) "Performance estimation of machine learning algorithms in the factor analysis of covid-19 dataset," *Computers Mater. Continua*, 66(2):1921–36.

[14] Chan H.P., Hadjiiski L.M., and Samala R.K. (2020). "Computer-aided diagnosis in the era of deep learning," *Med. Phys.*, 47(5):e218–27.

[15] Van Tulder G. and de Bruijne M. (2016) "Combining generative and discriminative representation learning for lung CT analysis with convolutional restricted Boltzmann machines," *IEEE Trans. Med. Imaging* 35:1262–72.

[16] Chakraborty S. and Mali K. (2020) "An overview of biomedical image analysis from the deep learning perspective," *Appl. Adv. Mach. Intell. Comput. Vis. Object Recognit.: Emerg. Res. Opportun.*, 197–218.

[17] Bhatt C., Kumar I., Vijayakumar V., Singh K.U., and Kumar A. (2021) "The state of the art of deep learning models in medical science and their challenges," *Multim. Syst.*, 27(4):599–613.

[18] Dai Y., Wang G., Dai J., and Geman O. (2020) "A multimodal deep architecture for traditional Chinese medicine diagnosis," *Concurr. Comput. Pract. Exp.*, 32(19):e5781.

[19] Krizhevsky A., Sutskever I., and Hinton G.E. (2017) "ImageNet classification with deep convolutional neural networks," *Commun. ACM*, 60(6):84–90.

[20] Atienza R. (2018) *Advanced Deep Learning with Keras: Apply deep learning techniques, autoencoders, GANs, variational autoencoders, deep reinforcement learning, policy gradients, and more*. Packt Publishing Ltd., pp. 1–10.

[21] Prabhu P., Karunakar A.K., Anitha H., and Pradhan N. (2020) "Classification of gait signals into different neurodegenerative diseases using statistical analysis and recurrence quantification analysis," *Patt. Recognit. Lett.*, 139:10–6.
[22] Zeng W., Yuan C., Wang Q., Liu F., and Wang Y. (2019) "Classification of gait patterns between patients with Parkinson's disease and healthy controls using phase space reconstruction (PSR), empirical mode decomposition (EMD) and neural networks," *Neural Netw.*, 111:64–76.
[23] Zhao A., Qi L., Li J., Dong J., and Yu H. (2018) "LSTM for diagnosis of neurodegenerative diseases using gait data," *Ninth Int. Conf. Graphic Image Process (ICGIP 2017)*, vol. 10615, p. 106155B. International Society for Optics and Photonics.
[24] Grover S., Bhartia S., Yadav A., and Seeja K.R. (2018) "Predicting severity of Parkinson's disease using deep learning," *Procedia Comput. Sci.*, 132:1788–94.
[25] Priya S.J., Rani A.J., Subathra M.S.P., Mohammed M.A., Damaševičius R., and Ubendran N. (2021) "Local pattern transformation based feature extraction for recognition of Parkinson's disease based on gait signals," *Diagnostics*, 11(8):1395.
[26] Zeng W., Yuan C., Wang Q., Liu F., and Wang Y. (2019) "Classification of gait patterns between patients with Parkinson's disease and healthy controls using phase space reconstruction (PSR), empirical mode decomposition (EMD) and neural networks," *Neural Netw.*, 111:64–76.
[27] Lahmiri S. and Shmuel A. (2019) "Detection of Parkinson's disease based on voice patterns ranking and optimized support vector machine," *Biomed. Signal Process. Control*, 49:427–33.
[28] Kim I., Rajaraman S., and Antani S. (2019) "Visual interpretation of convolutional neural network predictions in classifying medical image modalities," *Diagnostics*, 9(2):38.
[29] Saha R. (2019) *Classification of Parkinson's Disease Using MRI Data and Deep Learning Convolution Neural Networks*. Creative Components. 241. https://lib.dr.iastate.edu/creativecomponents/241.
[30] Pavithra M., Rajmohan R., Ananth Kumar T., and Ramya R. (2021) "Prediction and classification of breast cancer using discriminative learning models and techniques," *Mach. Vis. Inspect. Syst.: Mach. Learn. Based Appr.*, 2:241–62.
[31] Mamun K. (2018) *Deep Brain Stimulation: A Light of Hope for Parkinson's Patients,* https://theindependentbd.com.
[32] Noor M.B.T., Zenia N.Z., Kaiser M.S., Al Mamun S., and Mahmud M. (2020) "Application of deep learning in detecting neurological disorders from magnetic resonance images: a survey on the detection of Alzheimer's disease, Parkinson's disease and schizophrenia," *Brain Inform.*, 7(1):1–21.
[33] Albawi S., Mohammed T.A., and Al-Zawi S. (2017) "Understanding of a convolutional neural network," *2017 Int. Conf. Eng. Technol. (ICET)*, IEEE, pp. 1–6.
[34] Dakka J., Bashivan P., Gheiratmand M., Rish I., Jha S., and Greiner R. (2017) *Learning Neural Markers of Schizophrenia Disorder using Recurrent Neural Networks*. arXiv:1712.00512.

[35] Wang T., Tian Y., and Qiu R.G. (2019) "Long short-term memory recurrent neural networks for multiple diseases risk prediction by leveraging longitudinal medical records," *IEEE J. Biomed. Health Inform.*, 24(8):2337–46.

[36] Rastgoo R., Kiani K., and Escalera S. (2018) "Multi-modal deep hand sign language recognition in still images using restricted Boltzmann machine," *Entropy*, 20(11):809.

[37] Kaur M. and Singh D. (2020) "Fusion of medical images using deep belief networks," *Cluster Comput.*, 23(2):1439–53.

[38] Zhao A., Qi L., Dong J., and Yu H. (2018) "Dual channel LSTM based multi-feature extraction in gait for diagnosis of neurodegenerative diseases," *Knowledge-Based Syst.*, 145:91–7.

[39] Gokulan S., Narmadha S., Pavithra M., Rajmohan R., and Ananthkumar T. (2020) "Determination of various deep learning parameter for sleep disorder," *2020 Int. Conf. Syst. Comput. Autom. Netw. (ICSCAN)*, IEEE, pp. 1–6.

[40] Narmadha S., Gokulan S., Pavithra M., Rajmohan R., and Ananthkumar T. (2020) "Determination of various deep learning parameters to predict heart disease for diabetes patients," *2020 Int. Conf. Syst. Comput. Autom. Netw. (ICSCAN)*, IEEE, pp. 1–6.

[41] Rajmohan R., Pavithra M., Ananth Kumar T., and Manjubala P. (2021) "Exploration of deep RNN architectures: LSTM and Gru in medical diagnostics of cardiovascular and neuro diseases," *Handbook of Deep Learning in Biomedical Engineering and Health Informatics*, Apple Academic Press, pp. 167–202.

[42] Wang L., Chen K.C., Gao Y., *et al.* (2014) "Automated bone segmentation from dental CBCT images using patch-based sparse representation and convex optimization," *Med. Phys.*, 41:043503.

[43] Usharani S., Prithivi S., Sharmila S., Bala P.M., Ananth Kumar T., and Rajmohan R. (2021) "Mobile Application for Doctor Appointment Scheduling," *2021 Int. Conf. Syst. Comput. Autom. Netw. (ICSCAN)*, pp. 1–6, doi: 10.1109/ICSCAN53069.2021.9526398.

Chapter 3

Intelligent approach for network intrusion detection system (NIDS) utilizing machine learning (ML)

Shubham Sharma[1], Pronika Chawla[2], Naincy Chamoli[3], Disha Pahuja[2] and Maanya Mocha[2]

In this vociferous twenty-first century, we are surrounded by technology which has a huge impact on the lives of every generation from a small kid to an old-aged man, all are busy using technology. This amount of human involvement in technology motivates rapid advancement in technology where researchers and engineers put their all efforts into solving problems that will impact the lives of billions of users. These problems are either conspicuous or abstract. Any new advancement for conspicuous problems can be noticed by end-users but the advancement for abstract technology may not be noticed by end-users with such ease but they also play a major role in solving problems that need to be addressed. The diverse concept of future virtual systems, which includes various access systems, recurrence groups, and cells – all of which include inclusion areas in order to help remote administrators with the task of organizing and transmitting. Artificial intelligence (AI) and machine learning (ML) can help remote administrators to beat these difficulties by examining the geographic data, designing boundaries and memorable information to figure the pinnacle traffic, asset use and application types, enhancing and calibrating organize boundaries for limit extension and dispense with inclusion gaps by estimating the obstruction, and utilizing the between site separation data fifth-generation (5G) can be a key empowering influence to drive the ML and AI incorporation into the system edge. The figure underneath shows how 5G empowers synchronous associations with different Internet of Things (IoT) gadgets creating gigantic measures of information. The coordination of ML and AI with 5G multi-get to edge processing empowers remote administrators to offer an elevated level of robotization from

[1]Tata Consultancy Services
[2]MRIIRS
[3]UTU

the conveyed ML and AI engineering at the system edge, application-based traffic guiding and accumulation across heterogeneous access systems, dynamic system cutting to address shifted use cases with various quality of service necessities and ML/AI as an administration offering for end clients. In this chapter, we address all 5G technology, challenges associated with 5G technology, and the role of AI and ML in 5G technology. IoT will become more fruitful with AI, ML and 5G technology. All these technologies work in an interconnected system over a network because of this their reliability on the network increase. Therefore, in this chapter, we will also discuss how network intrusion detection can be done with the help of AI in a much faster way with better accuracy.

3.1 Introduction

The escalation of people working, learning and socializing on the web is at an outright top as we explore the current COVID-19 pandemic. Presently, like never, it is progressively unmistakable how indispensable availability, correspondence and solidarity are despite misfortune. Building superfast web associations, fibre to the reason and fifth-generation (5G) wireless communications technologies can make more extensive the associated networks and economies across neighbourhoods, urban areas, nations and the world. 5G can change our lives, including giving us greater limits and availability for huge quantities of gadgets – and specifically extends our compass to new, energizing advancements including augmented reality, the Internet of Things (IoT), and artificial intelligence (AI). With 5G, you can get to better than ever encounter including moment admittance to cloud administrations, broad cloud gaming, consistent video correspondence, increased augmented reality shopping and significantly many more things. With 5G we truly can 'plan for an impressive future' with the office for additional individuals to the interface on the double, more downloads and monstrous organization limit.

The IoT – the organization associated with 'smart' gadgets that convey effortlessly is changing how we live and work. At ranches, remote IoT sensors can communicate data about soil dampness and supplements to horticultural specialists the nation over. IoT caution frameworks, outfitted with batteries that keep going for quite a long time, and give property holders long haul security. Wearable wellness gadgets – for the two individuals and pets – can screen action levels and give input on pulse and breath. Even though these applications fill various needs, they all offer one trademark: reliance on solid connectivity [1]. Connectivity and computing ways to deal with the influence of the IoT from an information and communication point of view. With the ascent of the IoT, there is clearly a torrential slide of data being added to the organizations and data handling, stockpiling and investigation foundation of this advanced world. They are identified with security issues and mean unapproved admittance to get profoundly classified data. A data breach is a major issue that can hurt the operational cycles of any business, just as contrarily swaying the general standing, and causing a monetary burden. There are various sorts of data breaches, like taken data, stolen information, ransomware, phishing, password guessing, viruses, malware, denial-of-server attacks etc. Considering that practically all the data that organizations have is put away on the web, online protection is a consuming issue that should be tended to conveniently and successfully. In addition, it is obligatory to upgrade the security level to forestall possible dangers and ensure your secret data.

Network security shields your network and information from penetration, intrusions and different dangers [2]. This is a tremendous and all-encompassing term that portrays equipment and programming arrangements just as cycles or rules and setups identifying with network use, openness and generally speaking danger insurance. Network security includes access control, infection and anti-virus programming, application security, network examination, kinds of network-related security (endpoint, web, remote), firewalls and VPN encryption, and limitless options are present from there. A network-based intrusion detection framework (NIDS) recognizes noxious traffic on a network. NIDS for the most part requires indiscriminate network access to investigate all traffic, including all unicast traffic. NIDS are latent gadgets that do not meddle with the traffic they screen [3]. The NIDS sniffs the inner interface of the firewall in read-just mode and sends alarms to a NIDS management worker by means of an alternate (i.e., read/compose) network interface. NIDS track traffic to and from all network devices and are located at a strategic point or points within the network. It plays out an investigation of passing traffic on the whole subnet and matches the traffic that is given to the subnets to the library of known assaults. When an assault is recognized, or strange conduct is detected, the alarm can be shipped off the executive. NIDS, for example, will be installed on the subnet where firewalls are located to check whether anyone is trying to breach the firewall. Attacks on network security are basically actions against IT assets that are targeted to destroy, modify or steal sensitive information. In the current scenario, data is very vulnerable as many enterprises allow their employees to access data from their mobile/cellular devices, leading to theft or complete destruction of data.

Common cyber-attack types are as follows:

- Denial-of-service (DoS) and distributed denial-of-service (DDoS) attacks
- Man-in-the-middle (MitM) attack
- Phishing and spear-phishing attacks
- Password attack
- Eavesdropping attack
- Malware attack

3.1.1 DoS and DDoS attacks

In a DoS attack, a system is sent multiple requests, overwhelming the server, which makes it difficult for the system to respond to the requests. A DDoS attack is also similar in the sense that it also attacks the system's resources, but the difference is that instead of a single machine the requests are launched from a large number of host machines that are controlled by the attacker.

3.1.1.1 Ping of death attack

In this attack, Internet protocol (IP) packets are used to 'ping a target system' with an IP size over 65,535 bytes. As IP packets of this size are not allowed, the attacker fragments the IP packet. Now once the target system reassembles the packets, it then may experience buffer overflows and other crashes.

3.1.1.2 Botnets

Botnets usually refer to the systems infected with malicious software which are under the hacker's control to perform DDoS attacks. These may also be referred to as zombie systems and are mostly used to carry out attacks. This makes it difficult to locate these systems in a DDoS attack as these bots are usually located in varying geographic locations.

3.1.2 Man-in-the-middle (MitM) attack

A man-in-the-middle attack or MitM attack is when an attacker positions itself between the communication of two systems (such as a client and a server) and thus can capture their communication for its own benefit.

Common MitM attacks are as follows:

Session hijacking: In this attack, the attacker basically tries to hijack a session between a client and server. The IP address of the client is substituted with the IP of the attacking server. The communication with the server continues with this false IP.

IP spoofing: IP spoofing is used to convince a system that it is communicating with a trusted system, eventually providing access to the system to the attacker. The attacker then sends a packet with the IP address of a trusted host so that the target might accept the packet and then act upon it.

Replay: This occurs when the attacker intercepts a communication and saves old messages and then tries to send them later; basically, the attacker tries to impersonate one of the participants.

3.1.3 Phishing and spear-phishing attacks

In a phishing attack, people are sent emails that appear to be from legitimate sources and often try to gain their personal information of the people. They usually contain deals that are too good to be true and often attract people and trick them to give their private data such as credit/debit card information, social security number etc. **Spear phishing** is a targeted phishing activity. In this, an email that is specifically targeted towards a person is sent to him, which makes it easier for the people to follow along with it as it appears to be genuine.

3.1.4 Password attack

As passwords are the most common method of authenticating users, it is quite common and effective to try and crack the password to gain access to the data. The passwords can be easily acquired by 'sniffing' which is looking around the person's desk, social engineering or brute force or dictionary attacks.

- In a brute force attack, the attacker tries to get access to the system by guessing the password.

- In a dictionary attack, a dictionary of common passwords is used to attempt to gain access to the target system.

3.1.5 Eavesdropping attack

An eavesdropping attack occurs when network traffic is intercepted. By this, the attacker may gain access to the passwords, credit/debit card information and other sensitive data.

3.1.6 Malware attack

Malware is basically malicious software that may get installed in the system without the consent of the user. This happens when the user tries to download some software or application via the Internet, or untrusted resources. To sum up, the one question that arises is why network security is so important and why it has become an important parameter for any form of communication. There has been a significant uptick in Internet-based network attacks because of new technologies, hacker intelligence and usable resources. Network security is critical for both personal and business networks. Any households with high-speed broadband have one or two wireless routers, which can be hacked if not properly protected. These cyber-attacks have two major consequences for businesses. To begin with, they could cause significant financial harm to the organization as well as data loss. Second, they could lose more money because of the damage to their image.

3.2 Related work

This section discusses some recent accomplishments in this area [4]. The importance of network security has risen dramatically, and many devices have become aware of the need for network security. The most widely used methods are NIDS. Although new challenges can be more dangerous, many master dynamic plans are expected. These technologies complete the job by creating a profile of a typical Internet user. Then, using this profile, continuously monitor device activity for malicious activity. When the framework senses an irregularity, hazard or irregular fluctuation in traffic qualities or properties, it takes the necessary steps, such as raising an alarm. The primary goal of IDS is to detect whether a malefactor is attempting to disrupt the functioning of a system. Hosts, servers, IoT computers, routers and other intermediary devices are examples of systems. Anomaly-based and signature-based structures are the two types of NIDS. Signature-based systems are the most common form of commercial NIDS, although anomaly-based systems are still under development. Human operators were almost entirely responsible for detecting network threats in the past. They foresaw network irregularities in front of consoles, based on their extensive experience. With the rapid development in network capacity, this mission gradually required significant speed and accuracy upgrades. AI, a progressive and exciting computer science branch, and one of its subfields, ML, where the key concept is learning from results, is one suggested way to accomplish this. The authors of this paper would attempt to include a general overview of AI algorithms, with a focus on their usage for network intrusion detection.

AI is basically the study of mental faculties with the help of computational models. The main fields of AI research are ML, natural processing language, computer vision, robotics, reasoning, knowledge representation and general intelligence [5]. ML deals with forming algorithms that can learn from the provided datasets. These algorithms work by constructing models by providing them input and then using the existing models to predict the outcomes. The use of such algorithms makes them the perfect candidates for tasks such as intrusion detection. Nonetheless, there are other methods that do not use ML algorithms. The main areas where ML is usually applied involve classification, regression and clustering. Classification is used for identifying group membership; regression is used for predicting response and clustering is the set of inputs further divided into groups such that the group members have homogenous characteristics [6].

We also observed that once an algorithm has learned from the provided data, it can be used to predict the outputs of other similar data. This happens in the following manner: First, the given input is read and classified, to start the training phase of the data. Then the specific algorithm is applied to the dataset. In the third step, the output is predicted using the already available datasets, and all the results are gathered. The most common use of this is to track the heart rate of patients. Also, the performance of an ML algorithm is calculated using its F-score. This, however, is not enough on its own for IDSs. This also implies that it is necessary to apply a layered approach for security. This can be helpful in a case if a layer does not detect an intrusion, the other layer might detect it. This kind of layered approach may also help in cases when the first layer detects some intrusions, it can pass that data to the other layers. In some other algorithms, we can use the library scikit-learn, which is a robust ML library for Python which is built upon other libraries such as NumPy, SciPy and Matplotlib. The scikit-learn library provides us with some of the most important algorithms and provides ways to calculate the F-score [7].

The development of a NIDS using ML and deep learning (DL) methods usually includes three main steps: (i) data pre-processing, (ii) training and (iii) testing. For any of the possible solutions, the dataset is first pre-processed to convert it into a form that the algorithm can understand. Encoding and normalization are usually involved in this step. Cleaning the dataset, such as deleting entries with missing data and redundant entries, is often necessary, and this is done during this process. After that, the pre-processed data is uniformly split into two parts: the training dataset and the testing dataset. Traditionally, the research dataset accounts for about 80% of the total dataset size, with the remaining 20% constituting the study dataset. After the model has been conditioned, it is validated with the testing dataset and analysed based on its predictions. The network traffic instance will be expected to belong to either the normal or attack category in NIDS models. The performance of AI-powered NIDS is strongly dependent on training with a viable dataset. To improve the results of ML models, the model can be trained with a limited dataset. However, unless the dataset is named in nature, ML is not optimal for larger datasets. For larger databases, DL approaches are chosen because labelling is a costly and time-consuming operation. From raw datasets, these methods will learn and find relevant patterns. To make NIDS effective at detecting zero-day attacks, it must be updated daily with new data collected from network traffic monitoring. Because of the huge dataset and the deep design of the DL algorithms, the learning process can use more computing resources and time. The more the NIDS model is practised, the better it will be at detecting anomalies.

3.3 Cloud computing

Cloud computing is a service delivery model where it is classified into three categories: Infrastructure-as-a-service (IaaS), platform-as-a-service (PaaS) and software-as-a-service (SaaS). Cloud computing changes the way we are accessing and using computing services. It provides ubiquitous and on-demand service delivery of computing services and is charged based on usage. Cloud resources can be utilized in an efficient manner to implement this framework. Cloud computing has lots of advantages like available on-demand, scalability, rapid provisioning, elasticity and many more, but with all these advantages it also has a few disadvantages like latency and security because after all we are using someone else's infrastructure for our work. Therefore services should be availed only from trusted cloud providers. This NIDS framework can be put on the cloud so that every inbound traffic should pass through this framework, and it helps to find whether the traffic is harmful or not. By putting this framework over the cloud, we can easily scale up and scale down it on demand and link this with multiple services, applications and devices because of which we do not have to design this framework individually for each application. As discussed, cloud security and latency are two of the limitations of cloud computing because of which, maybe, this is not useful for the real-time applications like IoT devices or for confidential, red-classified information but this can also be mitigated using fog computing where an additional layer on cloud is added which is nearer to the edge devices [8].

As shown in Figure 3.1, for Organization B, all the inbound traffic is configured with this network intrusion detection framework then the framework has the capability to classify normal and anomaly in inbound traffic. On that basis if traffic is normal then it can hit the Org B server, otherwise, an error message will pop up on the screen of 'Access Denied'. In Figure 3.1, let us suppose there are two people in Org A and one has malicious intent who is using a different network, otherwise, the other one is just doing his work. If both hit the Org B network then the person having malicious intent will not be able to access the servers of Org B because of this framework.

3.3.1 Machine learning

ML is one way of making a machine intelligent by feeding processed data over algorithm and those algorithms have the capability of learning from an experience as a human does, and once the machine learns from the previous data, it will be able to recognize, classify and recommend things based on the requirement [9]. For this framework, we used ML to predict anomalies in inbound traffic within no time and inbound traffic with anomalies is not allowed to access the network. As discussed

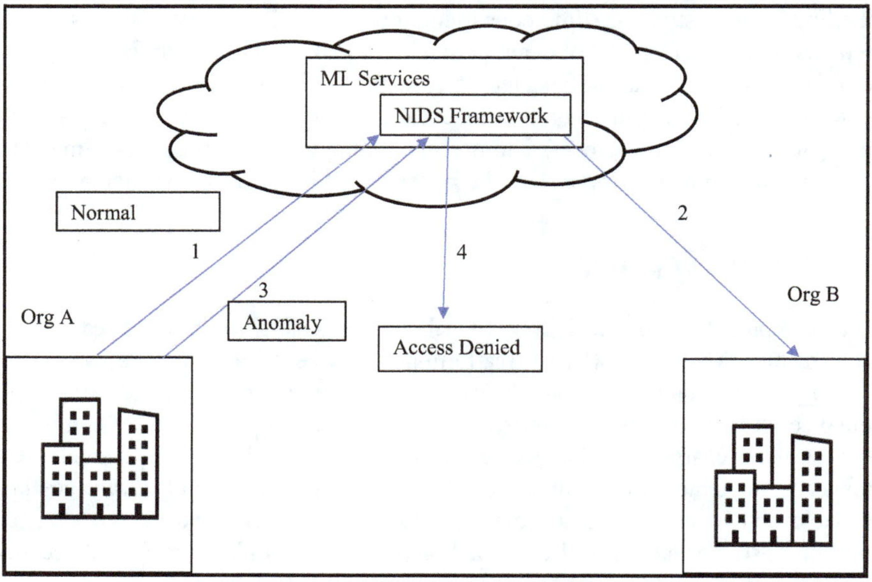

Figure 3.1 Implementation of NIDS framework over the cloud

before, for training an ML model, we need to feed data into ML algorithms. Therefore, we are also using one dataset for this.

3.3.2 Exploratory data analysis

The local area of network was engaged in a real-world environment and impacted by various assaults. An association is an arrangement of transmission control protocol (TCP) parcels beginning and finishing eventually span between which information streams to and from a source IP address to an objective IP address under some distinct convention. Likewise, every association is named as one or the other ordinary or as an assault with precisely one explicit assault type. Every association record comprises around 100 bytes. For every TCP/IP association, 41 quantitative and subjective elements are gotten from ordinary and assault information (3 subjective and 38 quantitative elements). The class variable has two classifications: Normal and anomaly (Figure 3.2).

From Figure 3.3 it can be concluded that the maximum number of TCP protocol type is around 81.3% and then the UDP protocol type is around 12.1% and the ICMP protocol type is around 6.56% (Figures 3.4 and 3.5).

Feature engineering is one of the most important areas in machine learning because each feature plays a different role in this, where if we missed any important feature and use only the least important features to predict an anomaly, then maybe some normal traffic can also be predicted as an anomaly. Feature importance is shown in Figure 3.6 for the dataset we are using to build this NIDS framework (Figure 3.7).

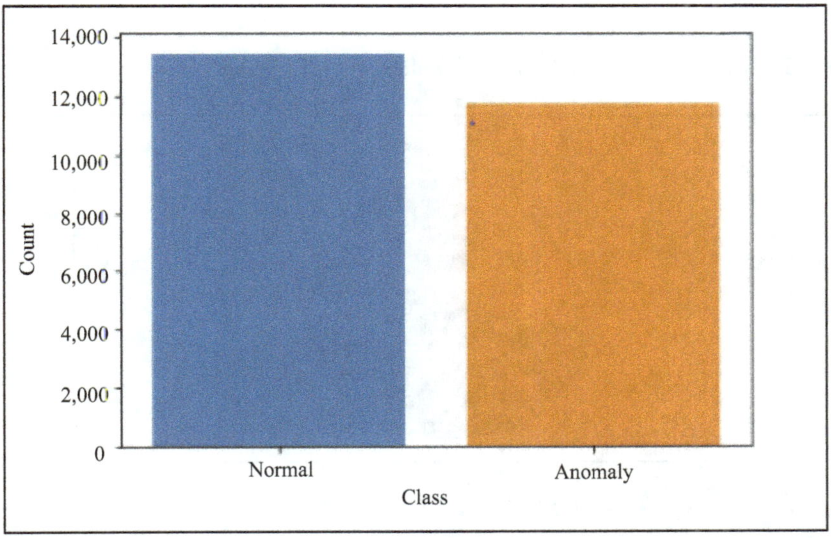

Figure 3.2 Count of class classification

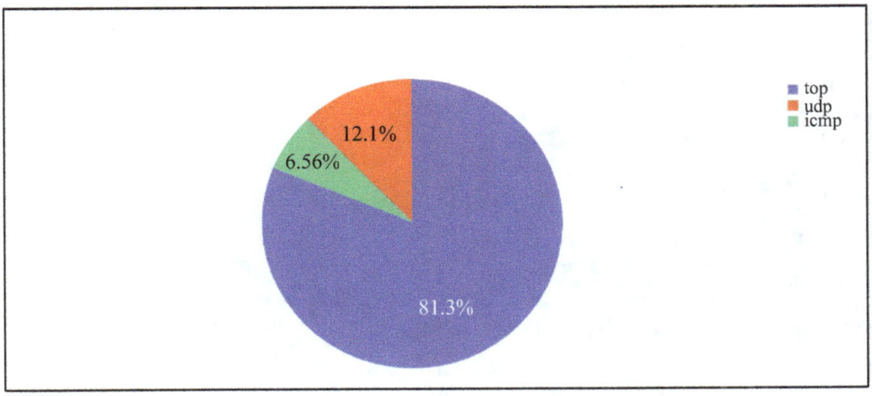

Figure 3.3 Pie chart of protocol type

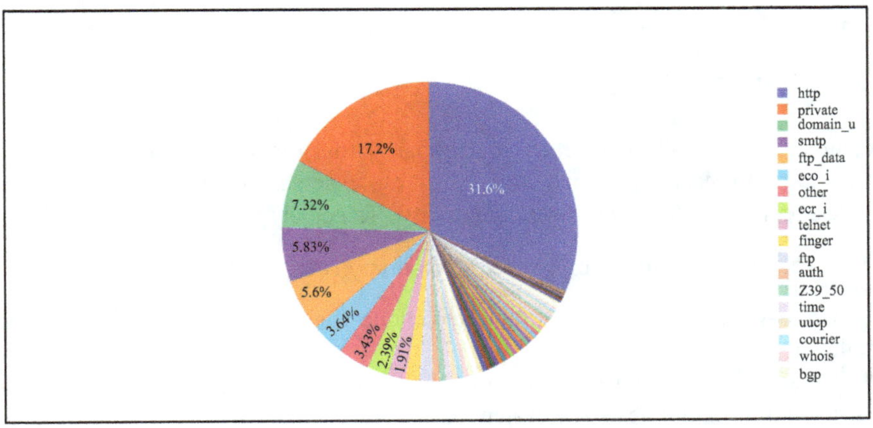

Figure 3.4 Pie chart of service

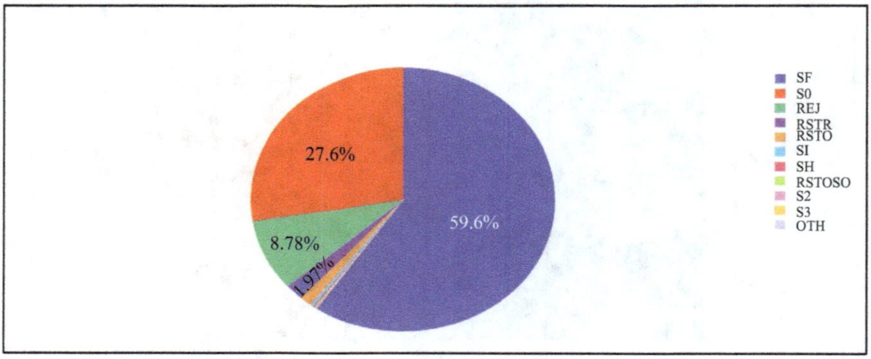

Figure 3.5 Pie chart of flag

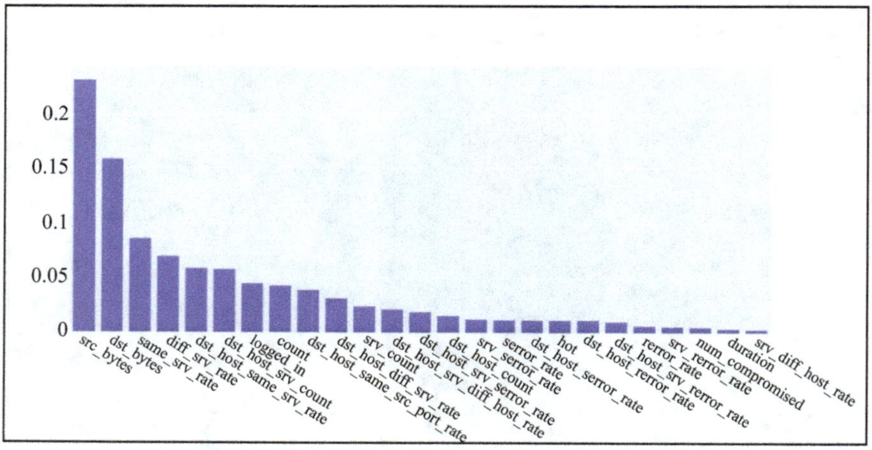

Figure 3.6 Feature importance

3.4 Results

Accuracy is the level of closeness to genuine worth. It is how much an instrument or cycle will rehash a similar worth. All in all, exactness is the level of veracity, while accuracy is the level of reproducibility (Table 3.1).

$$\text{Accuracy} = \text{Number of correct predictions}/\text{Total number of predictions}$$

Cross-validation is a measurable technique used to gauge the ability of AI models.

Cross-validation is a resampling methodology used to assess AI models on a restricted information test. The system has a solitary boundary called k that alludes to the number of gatherings that a given information test is to be parted into. Thus,

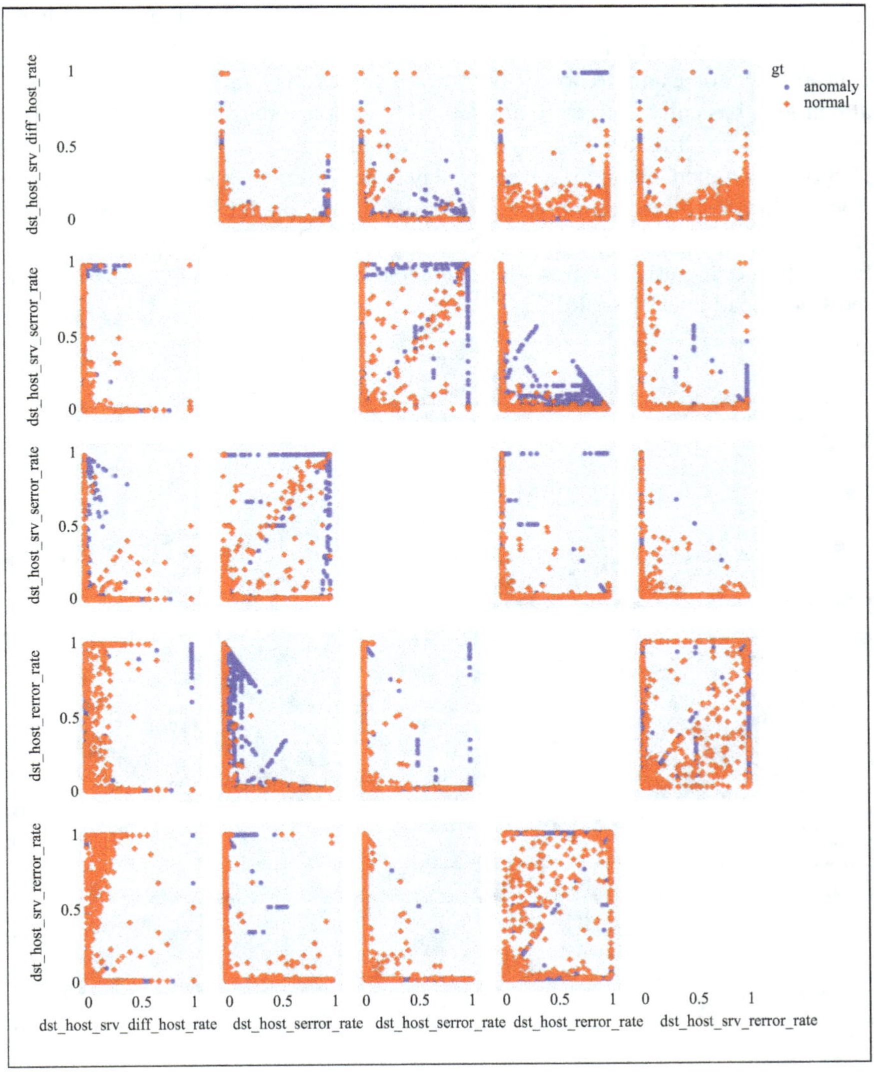

Figure 3.7 Bivariate distribution of numeric feature pair

Table 3.1 Accuracy result of various machine learning algorithm

ML algorithm	Accuracy (%)
Logistic regression	97.27
K-nearest neighbour	99.36
Support vector machine	97.62
Decision tree	99.43
Random forest	92.37

the method is frequently called k-overlap cross-validation. At the point when a particular incentive for k is picked, it could be utilized instead of k in the reference to the model; for example, k=10 becomes a ten-overlay cross-validation. Cross-validation is basically utilized in applied AI to appraise the ability of an AI model on concealed information. That is, to involve a restricted example to appraise how the model is relied upon to act every day when used to make expectations on information not utilized during the preparation of the model. For this sample test, we use the value of k as 5, and Figures 3.8 and 3.9 show the classification report after each fold. Figure 3.10 shows the average evaluation metrics over five folds for precision, recall and F1-score.

```
=====Evaluation of fold0 starts=====
=====Classification Report=====
              precision    recall  f1-score   support

           0       1.00      1.00      1.00      1872
           1       1.00      1.00      1.00      2159

    accuracy                           1.00      4031
   macro avg       1.00      1.00      1.00      4031
weighted avg       1.00      1.00      1.00      4031

=====Evaluation of fold0 finishes=====

=====Evaluation of fold1 starts=====
=====Classification Report=====
              precision    recall  f1-score   support

           0       1.00      0.99      1.00      1894
           1       0.99      1.00      1.00      2137

    accuracy                           1.00      4031
   macro avg       1.00      1.00      1.00      4031
weighted avg       1.00      1.00      1.00      4031

=====Evaluation of fold1 finishes=====

=====Evaluation of fold2 starts=====
=====Classification Report=====
              precision    recall  f1-score   support

           0       1.00      0.99      1.00      1893
           1       0.99      1.00      1.00      2138

    accuracy                           1.00      4031
   macro avg       1.00      1.00      1.00      4031
weighted avg       1.00      1.00      1.00      4031

=====Evaluation of fold2 finishes=====
```

Figure 3.8 Classification report for Fold 0, Fold 1 and Fold 2

```
=====Evaluation of fold3 starts=====
=====Classification Report=====
              precision    recall  f1-score   support

           0       1.00      0.99      1.00      1878
           1       0.99      1.00      1.00      2152

    accuracy                           1.00      4030
   macro avg       1.00      1.00      1.00      4030
weighted avg       1.00      1.00      1.00      4030

=====Evaluation of fold3 finishes=====

=====Evaluation of fold4 starts=====
=====Classification Report=====
              precision    recall  f1-score   support

           0       1.00      1.00      1.00      1841
           1       1.00      1.00      1.00      2189

    accuracy                           1.00      4030
   macro avg       1.00      1.00      1.00      4030
weighted avg       1.00      1.00      1.00      4030

=====Evaluation of fold4 finishes=====
```

Figure 3.9 Classification report for Fold 3 and Fold 4

```
=====Average evaluatin metrics over 5 folds=====
=====Metrics Macro=====
Precision = 0.9962987509772029 | Recall = 0.9960304507629921 | F1-score = 0.9961597875703376
=====Metrics Micro=====
Precision = 0.9961792263801691 | Recall = 0.9961792263801691 | F1-score = 0.9961792263801691
=====Metrics Weighted=====
Precision = 0.9961876816931637 | Recall = 0.9961792263801691 | F1-score = 0.9961786846657901
```

Figure 3.10 Average evaluation metrics over five folds

References

[1] Li, S., Da Xu, L., and Zhao, S. (2018). 5G Internet of Things: a survey. *Journal of Industrial Information Integration*, 10, 1–9.

[2] Burton, J. (2003). Introduction to intrusion detection systems. In: *Cisco Security Professional's Guide to Secure Intrusion Detection Systems* (pp. 1–38). Rockland, MA: Syngress Publishing.

[3] Rubin, S., Jha, S., and Miller, B.P. (2004). Automatic generation and analysis of NIDS attacks. *20th Annual Computer Security Applications Conference*, pp. 28–38, doi: 10.1109/CSAC.2004.9.

[4] Ahmad, Z., Shahid Khan, A., Wai Shiang C., Abdullah, J., and Ahmad F. (2021). Network intrusion detection system: A systematic study of machine learning and deep learning approaches. *Transactions on Emerging Telecommunications Technologies*, 32(1), e4150.

[5] Bhatia, M., Sharma, S., Hooda, M., and Debnath, N.C. (2020). A machine learning approach toward meal classification and assessment of nutrients value based on weather conditions. In: *Big Data Analytics and Intelligence: A Perspective for Health Care*. Emerald Publishing Limited.

[6] Stampar, M. and Fertalj, K. (2015). Artificial intelligence in network intrusion detection. In: *2015 38th International Convention on Information and Communication Technology, Electronics and Microelectronics (MIPRO)*, IEEE, pp. 1318–1323.

[7] Repalle, S.A. and Kolluru, V.R. (2017). Intrusion detection system using ai and machine learning algorithm. *International Research Journal of Engineering and Technology (IRJET)*, 4(12), 1709–1715.

[8] Bhatia, M., Sharma, S., Bhatia, S., and Alojail, M. (2020). Fog computing mitigate limitations of cloud computing. *International Journal of Recent Technology and Engineering (IJRTE)*, SCOPUS.

[9] Sharma, S. and Chamoli, N. (2021) Real-time patient health monitoring. In *Computational Intelligence and Predictive Analysis for Medical Science: A Pragmatic Approach* (pp. 193–216). Berlin: De Gruyter.

Chapter 4

Classification methodologies in healthcare

Adri Jovin John Joseph[1], *Ferdin Joe John Joseph*[2], *Oswalt Manoj Stanislaus*[3] *and Debashreet Das*[4]

Predictive analytics in healthcare has attained peculiar growth due to the advent of machine learning (ML) techniques. ML and the capability to handle veracious and voluminous datasets produced from different sources contribute to the same. The data is acted upon by various techniques to get it cleansed, normalized, and converted into a form that can be inferred. Once the data is ready for inference, it needs to be classified to get relevant results. On this ground, the classification of medical data is essential for the diagnosis of medical conditions and for treating them appropriately. From a narrow perspective, it could be understood that the most used classification methodologies are logistic regression, naïve Bayes, stochastic gradient descent, K-nearest neighbor, decision tree, random forest, support vector machine, and kernel approximation. Many variants in classification methodologies have also been introduced in recent days by hybridizing them with other optimization techniques. This chapter provides a brief insight into the application of classification algorithms and their relevant variants in classifying diverse medical data based on requirements. This chapter also recommends which classification method shall facilitate a good diagnosis and decisive results in healthcare scenarios. It also provides information on how voluminous data are handled by healthcare applications, which are acted upon by ML approaches. It also discusses the pros and cons of using ML approaches in medical diagnosis. It is intended to discuss specific case studies derived from various reputed research organizations so that the reader shall receive a better understanding of the algorithms discussed in this chapter.

[1]Department of Information Technology, Sri Ramakrishna Institute of Technology, India
[2]Faculty of Information Technology, Thai-Nichi Institute of Technology, Thailand
[3]Department of Computer Science and Business Systems, Sri Krishna College of Engineering and Technology, India
[4]Department of Computer Science and Engineering, Karunya Institute of Technology and Sciences, India

4.1 Introduction

Artificial intelligence (AI) has become a new game-changer in most industries today. The healthcare domain is no exception to the intervention of AI. Healthcare has become data-driven, and all the data extracted from the patients can be stored in large storage. These data carry essential information about the status of the health of patients. They are highly voluminous and veracious. These data may be from different devices using different calibrations as well. Though some data do not directly contribute to certain decisions, they may be necessary for the process of diagnosis. Reducing the overheads in diagnosis is not an easy task. AI has dramatically reduced the overheads in diagnosis and data management. It requires lots of studies and repeated experimentation. These types of experimentations started in the 1980s using data mining techniques over extensive databases. Later, a blend of conventional data mining techniques and elementary machine learning (ML) algorithms was used [21,22]. In recent days, ML algorithms have been widely used in data mining to ease the extraction of valuable data. As already stated, the task is nevertheless so easy due to the dimensionality of data. Healthcare data is too complicated and has several parameters, which may be redundant or irrelevant to the context of research. It is often termed the "curse of dimensionality" by most researchers [1,2].

Identification of valuable parameters for the diagnosis of a particular disease or a medical condition is highly challenging. A parameter, if selected wrongly, may not produce a fruitful result. Parameters may also be termed as features by some of the researchers. There are substantial shreds of evidence that support the argument that parameter identification has a considerable impact on the classification and prediction of healthcare data [3–6]. Therefore, parameter identification or feature identification plays a vital role in healthcare diagnosis. The curse of dimensionality cannot be prevented, unlike in most systems, but can be mitigated. Two ways can be used to mitigate the problem [7]:

(1) Include prior knowledge about the unknown function, which requires to be approximated. It requires expert support, and the knowledge should be problem-specific. It might not be possible to provide a generic solution in most cases. It is a solution-specific approach.

(2) Design a learning system that provides increased smoothness in dealing with the problem with increased input dimensionality. The process is automated and provides a generic solution. However, this requires an expert review to validate the results. It is a generic approach. Though stated to be a generic approach, the high-dimensional input is validated for the respective outputs.

The critical challenge that is faced concerning data is that, unlike at the end of the 1990s, nowadays, data produced by healthcare devices are terabyte and zettabyte scalable. The volume of data is not in scales of gigabytes. The more the data, the more accurate the results will be. Further, these data may or may not be structured or follows the same scheme. Also, it is expected that the data does not follow the same scheme when it originates from various devices. Though most of the data produced by a device used for a specific diagnosis may have standard data formats, we cannot universally state that the device may not collect other data that the manufacturer feels to the effective diagnosis.

Moreover, the calibration of the devices also contributes to variation in the data that is being collected. The data which are collected from various data sources need to be cleansed. The missing values need to be filled or eliminated if required or if it does not have a significant role in the problem that has been dealt with clearly. Further normalization of data needs to be done. If normalization is not done appropriately, the system may produce anomalous results. The success in classification depends on the quality of data that is fed to the system [8]—better classification results in better prediction.

We have so far discussed the role of AI in prediction and diagnosis in healthcare. We had also discussed the role of data cleansing and calibration of devices in decision-making. In the forthcoming sections, we discuss the various classification algorithms, parameter or feature selection methods, and various real-time scenarios where classification played a pivotal role in healthcare.

4.2 Classification algorithms

Healthcare encompasses the comprehensive processes of disease, injury, and other physical and mental impairments in humans. Most countries' healthcare industries are evolving at a rapid pace. The healthcare industry can be considered a data-rich environment because it generates massive amounts of data such as electronic medical records, administrative reports, and other benchmarking findings. However, these healthcare data are underutilized. However, data mining can search these massive amounts of data for new and valuable information. Data mining in healthcare is primarily used to predict various diseases and assist doctors in making clinical decisions. The following is a discussion of the various methods used in the healthcare industry.

The discovery of a predictive learning function that classifies a data item into one of several predefined classes is called classification. The related work in classification will be discussed in the subsections that follow.

4.2.1 Statistical data

Time series analysis is widely used for healthcare data, including the vital stats of the patients. These vital stats are analyzed using multivariate time series (MvTS). This MvTS is normally used for multivariate analysis for the progression of temporal data using multivariable statistical analysis. The Mahalanobis distance is used to distinguish one cluster of data against another to obtain inferential analysis and predictive analytics. In parallel, the Mahalanobis space is used to represent a metric, namely, the degree of deviation from the known reference group. Some studies and methodologies used the Mahalanobis Taguchi system (MTS) for ulcer prediction with these statistical classifiers. The imbalance of samples within classes is widely prevalent in healthcare datasets. When using skewed or imbalanced datasets, data mining algorithms are frequently affected by skewed distribution.

This problem frequently results in highly predictive classification accuracy over the majority class and poor accuracy over the minority class. This method would be an excellent method to test accurate dataset pressure ulcers because it can distinguish the degree of abnormality of observations. This method is also used because the MD is appropriately scaled. The tests were carried out in four phases, with scaled datasets ranging from 14 to 8, 5, and 2. Based on the significant difference between the standard and abnormal examples, the paper results show that the measurement scale for this algorithm has a good performance. The MTS proves to have higher sensitivity and g-means values in the testing stage due to its suitability for scaling. In terms of sensitivity, the MTS has improved.

4.2.2 Discriminant analysis

Linear discriminant analysis (LDA) is an unsupervised feature reduction-based technique to classify data based on measurements with unlabeled observations [16]. This is used in the classification of many medical data from the structural format of clinical observations. Parkinson's disease based on non-motor nerve symptoms for predicting the severity was done by LDA-based distancing. This research aimed to study the relationships and correlation between non-motor and motor nerve-based symptoms. This algorithm captures the ability of statistical dependencies for the predictor variables and reduces the features. These variables increase the ability of statistical dependencies among predictor variables and it is used to explore the linear constraint to discover the synergy among the symptoms ranging from non-motor to motor nerves. This model using LDA is found better than many algorithms and it was studied among a sample set of patients with an accuracy ranging to an estimate of around 70% as the performance of the algorithm improved. This improvement was due to the dependencies associated with the data in linear time series format. Some studies indicate that there are many methodologies proposed but the algorithm based on LDA for Parkinson's disease is used to evaluate the metrics of classification accuracy. This evaluation was done to find the most significant risk factor and to predict the initial set of significant risk factors in advance. This analytics is used for warning about the early stages of illness. The studies conducted by various sources and the experimental results based on the empirical study of motor and non-motor symptoms are suitable enough to identify the diseases and provide with a product that issues early warning about prolonged illness in a curable state and timeline which is proven like a linear relationship [22].

4.2.3 Decision tree

Decision tree is also used in clinical data for making a classifier based on multiple tuples in the structured data. Clinical data are mostly structured data with numerous varieties of tuples, and they are analyzed for feature importance. The important features identified were subjected to the classifier model of decision trees. The selected tuples are parsed using the correlated subspaces within the defined decision trees. This decision tree is used on the datasets made available from the clinical data collected. The prognostic analytics obtained from this classifier model was found to be high enough to justify the classification-based evaluation metrics. The datasets used in this classifier were examined for a balance in the sampling techniques to obtain an overall coverage of the whole set of patients in the given scenario. There is a limitation that these decision trees cannot be used for the branches resulting from recursively separate observations.

4.2.4 K-nearest neighbor (KNN)

KNN is one of the simplest classification algorithms used in clinical data for the classification of patients. This is an instance-based classifier that keeps updating when a new instance of data gets updated in the database. This algorithm uses all the instances available in an n-dimensional space Rn. This space of Rn involves the training set of data and depends on the *n* number of instances available in the data at that moment. Owing to this retention of data, the performance shows an increase in its efficiency. Since new instances of data keep adding to the database and training happens again and again for every update in transactions, this algorithm is found to be time-consuming but the performance for certain clinical automation is found to be remarkable. As a result of this performance, many clinical products used to identify some disease progression use this algorithm.

4.2.5 Logistic regression (LR)

Contrary to the name, LR is a classification algorithm. This algorithm is used on a given data consisting of a set of features that are continuous or discrete or in combination. This is strictly a binary classification algorithm that provides output in Boolean format. The data input in a linear format is passed through the logistic function and computes a linear combination for inputs. This classifier is used for binary classification-based problems with clinical data and in some experiments, the outcome is found to be competitive. The tuples in the structured data are taken as feature sets and using some feature reduction techniques, most of the tuples are pruned to suit the solution needed for the problem. The dataset with a prominent size makes the boundaries of accuracy more significant and thereby leading to a rise in the performance evaluation with this algorithm.

4.2.6 Bayesian classifier

Bayesian classifiers based on the naive Bayes algorithm are used to make predictive models. This works based on the probabilistic function of the Bayesian theorem. These classifiers are known for their computational efficiency in terms of time and space while dealing with a huge corpus of clinical data. The ability to deal with missing values in the data frames is done naturally by default and more efficiently. Most of the studies done on clinical data with this algorithm reported high performance in terms of classification accuracy based on the models designed on the sample of data trained. This is due to the averaging approach of the probabilistic calculation of tuples, and the classifier model was used to extract more features with a drastic decline in the magnitude of overfitting between training and testing accuracy. This algorithm is ideal for datasets with a considerable amount of missing data as well as sparse tables [14].

Healthcare is the spine of human civilization, and it is witnessed predominantly during the management of pandemics. The business associated with healthcare has a large magnitude of stake with medical science. The data associated with this industry is huge beyond the magnitude of normal storage and are termed as classified and confidential under some non-disclosure policy with the health providers. Many data mining algorithms and techniques are used to arrive at practical decisions. For detecting cardiovascular diseases, many tests are done on patients, whereas the involvement of data mining reduces the need for some of the tests. The tools needed to analyze the hidden information are not sufficient enough. This led to the development of data mining algorithms to classify cardiovascular diseases using data mining algorithms. Naive Bayes algorithm is one of the most widely used methods, specifically for cardiovascular diseases.

4.2.7 Support vector machine (SVM)

SVM is a statistical algorithm found to be most suitable for disease diagnosis. Many SVM-based variants like twin SVM [30] and Pinball SVM [31] are used to predict diseases associated with the brain, like Alzheimer's and Parkinson's diseases. It has various kernel functions of linear, polynomial, radial basis function, and sigmoid. Each kernel function performs well for different problems. The parameters of SVM are fine-tuned and reconfigured to obtain the required performance level. This algorithm comes from the basis of a support vector classifier (SVC), and it helps in handling classification problems that need high generalization performance. This leads to the minimization of structural risk in terms of generalization error. As it works with various kernel functions, it resembles a quadratic programming problem based on linear constraints. With this functionality, SVM is used widely in medical diagnosis. The heart disease dataset from UC Berkeley and the extensive research done in this regard form a substantial base for the usage of SVM in clinical data. Some studies show that SVM and its variants are used for a comparative analysis of the proposed methodology.

The ability of generalization is controlled in SVM by the factors of error occurring during training and the threshold of learning capacity exhibited by the classifier in the learning machine. Varying the features meticulously edited with the inclusion and pruning of tuples entails a reduction in the error rate. As a result of this computational advantage, SVM sounds like a promising option for solving problems related to clinical diagnosis using ML. This option is advantageous to data with a phenomenal increase in the dimensional space.

4.3 Parameter identification

Over the recent years, the dimensionality of the data handled by data scientists and data analysts has increased exponentially. This high dimensionality has posed a lot of problems to the learning algorithms that are available today [17], which is usually termed the "curse of dimensionality" [15]. Whenever a real-world problem is taken, the problem is engulfed by several parameters which result in overfitting outcomes that degrade the overall learning. Since the performance pertaining to learning is highly impacted by the dimensionality of the data, dimensionality reduction techniques play a major role in learning. Hence, dimensionality reduction has vital application in ML and data mining. One of the most prevalent techniques followed by most researchers to reduce the dimensions in each real-world problem is feature selection. The underlying principle of feature selection is selecting a small set of parameters or features based on the criteria fixed for evaluation. There is evidence of the contribution of feature selection to improved learning performance. This also has an indirect effect on lowering computational cost and interoperability. Feature selection methods can be broadly classified into

- supervised feature selection
- unsupervised feature selection
- semi-supervised feature selection

The above-mentioned feature selection techniques are based on the labeling of the training set. The supervised ones have labeled training sets, whereas the unsupervised ones will not have labeled training sets. The supervised feature selection technique is categorized as

- filter models
- wrapper models
- embedded models

There will be an influence of the learning bias over the bias of the feature selection algorithm. These can be isolated by means of filter models. Filter models delimit the influence of the bias of learning algorithms over the bias of feature selection. This depends on various factors like distance, dependency, attribution, and correlation. This provides outputs with the highest form of accuracy [18].

The wrapper model does not function like that of the filter model. It takes into account the accuracy of the learning algorithm to ascertain the validity of the features that are selected. However, the memory overhead will be high on utilization of the wrapper model when the dimensionality is high.

The embedded model was introduced to overcome the shortcomings of the wrapper model. This combines the advantages of the filter model and wrapper model, thereby improving the accuracy in most cases where the dimensionality of data is very high.

Many researchers were also interested in increasing the number of unsupervised features. Unsupervised feature selection is so far limited to search problems without classy labels and relies on good measurement clustering [12], which can also result in a large subset of non-prominent features. There are additional restrictions that make the oversized facts to be impossible to a huge magnitude for the improvement of applicable skills. The effectiveness of feature selection follows as the next major issue.

Feature selection is rarely monitored but a complete assessment is done and observed to be effective in [9]. For supervised feature selection to happen in an effective manner, the label information needs to be specified in an appropriate and accurate form, failing which the supervised feature selection may fail [23].

Unsupervised feature selection works with unmarked facts but assessing skill relevance is miles of difficulty [19]. It is not uncommon for subjects to have large dimensions but small classified pattern lengths. For high-dimensional data with small, classified samples, the hypothetical space is too large, but the limits are too small (classified instances). Combining the two de facto features poses a whole new challenge in research. Based on the idea that classified non-labeled data are selected from the same population generated by the target idea, semi-supervised feature selection uses each classified unlabeled data to estimate the importance of features which is termed feature weighing. Feature weighing is understood as a generalization of feature selection. When a feature is selected, binary weights are assigned to the feature. If a valid feature is encountered one is selected, else it will be zero.

Feature weighing assigns a weight to each feature, which is usually set within the range [0,1] or [-1, 1]. The importance of features is directly proportional to the increase in weight. A global uniform weight in the form of a scalar is assigned by most of the feature weighing algorithms for the features contained in all instances of data. Various factors contribute to this process depending on the location of the data in the search space. The factors that affect include noise, relative importance, degree of relevance, etc. There are neighborhood characteristic selection algorithms in which the functional neighborhood selection is achieved specifically for test examples, but this is not uncommon in lazy learning algorithms such as KNN [3,9]. Feature selection or feature weighing usually happened at the time of classification, as opposed to the idea that it happens during training. It is the knowledge of the test instances that enhances the ability of feature selection rather than that of the contribution of the training instances. The approach to feature selection usually consists of four simple steps:

- subset generation
- subset evaluation
- stopping criteria
- result validation

During the subset generation, the feature subset selection is primarily based on the specific search strategy which is evaluated based on a certain evaluation criterion. The subset having the best fit is chosen from all the candidates available in the search space until the stopping criteria are met. In the final step, the chosen subset can be validated for correctness using domain expertise or a predetermined validation set [13].

Considering any real-world problem, it may not be possible to provide a class label for the parameters or features related to the problem without appropriate knowledge about the problem. However, to solve the problem, each feature needs to be labeled appropriately. There are a lot of candidate features associated with the real-world problem, which may be redundant or insignificant with respect to the area of interest of a researcher. Hence, feature selection plays a major role in the process of classification. It helps to identify the significant features and ignore the unimportant features relevant to a problem.

4.3.1 Feature selection for classification

Supervised learning is used to solve most real-world problems. Each instance is related to a class label and the class probabilities and class-conditional probabilities are unknown. It is an inevitable fact that the required knowledge about the most relevant features is very less while dealing with real-world situations. To overcome this shortcoming in the required knowledge, many candidate features are introduced to represent a domain which may result in irrelevant or redundant features. It is very hard to find out the actual classifiers due to the huge volume of data that is required to be classified. However, it is essential to remove the irrelevant or redundant features during the classification process because the presence of more irrelevant features may increase the running time of the learning algorithm. This may also result in a general classifier that may not produce appropriate results for a specific problem. Hence, it is required to identify the parameters or features with utmost care since it provides a better insight into the concept of the real-world classification problem. Figure 4.1 illustrates the general feature selection strategies deployed for a classification framework.

Feature selection has a high impact on the training phase of any classification technique. There are various approaches to improve the efficiency of feature selection. One such approach is the divide-and-conquer technique, where the whole features of the data are not directly taken into the learning algorithm. Instead, feature selection is done to identify a subset of features followed by the processing of data pertaining to the selected feature. In most cases, the feature selection and learning algorithm will be independent of one another or the feature selection will depend on a predetermined set as in the filter model and wrapper model, respectively. The classifiers which are identified for the prediction phase are based on the features that are selected. The downsized subset selection of features depends on the following criteria:

- Lack of significance in the magnitude of classification accuracy
- When selected features are taken alone with the resulting values, it results in a distribution of classes that are as close as possible.

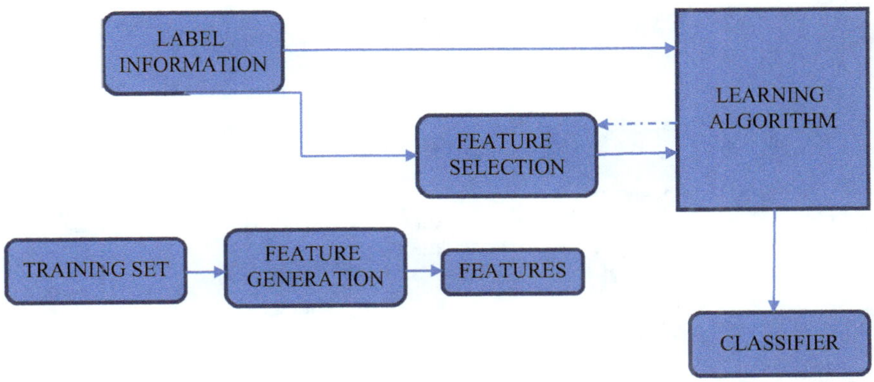

Figure 4.1 A general framework of feature selection for classification

The procedure of finding the ideal features is an exhaustive process to find the best one. This is due to the fact that the feature selection methods search and find the best ones among 2 million candidate subsets according to some methodologies studied. The search for features among the subsets of data is done for this purpose. This is computationally expensive and practically negligible for small and medium number of features (m). Heuristic or random search methods are employed to perform feature reduction. This feature reduction results in the decrease of computational complexity and compromises the performance of the methodology. To prevent the subset search to become exhaustive, feature reduction is used as a stopping criterion.

There are two major problems that any ML algorithm encounters with respect to the data being used for processing. They are:

- overfitting
- curse of dimensionality

Overfitting happens when there are significant features of the data which are required to be generalized but have not done so. This may be due to the fact that the parameters or the feature may be beyond the scope of the dataset. The curse of dimensionality, as we previously discussed, is the result of the high dimension of data which is unintuitive in nature and has varied properties. The role of feature selection is not just restricted to the number of features but also toward the optimization of the performance of the model (Figure 4.2). An additional advantage that feature selection offers is model interpretation. The reduction in the number of features helps in providing direct results. The proposed model for the specific real-world problem acts as a model that can be used for further predictions of the same kind.

The correlation coefficient is used to get the statistics between the input and output tuples, and it forms the basis for filter feature selection. The statistical

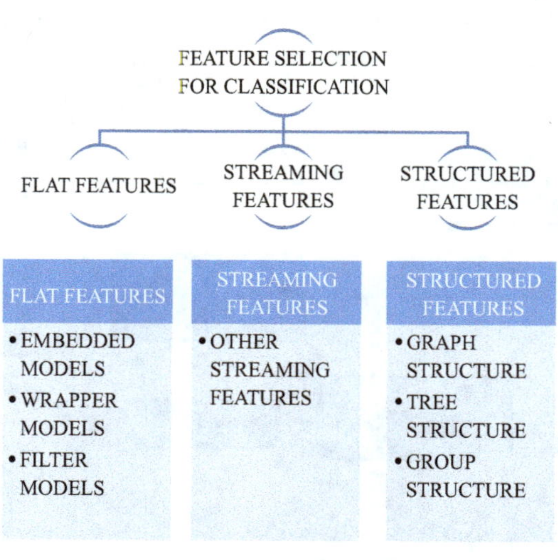

Figure 4.2 Classification of algorithms of feature selection

measures are chosen on the measures which are highly dependent on the data types concerned with the variables.

The data types involved in features include numerical and categorical aspects. The numerical data is in the integer format with variables like height, weight, etc., and are in integer or floating-point types. The categorical data are represented in Boolean or ordinal values, like the picket scale, for the data categories in labels.

In general, the most common numerical variables that are used for feature selection are

- integer variables
- float variables

Apart from the above-mentioned numerical variables, some of the categorical variables that are used are

- Boolean variables
- ordinal variables
- nominal variables

It is easier to choose a statistical measure for a filter-based feature selection technique if one possesses an expertise in the data type of the variables that are used in the dataset (Figure 4.3).

There are also two groups of variables that are not related to the data type, namely, the input variable and the output variable. The input variable is one that is provided as input to a model whereas the output variable is the one that is sought out of the model.

The datasets in structural form have numerous tuples and not all of them are needed. So, we use feature selection to reduce the size of features in the group of variables. A response variable is created to get the output variables with the intention of prediction. Predictive modeling is done with algorithms to get the output variables

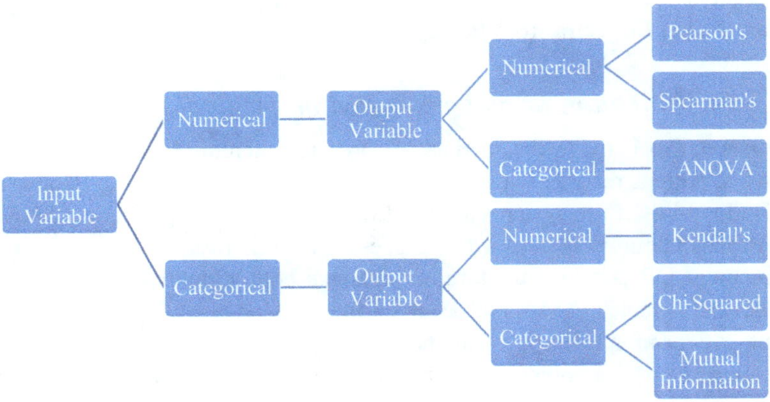

Figure 4.3 Choosing the feature selection method

in the desired way, and it indicates the type of problem addressed. If the output variable is numerical, it is a clear indication that it is a regression-based problem based on a set of numeric variables. If the output variable sets the category of labels based on the probabilistic values, then it is a classification-based prediction.

Numerical output: Regression analysis, time series analysis, etc.

Categorical output: Classification and categorization of data.

Various filter-based feature selection methods are employed to obtain the statistical measures and calculate the target variable at the rate of one variable at a time. This methodology is known as univariate statistical measures. Any interaction between the input variables in the set of features is not filtered as ideal as the perfect fit of the classifier model [20].

4.4 Real-time applications

Medical data classification starts with a patient data monitoring system until the forecast of adverse medical conditions, which leads to diseases like Parkinson's disease and Alzheimer's. With the necessity of social distancing and more sanitation, COVID-19 has paved the way for the automation of healthcare and disease diagnosis using various classifiers. These tasks are done using various ML and deep learning systems. While the population of a few countries is increasing, certain countries in the eastern hemisphere are going toward aging and super-aging societies. These aging communities are badly in need of automation systems aided by classification systems. The applications of classification problems in healthcare are explained as follows:

- Classification of patients based on medical record
- Predictive analytics and diagnostic analytics based on medical records
- Classification of diseases based on medical imaging
- Mixed reality-based automation to help aid aging society
- Tiny ML-based classification systems for medical gadgets
- Classification systems for insurance claim management
- Case study: Inspectra from Perceptra

4.4.1 Classification of patients based on medical record

For a hospital to manage the medical records is a herculean task. There will be millions of patient records with billions of entries. Though flat file databases are used, Hadoop-based file systems are used with cloud-based applications. Polar DB from Alibaba Cloud is seen as a fast processor for this massive corpus of medical records. Polar DB is robust to store and manage both structured and unstructured data like sparse tables, XML, and JSON.

The patient monitoring systems hold a vital role in the administration of hospital management. It involves the operations of allocation of specialty to be treated, pharmacy management, insurance claim systems, and legal affairs. Any mishap in these systems leads to severe consequences like the loss of money and, more importantly, the hospital's reputation. Gone are the days of enterprise resource planning, and now it is the dawn of cloud-based systems, which reduces the need for physical infrastructure in terms of information technology IT systems.

The classification of patients based on their history is normally done to group the patients based on specialty. It helps in managing the supply chain of medical services. Usually, the medical data in hospitals are managed in the form of No SQL. The data requires much processing, and it is done using complex data cleaning algorithms and then converted to a format that is efficient enough for the classification algorithms to handle. The classification algorithms will process this data and then categorize the patients based on priority levels ranging from emergency to casual visits to physicians.

4.4.2 Predictive analytics and diagnostic analytics based on medical records

Medical records consist of various columns, which are typically numerical or categorical data. Numerical data are normalized within a range, and categorical data are labeled with a numeric value to obtain Likert scale values to enumerate the categorical data. These data columns are defined as features, and they are fed into unsupervised algorithms for clustering and supervised learning algorithms for classification. Clustering is done to group the patients during chaotic situations like pandemics and manage the crowd of people visiting healthcare facilities more effectively. Classification is done on data like vital signs and history of medical observation to predict if a patient who got treated in the past may get any health issues again in the future or not.

Blockchain-based medical record management systems have been developed nowadays, promising healthcare development beyond cryptocurrencies. This technology will help secure large volumes of highly confidential and sensitive patient information to a greater extent. The confidentiality, privacy, and integrity of patient data are assured. Healthcare providers have chains of hospitals that get full access to patient records within the hospital network or any authorized remote access.

4.4.3 Classification of diseases based on medical imaging

Disease classification from medical images is one of the most sought-after avenues of research and development in healthcare. X-ray, computed tomography (CT) scan, and magnetic resonance imaging (MRI) are the most widely used imaging techniques for disease diagnosis in non-invasive and non-destructive medical procedures. The applications of medical imaging technology include mostly segmentation and classification problems.

Segmentation is used to detect the abnormalities and then localize the boundaries of the affected body parts. Segmentation coupled with classification algorithms is used to detect and localize the presence of abnormalities in the internal organs [10].

As the case study of the Inspectra app is discussed in this chapter, there are a lot of real-time applications to detect diseases related to viral pneumonia, cysts, and tumors. These applications helped the medical community to save lives by preventing people from getting into advanced levels of cancer, life-threatening conditions, and early diagnosis of cardiopulmonary systems. Nowadays, deep learning algorithms involving convolutional neural networks and transformers [25] develop such applications. Convolutional neural network-based architectures using a variety of layers from the existing architectures [26,27] are used to improve classification performance in medical images.

4.4.4 Mixed reality-based automation to help aid aging society

Countries except those from South Asia and Africa are experiencing a decline in the birth rate, leading to most countries entering an aging or super-aging society. Countries like Japan, Thailand, Korea, Hong Kong, and Singapore are going toward an aging society. Owing to this factor, there is a spike in the need for elderly care to an exponential level apart from the surge in issues related to human resource management (Figure 4.4).

Robotics is key to addressing this issue on various fronts. Just like the need for radiologists in Thailand led to the rise of Perceptra in Thailand, many AI-based healthcare systems are getting developed. Virtual reality (VR) or augmented reality (AR), or the combination of both as mixed reality, is giving solutions to address the needs of older adults from being in a remote location. Some Japanese AI researchers develop Exoskeleton-based gadgets at the research level to address the issue of older adults affected with mobility-based diseases or conditions.

With the help of AI and advanced technologies, surgeries are performed by specialist doctors from being in a remote location. Currently, they require some assistance from paramedical staff and junior resident surgeons from being in the

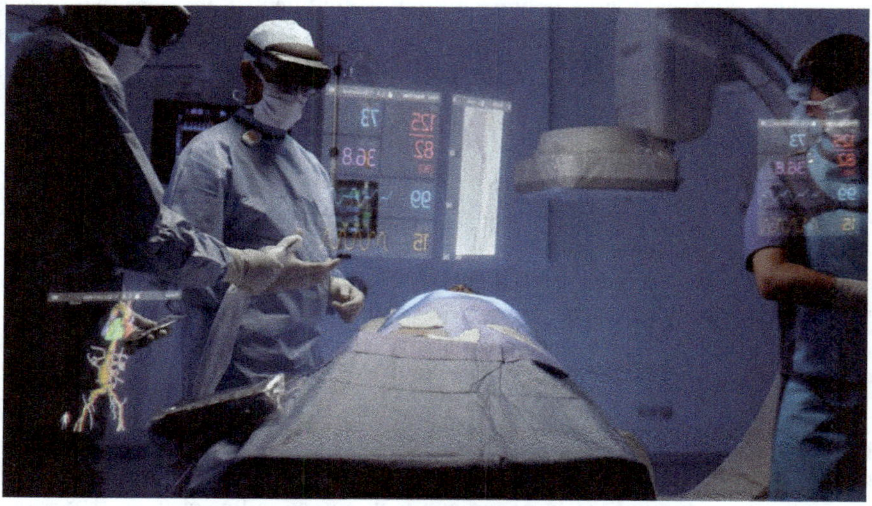

Figure 4.4 Phillips Microsoft augmented reality system used by surgeons (Source: philips.com)

operation theatres. Decision tree-based algorithms are used in classifying the actions. In combination with classification algorithms, AR, VR, and mixed reality make the level of medical robotics in remote locations feasible and reduce the need for the workforce in terms of healthcare workers to a greater extent.

4.4.5 Tiny ML-based classification systems for medical gadgets

Tiny ML deals with ML done at gadgets dealing with the Internet of Things (IoT). IoT devices are generally microprocessors like Raspberry Pi or NVIDIA Jetson Nano, and microcontrollers from the ESP family IoT are used mainly for remote monitoring systems for patients with limited mobility. These devices have limited RAM and ROM. This storage limitation led to the development of embedded systems to collect data and transmit data through edge [29] or fog-based systems. Fundamental decisions may be taken through edge servers, but data processing and data-driven decision-making is the least explored solution for medical IoT. ML algorithms are developed in IoT microcontrollers too. Since most of the IoT microcontrollers support C and C++, they are good enough to support ML at the level of embedded systems at a micro-level. So, this way of developing IoT applications powered by ML is transforming the development in medical IoT (Figure 4.5). It includes predicting adverse conditions based on real-time patient data to emotion recognition-based patient monitoring systems.

4.4.6 Classification systems for insurance claim management

Insurance claims play an essential role in the financial management of healthcare facilities. It has been observed that there are many fraudulent claims processed

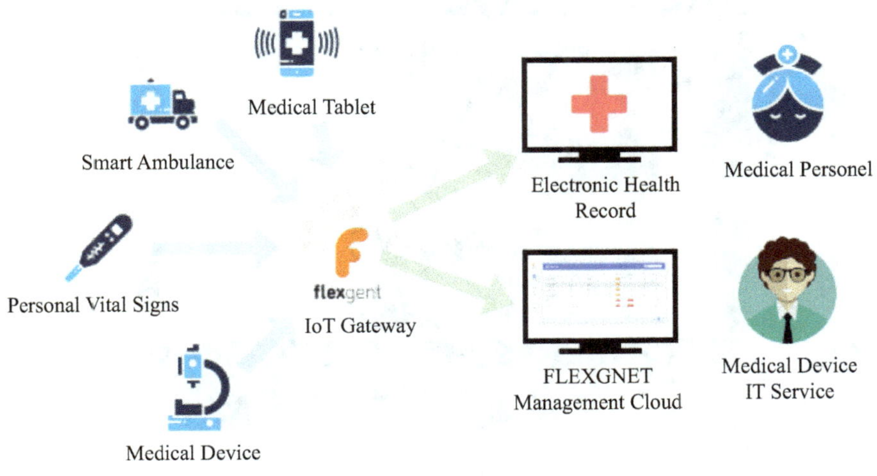

Figure 4.5 Flexgent IoT system architecture (Source: softgent.com)

without having enough evidence to reject those. There is a history of claims, and they all come under a particular pattern in statistical analysis [11]. ML algorithms are implied on the data obtained from the prospective beneficiaries, and a credit score is obtained from a series of features which brings uniqueness to the data to be classified. Classification algorithms like decision trees on text data help to identify the subjectivity and polarity scores. These scores will add up to make a decision. Many other algorithms like LR, naive Bayes, artificial neural networks, and SVCs are used to identify genuine claims and avoid fraudulent ones.

The use of blockchain technology in insurance management systems and the implication of ML-based classification for the process of consensus in the data logs from the blockchain system helps improve the safety and security of insurance claims.

4.4.7 Case study: Inspectra from Perceptra

Perceptra Co. Ltd is a healthcare-based AI solution provider from Thailand. Perceptra mainly focuses on deep learning in computer vision. This company is steering an initiative of Thai AI in healthcare, especially in the fight against COVID-19 [24]. They have developed a chest X-ray application named "Inspectra," an AI system that assists physicians in diagnosing abnormalities in the chest from X-ray images.

It is equipped with an algorithm that measures the ratio of the width of the heart to the lungs. It is a cardiothoracic ratio (CTR) from chest X-ray photographs and is used to diagnose abnormal heart conditions or enlarged heart (Figure 4.6).

They applied deep learning techniques based on the U-Net model to first segment the lungs and the heart from chest X-ray images. The ratio of the width of the heart to the lungs is calculated as CTR (Figure 4.7).

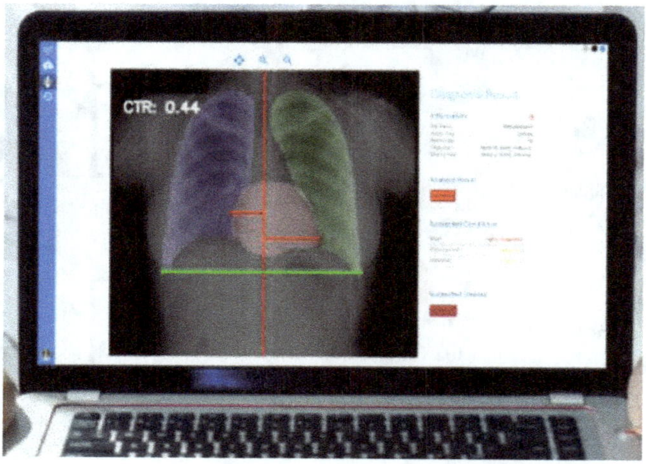

Figure 4.6 Cardiothoracic ratio (Source: https://inspectra.perceptra.tech/publications/)

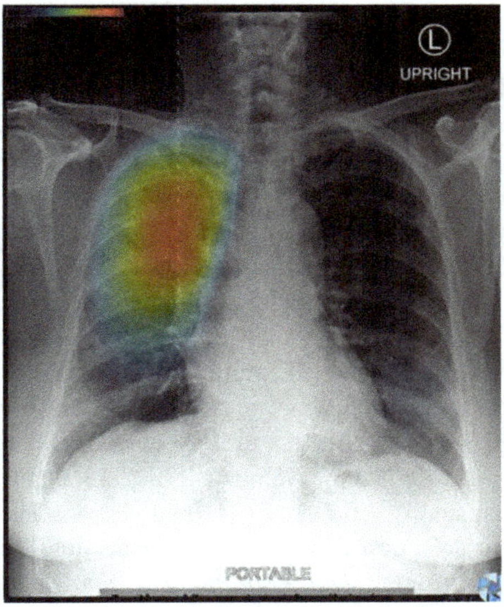

Figure 4.7 Inspectra's sample outcome

This product was developed in collaboration with Siriraj Hospital, Bangkok. Today, Perceptra is helping hospitals across Thailand with disease diagnoses. They are mainly working on COVID-19 X-ray-based automated diagnostic systems. Anyone from any country can upload an X-ray image into the Inspectra application and wait for AI-based diagnosis results. Thailand has experienced the outbreak of COVID in three waves as it became the second country to get affected after China. The Delta variant spiked the number of infections in the third wave compared to the previous two waves. The severity of lung infection is more in the third wave and results in an exponential mortality rate. If someone is infected with COVID-19, early detection and early treatment for lung infections will reduce mortality. It is the problem addressed by Inspectra, and it uses tomography computation. Inspectra used 1.5 million chest X-ray images from all over the world, including 400,000 images from hospitals in Thailand. The classification accuracy was found to be par in excellence compared with the classification done by trained radiologists. Perceptra has a team of academic researchers and efficient ML engineers who made this product possible enough to make it a reality. The main intention of Perceptra is to provide AI-based solutions to solve the problem of shortage in the number of radiologists in Thailand. Apart from COVID-19, they are working on applications that can detect diseases from medical images like CT scan, X-ray, and MRI.

4.4.8 Deep learning for beginners

A detailed user manual is presented in [28] for anyone who wants to learn to implement deep learning algorithms. This material consists of GitHub links to support with necessary code and is compatible with Keras and TensorFlow.

References

[1] Guyon I and Elisseeff A. An introduction to variable and feature selection. *Journal of Machine Learning Research*, 2003;3:1157–82.

[2] Chen Y, Wang Y, Cao L, and Jin Q. An effective feature selection scheme for healthcare data classification using binary particle swarm optimization. In: *2018 9th International Conference on Information Technology in Medicine and Education (ITME)*, 19 October 2018, IEEE, pp. 703–7.

[3] Dash M and Liu H. Feature selection for classification. *Intelligent Data Analysis*, 1997;1(1–4):131–56.

[4] Tang J, Alelyani S, and Liu H. Feature selection for classification: a review. *Data Classification: Algorithms and Applications*, 2014;37.

[5] Kwak N and Choi CH. Input feature selection for classification problems. *IEEE Transactions on Neural Networks*, 2002;13(1):143–59.

[6] Janecek A, Gansterer W, Demel M, and Ecker G. On the relationship between feature selection and classification accuracy. In: *New Challenges for Feature Selection in Data Mining and Knowledge Discovery*, 2008, PMLR, pp. 90–105.

[7] Haykin S. *Neural Networks and Learning Machines, 3/E*, 2010, Pearson Education, India.

[8] Singh D and Singh B. Investigating the impact of data normalization on classification performance. *Applied Soft Computing*, 2020;97:105524.

[9] Alelyani S, Tang J, and Liu H. Feature selection for clustering: a review. In *Data Clustering: Algorithms and Applications*, 2014, Chapman and Hall/CRC, pp. 29–60.

[10] Domeniconi C and Gunopulos D. Local feature selection for classification. In: *Computational Methods of Feature Selection*, 2007, Chapman and Hall/CRC, pp. 227–48.

[11] Duda RO and Hart PE. *Pattern Classification*, 2006, John Wiley & Sons.

[12] Dy JG and Brodley CE. Feature subset selection and order identification for un-supervised learning. In: *Proc. 17th International Conference on Machine Learning*, 2000, Morgan Kaufmann, pp. 247–54.

[13] Dy JG and Brodley CE. Feature subset selection and order identification for unsupervised learning. *ICML*, 2000;247–54.

[14] Cawley GC, Talbot NL, and Girolami M. Sparse multinomial logistic regression via Bayesian L1 regularisation. *Advances in Neural Information Processing Systems*, 2007;19:209.

[15] Hastie T, Tibshirani R, and Friedman J. *The Elements of Statistical Learning*, Springer, 2001.

[16] Hastie T and Tibshirani R. Discriminant adaptive nearest neighbor classification and regression. *Advances in Neural Information Processing Systems*, 1996;409–15.

[17] Liu H and Motoda H (eds.). *Computational Methods of Feature Selection*, CRC Press, 2007.

[18] Liu H and Yu L. Toward integrating feature selection algorithms for classification and clustering. *IEEE Transactions on Knowledge and Data Engineering*, 2005;17(4):491–502.

[19] Mitra P, Murthy CA, and Pal SK. Unsupervised feature selection using feature similarity. *IEEE Transactions on Pattern Analysis and Machine Intelligence*, 2002;24(3):301–2.

[20] Peng H, Long F, and Ding C. Feature selection based on mutual information criteria of max-dependency, max-relevance, and min-redundancy. *IEEE Transactions on Pattern Analysis and Machine Intelligence*, 2005;27 (8):1226–38.

[21] Quinlan JR. *C4.5: Programs for Machine Learning*, 1993, Morgan Kaufmann.

[22] Robnik-Šikonja M and Kononenko I. Theoretical and empirical analysis of ReliefF and RReliefF. *Machine Learning*, 2003;53(1):23–69.

[23] Song L, Smola A, Gretton A, Borgwardt KM, and Bedo J. Supervised feature selection via dependence estimation. *Proceedings of the 24th International Conference on Machine Learning*, 20 June 200, pp. 823–30.

[24] https://inspectra.perceptra.tech/covid-19/. Retrieved on 18 October 2021.

[25] Han K, Wang Y, Chen H, *et al.* A survey on visual transformer. arXiv preprint arXiv:2012.12556. 23 December 2020.

[26] Sasubilli SM, Kumar A, and Dutt V. Machine learning implementation on medical domain to identify disease insights using TMS. In: *2020 International Conference on Advances in Computing and Communication Engineering (ICACCE)*, 2020, pp. 1–4, doi: 10.1109/ICACCE49060.2020.9154960.

[27] Sasubilli SM, Kumar A, and Dutt V. Improving health care by help of Internet of Things and bigdata analytics and cloud computing. In: *2020 International Conference on Advances in Computing and Communication Engineering (ICACCE)*, 2020, pp. 1–4, doi: 10.1109/ICACCE49060. 2020.9155042.

[28] Joseph FJ, Nonsiri S, and Monsakul A. Keras and TensorFlow: a hands-on experience. In: *Advanced Deep Learning for Engineers and Scientists*, 2021, Springer, Cham, pp. 85–111. https://doi.org/10.1007/978-3-030-66519-7_4.

[29] Chertchom P, Jiteurtragool N, Joseph FJ, *et al.* The development of IoT system for data analytics on students' behavior. Available at: http://library.tni.ac.th/thesis/upload/files/Res%20Pim%2024%20July%202021/Prajak%20Chertchom%20Res%20IT%202019.pdf.

[30] Kotturu PK and Kumar A. Comparative study on machine learning models for early diagnose of Alzheimer's disease: multi correlation method. In: *2020 5th International Conference on Communication and Electronics Systems (ICCES)*, 2020, pp. 778–83, doi: 10.1109/ICCES48766.2020.9137872.

[31] Dubey AK, Narang S, Kumar A, Sasubilli SM, and García-Díaz V. Performance estimation of machine learning algorithms in the factor analysis of covid-19 dataset. *Computers, Materials & Continua*, 2021;66(2):1921–36.

Chapter 5
Introducing deep learning in medical domain

Prithi Samuel[1], Anusha Bamini[2], P. Nancy[3], S. Oswalt Manoj[4] and Maruti Perumal[5]

The buzz words for today, namely artificial intelligence (AI), machine learning (ML), and deep learning (DL), are slowly entering the medical industry over the past decade, which brings technologies and solutions to a change in the structure of the medical field. These technologies are connected, and each one offers something different in the medical field by showing a difference in how medical professionals treat the patient. DL provides the power to transform and deliver a more decadent layer of medical technology solutions. It is progressively available with innovative technologies that have broad applications in the natural world medical field. DL plays a vital role in providing insight to medical professionals, which helps identify diseases at an early stage, thus delivering better tailored and most effective patient care. DL has become a well-known initiative that everyone has an idea about. The AI–DL industry has been developing quickly, which provides some sufficient development opportunities to the medical industry to bring a significant change. According to Gartner, almost 37% of all sectors surveyed use DL in their profession. It has been foreseen that by 2022, around 80% of modern developments will use AI and DL. It has been observed that the year 2021 would bring the most powerful deep understanding and AI trends that might reshape the country's economic, social, and medical domains.

This chapter describes DL, the history of deep knowledge in the medical field, the barriers to deep understanding, and DL opportunities in the medical industry. It also focuses on the various methods or algorithms of DL such as convolutional neural network, deep autoencoder, deep Boltzmann machine, deep belief network in biological systems, medical imaging, and health record and report management. It also discusses the various applications of DL in healthcare and how deep understanding is used in medical image analysis.

[1]Department of Computational Intelligence, School of Computing, SRM IST, Kattankulathur Campus, India
[2]Department of Computer Science and Engineering, Karunya Institute of Technology and Sciences, India
[3]Department of Computing Technologies, School of Computing, SRM IST, Kattankulathur Campus, India
[4]Department of Computer Science and Business System, Sri Krishna College of Engineering and Technology, India
[5]Full Stack Developer, Brazil

5.1 Introduction

The invention and implementation of technology instruments have resulted in constant innovation in medicine, genomics, pharmaceuticals, and monitoring systems. It has made data collection, interpretation, and processing much more accessible. Conversely, technological advancements allow for the storage of vast amounts of data containing valuable information. As of now, cameras to detect patients' gestures, cordless necklaces, identification cards for bioacoustics signal and blood circulation acquirement, smartwatches fiber content intelligent material, superintendent heart rate monitor, and sensing technologies can generate massive quantities of data in the form of photographs, audio, message, infographics, and indicators [1]. The abundance of biomedical data presents both challenges and barriers for healthcare research. Discovering the relationships between all of the many snippets of data in these datasets, for instance, is a critical topic for developing reliable surgical equipment using data-driven techniques and ML.

Human specialists in a particular topic are well-versed in a particular domain subject. Perspective, wide variability among translators, accessibility, and exhaustion are all factors that restrict human specialists [2]. Artificial intelligence (AI) was created to assist humans in completing tasks and overcoming these limits by incorporating cognition behavior in humans into robots and automation (AI). ML is a branch of AI that allows machines to understand and act like professionals without being specifically trained [3]. Some examples are support vector machine, logistic regression, regression models, naive Bayes, K-means clustering, and other ML techniques.

Deep learning (DL) is now having a significant impact in various fields of healthcare. The outstanding results documented in healthcare have been made feasible by the rising accessibility of healthcare statistics and the quick advancement of variants in DL approaches [4]. DL approaches can identify medically useful information disguised in massive quantities of healthcare data, which in turn can be used for making decisions, diagnosis, monitoring, and avoidance of chronic disorders. Health behavior response [5], electronic health records (EHRs) handling and extracting scientifically supported diagnosis from message [6], vision-related evaluation and characterization [7], motion detection and robotic-assisted healing process [8], hearing disorder services [9], treatment for cancer [10], cardiac treatment [11], and neural activity investigation [12] are some of the areas where DL can be used. It enables therapy to be more uncomplicated for healthcare practitioners, more comfortable for patients, and more efficient and effective surveillance. The development of DL in medicine has transformed the use of minimal equipment like a temperature gauge and heart monitor into computerized tomography (CT), ultrasonographic diagnostic devices, radio fluoroscopy, radiosurgery, laser therapy, blood transfusions, ventilation systems, and other devices that have transformed traditional patient care into remarkably resilient diagnosis willing to challenge many feared diseases [4]. Here there is a small doubt that, in the next few years, medical support and technology will enhance in various ways, leading to improved performance and providing higher-quality services.

5.1.1 DL in a nutshell

When it comes to machine learning (ML), DL is a subset that employs techniques to evaluate data, replicate thoughts and opinions, and generate representations. DL uses layering algorithms to process data, comprehend voice sounds, and continuously recognize objects. In a system, the input layer is the lowest layer while the perceptron is the last. The layers that reside between both the two are alluded to by hidden units. Typically, each layer is a simple, responsible approach with just a single kind of activation function. Understanding "deep" in DL seems to be that of the human brain, and we will not always comprehend all of the factors that go into a choice or conclusion since it incorporates progressively complicated layers.

On the other hand, DL is striving to emulate the human brain's incredible complexity. A neural network (NN) of perceptrons or digital neurons is built and arranged in layers, with each layer's perceptrons attached toward the perceptrons of the next layer. Every layer transmits information toward the next, with the preceding layer's output serving as feed. The NN training data, such as photographs of things, are fed, and each object is predicted. Naturally, its initial estimations will be terrible. However, the NN modifies the connections between its perceptrons until it delivers high accuracy predictions when feedback is provided. A trained model is obtained that can accurately categorize things.

Data is processed through a hierarchy of layers upon layers in DL models, upon each subsequent layer using the information from the preceding one to determine its results. While more data is processed, deep NNs grow more precisely, effectively learning from prior results to improve their capability to make associations and relationships. DL is dependent on how nerve cells in organisms' brains link against each other to acquire knowledge. Each successive layer of nodes is engaged whenever it obtains impulses from its surrounding neurons, analogous to how electrical currents pass between the cells of biological creatures.

DL generally necessitates minimal preprocessing steps. Several of the cleaning and normalization processes that human developers must accomplish when utilizing standard ML approaches are handled by the network itself. "For centuries, developing a learning algorithm that converted original information into an appropriate intermediate representation or support vector through which a classifier can detect patterns in the input requires thoughtful innovation and tremendous domain knowledge."

5.1.2 History of DL in the medical field

For the past couple of years, concepts like ML, DL, and cognitive science have been gradually infiltrating the medical field, delivering services and innovation that are transforming the face of healthcare. These innovations are interconnected, offering things unique to the profession and altering the way healthcare experts handle their duties and clinical management.

A significant number of DL highlights are now focused on comparatively small projects or premarketed research initiatives. On the other hand, DL consistently makes its direction into novel technologies with robust implementations in the legitimate healthcare context. The most encouraging business incorporates creative patient-confronting applications and a few remarkably well-established approaches for enhancing health information technology professionals. Diagnostic imaging solutions, chatbots that could recognize patterns in patient conditions, DL algorithms that can detect critical cancer types, and scanning electron microscopy services that use DL to recognize serious illnesses or particular kinds of psychopathy are all examples of DL applications in the healthcare system. DL has proven crucial in giving health practitioners knowledge that enables doctors to detect problems at an early stage, allowing them to provide significantly more tailored and appropriate quality of healthcare.

5.1.2.1 DL in a myriad of health sectors

The numerous sectors of healthcare where DL is illustrated in Figure 5.1 and discussed below.

- **Genetics:** A DL technology is being used to comprehend a gene, which aids patients receiving therapy in understanding the condition that may impact them. DL algorithms are utilized to improve the speed and accuracy of medical sector researchers and clinicians. With the financial industry's support, genetics is a sector that is rapidly increasing, and DL is poised to have a bright future.
- **Cell scope:** A cell scope is a device that employs DL strategies to enable parents to keep track of their children's health issues and conditions. This technique may be viewed on almost any gadget and reduces the number of times parents visit hospitals or medical facilities to see their kids or chat with a physician.
- **Insurance fraud:** DL is applied in investigating health coverage claim forgery. DL employs a predictive modeling technology that could foresee potentially fraudulent claims in the long term. DL aids the financial industry in sending targeted promotions and deals to patients.

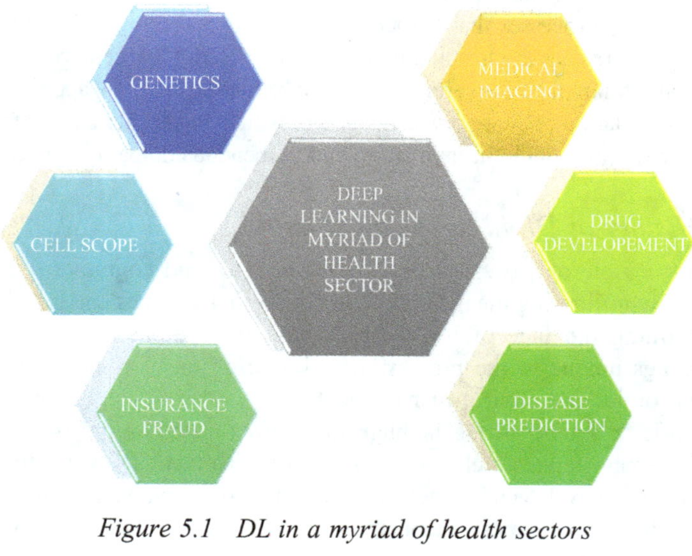

Figure 5.1 DL in a myriad of health sectors

- **Medical imaging:** A range of methodologies are employed in the healthcare sector to diagnose hazardous disorders, including CT scans, ECGs, and magnetic resonance imaging (MRI). Brain tumors, cardiac arrest, cancers, and many other ailments are among the dangerous diseases. As a result, DL could be used to counsel specialists who could analyze a surgeon's ailment and treat them appropriately.
- **Drug development:** Data learning aids in the invention of novel medicines as well as their development. The health records of a patient are examined, and medication is prescribed based on the findings. We can acquire understanding from the patient's diagnostics and records and disorder statistics about the symptoms, using DL in medical services.
- **Disease prediction:** Disease prediction systems have recently received much interest. The single-layer feedforward neural classifier is among the most commonly used ML approaches and was frequently used to forecast various diseases. In numerous real-life environments, DL techniques have been used to overcome healthcare challenges. Over recent years, NN ensembles have been successfully used in many scenarios, including diagnostic testing. By training a set number of NNs and then integrating their results, NN ensembles should significantly improve the streamlining abilities of instructional strategies.

5.1.3 Benefits of DL in the medical domain

Massive volumes of data-driven DL necessitate high-end workstations with strong graphical processing units to run in a reasonable length of time. Almost all of the application characteristics in conventional ML approaches must be defined by a specialist to reduce the dimensionality of the data and create correlations that are more evident to learning algorithms. DL techniques have the most significant benefit in that they aim to understand high-level patterns from images independently. It removes the necessity for domain experts and intensive semantic segmentation in theory. DL approaches are exploding to a never-seen-before degree of precision in complicated issues in which a significant type of technology is essential. However, there is a deficiency of domain awareness for features creation, for example, image processing techniques, speech processing, and automation.

DL algorithms can lower healthcare costs by offering accurate and automatic diagnosis predictions. It avoids essential and critical situations from being delayed in reporting and minimizes the administrative burden on healthcare providers to concentrate on more sophisticated diagnostics or allocate more resources to patient care. By reviewing medications and clinical information, we can reduce medication errors. It can provide faster diagnostics.

DL recognizes the significant correlations in the dataset and keeps track of former customers' knowledge, which can be utilized as a future analysis for patients on the basic similar signs or illnesses. When you demand a vulnerability index upon medication apart from discharging, DL enables us to build an approach based on whatever source of information is given. DL generates precise and quick forecasting models, allowing for trust and precise distribution of resources. DL methods result in cheaper costs and better outcomes.

5.1.4 Challenges and obstacles of DL in the medical domain

DL is still developing in pharmaceutical and biomedicine applications, despite some of its outstanding performance and potential. For DL to be able to tackle the fundamental health and primary difficulties, considerable challenges must be overcome. This section discusses some of the difficulties that come with implementing DL approaches.

- **Representation and transformation of medical data:** With the right type and quantity of data, DL algorithms can generate the most efficient observations and predictions. Real-world medical information is generally stored in unstructured configurations including sequence data (time series data, video and audio signals, DNA, and so on), trees (XML files, parse trees, RNA, and so on), textual information (ailments descriptions, tumor descriptions, and health records), or variations of these file types. Consequently, the DL technology's foundation can only handle quantitative input data because it is finally broken down into sequences of 0s and 1s for the computational machine to interpret. Certain subjective data is difficult to transform into a comprehensible form, and handling can be time-intensive.
- **Handling biomedical data stream:** Handling incredibly quickly and broadcasting data is another DL challenge. The healthcare business is undergoing a significant transition, with a massive amount of healthcare data being generated at a quick pace. The advantage is that physicians can use them in conjunction with the DL model to detect and administer public healthcare services for a variety of complex diseases. Such data can be obtained in multitudes of zettabytes of real-time physiological signals from a myriad of perspectives, which include diabetes management, neural activity, heart rate, and respiratory rate level; medical diagnostics from sonography, electrography, and MRI; and biomedical applications from sonography, electrography, and MRI.
- **Medical big data (MBD) analysis:** Disease control, therapy, and diagnosis are only several of the benefits of MBD analysis. Radiology devices, the web, fingerprint recognition, big clinical studies, biomarker data, professional archives, and organizational reimbursement records are all sources of MBD. In this discipline, DL provides a tool for doing intelligent and reliable statistical analysis, which will help specialists make more informed decisions, assess patient health conditions, and develop an effective AI. Unfortunately, there are a few obstacles that DL must overcome in this arena. The main hurdle is obtaining MBD due to the risk of unauthorized disclosure and lack of information sharing responsiveness, which might violate medical confidentiality and constitutional concerns, expensive items, and health researchers engaged. A further difficulty is the information gathering method performed via prescribed format and procedures, which can be time-consuming. The information is often modest compared to information from other contexts (e.g., social media), and they are derived from nonreplicable conditions or situations that are not often seen.

5.1.5 Opportunities of DL in the medical field

DL systems are currently facing issues such as confidentiality, comprehensibility of DL techniques, and developing effective frameworks to deal with varied and constantly evolving healthcare information. The use of DL technologies for patient care, such as voice and language processing, will remain to rise in the future. As a result, DL applications and services raise data privacy and security issues. According to investigators, DL models must also enhance in the following areas: data integration, comprehensibility of DL forecasts, and information protection.

In the healthcare field, DL is becoming more widely used, particularly in biomedical parameters like electrocardiogram, electroencephalogram (EEG), and electromyography (EMG). An electrocardiogram measures the heart's electrophysiologic activity, the brain's neural impulses activity is monitored by EEG, and EMG monitors the functional state of the body's nervous system.

Numerous physiological signs may now be recorded concurrently and constantly, thanks to the development of smart devices in recent years. Considering various purposes, multiple DL approaches may be needed to classify and predict these signals. Scientific investigations should concentrate on a specific generalized DL approach that can handle a variety of classifications. This strategy will save time and energy that otherwise would have been required to develop a unique solution for each classification.

5.2 DL applications in the medical domain

5.2.1 Drug discovery and medicine precision

The application of genomics to discover drugs generated a plethora of possible drugs, yet have stumbled upon numerous struggles in converting the findings into valuable drugs. Precision medicine symbolizes an innovative affiliation amid genomics and drug discovery, which affords cooperation of understanding into the contrivances and impending handling possibilities of ailment. The best example of precision medicine with gathering-focused dealings is oncology [13]. DL algorithms have accomplished exemplary enactment in big tasks. Convolutional neural network (CNN) finds its application in classifying images, detecting objects, and segmentations. Recurrent neural networks (RNNs) and their progenies like long short-term memory units (LSTMs) and gated recurrent units (GRUs) find their application in the translation of languages, recognition of speech, and generation of music and texts. Several of the fields which can probably benefit from DL's effectiveness is drug discovery. DL can dramatically improve and lower the price of drug discovery, which would be a lengthy and costly procedure. DL can be applied in the discovery of drugs under altered groups which include

- prediction of drug properties
- de novo drug design
- drug–target interaction (DTI) prediction

In drug properties prediction, the drug is given as input to the algorithm in the form of molecular fingerprint, text-based representation, and graph structure (2D or

3D graph), and the drug's property will be the output. In molecular fingerprint, input is given as a combination of binary digits (bits) sequences that characterize the occurrence or nonoccurrence of specific substructures in the molecule. In the graph structure method, the compound can be considered as a graph, where atoms act as vertices and edges are framed as the bonds between the atoms. DL libraries like Deep Graph Library, PyTorch-Geometric, and PyTorch-BigGraph have perceived a noteworthy accomplishment in this area.

De novo drug study aimed to design where certain newer compounds are designed using the receptor's 3D structure. It entails using molecular modeling techniques to assess the framework of the lead target compounds and to create lead changes. This can also be utilized to create new biological classes with equivalent substituents to the targeted by starting with a framework or structure that is genetically different from prespecified leads [14].

DTIs are vital to numerous applications like the discovery of drugs, repositioning of drugs [15], and prediction of side effects [16]. Nevertheless, the high cost involved in the validation of experiments outbreaks its achievement. Hence researchers have focused on the direction of prediction of DTIs computationally in the following ways:

- Finding whether a drug interacts with the target or not [17,18].
- Speculating the bondage of drugs with specific proteins [19,20].

5.2.2 Detection of diseases

DL algorithms have proved their efficiency in detecting diseases and their diagnosis compared to other traditional computational algorithms. It encompasses the usage of NNs, where modifications of the parameters are done if any first-hand data arrives. This feature makes it very efficient in providing the capability to acquire things by themselves without human intervention. The standard DL techniques for disease diagnosis/detection are elucidated below, which are used to train a model for effective diagnosis.

- **CNN:** It is a category of artificial NN with a minimum of three layers utilized for disease diagnosis. The three layers are convolutional, pooling, and fully connected. It works with the formulation of a feature extraction model specific to the disease. It is involved in the classification of diseases, segmentation of diseases and identification of diseases, prediction of diseases, and even drug discovery. CNN uses images for disease diagnosis. Skin cancer, breast cancer, and heart disease are the various diseases where CNN is applied with. Dilated convolutions is a variant of CNN used to diagnose diseases, which comprises a weight matrix with adjustable size (dilation). It predominantly finds its application in the classification and segmentation of diseases.
- **Fully convolutional networks:** A CNN category that does not necessitate segmentation and can be directly applied to disease diagnosis. It functions with textual data such as patients' medical records.
- **RNN/LSTM:** This technique is extensively used in language processing applications, and it comprises at least two stacks of LSTMs, which are RNN

cells. RNN is applied in the prediction of Alzheimer's disease, Parkinson's disease, and Crohn's disease.

- **Generative adversarial networks (GANs):** This method consists of two NNs. The first network produces samples, and the second evaluates the samples (discriminator). GANs predict leukemia and myocardial diseases. It uses MRI, CT, and X-ray to detect disease. Auxiliary classifier GAN (AC-GAN) is a variant of GAN that produces more exact results than GANs. Convolutional auxiliary classifier generative adversarial network (CAC-GAN) is another category of GAN which is more appropriate for multilabel classification problems such as disease diagnosis and disease prediction.
- **Attention-based deep neural network (attentional NN):** This neural technique network incorporates the context of input data into its output prediction via an attention mechanism, building it appropriate for multifaceted chores such as disease diagnosis and planning of treatment.
- **Adversarial autoencoders:** This technique builds feature extraction models specific to diseases. It has an encoder that transforms disease images or disease symptoms into disease-relevant features, and a decoder performs the reverse process from these features back to disease images or disease symptoms.

5.2.3 Diagnosing patients

DL and CNNs can be used to diagnose melanoma in images of skin lesions. Prediction of the presence of cancer tumors can be done by combining DL and image processing techniques. Data scientists use a CNN to detect coronary artery disease from images of CT or MRI. With the use of disease markers like the texture of the liver and the ability to reflect or transmit echoes, liver diseases like fatty liver can be identified by training the ultrasound images using deep NNs. Even lung diseases like tuberculosis, pulmonary nodules, and lung masses can be identified by applying DL techniques and image processing utilizing scan images. DL models can be used to develop disease diagnostic algorithms for psoriasis disease based on images of the skin lesions caused by this disease.

5.2.4 Healthcare administration

Because of its ability to provide précised suggestions in diagnosis without human intervention, DL algorithms can minimize the cost involved in providing care, avoid time delays in broadcasting emergency cases, and reduce errors in diagnosis by auditing prescriptions and diagnostic results and make the process quicker.

5.2.4.1 Medical imaging

Regarding image classification, ailment identification, and diagnosis, MR and CT techniques use computer vision and image recognition. By merging data images of cell size, density, and pattern, DL algorithms can make correct predictions. Such structures can be used to emphasize important parts of an image. Diabetic retinopathy, Alzheimer's disease, and breast nodules are all detected using DL algorithms. DL algorithms simplify complex data analysis, allowing anomalies to be identified and given preferential treatment more precisely.

5.2.4.2 Patient care: data analytics in healthcare and insurance

DL models can analyze EHRs, which include structured and unstructured data such as medical documentation, laboratory results, diagnoses, and prescriptions, faster and much more accurately than traditional models. Additional information is nowadays collected from smartphones and wearable devices. DL models can also convert data with mobile apps and monitor the risk factors. Apps related to mental health, such as Happify, Moodkit, Woebot, and Wysa, influence DL models for more genuine chats with patients. DL models also enable healthcare organizations to provide individual attention to patients after a thorough study of medical history, the symptoms experienced by the patients, and the results of various medical tests. Models are also used in performing audits of prescriptions to minimize the probable errors that might occur in diagnosis and prescriptions. DL models support insurance companies in making proposals to their clients through influential prognostic analytics. It is also utilized in recognizing fraudulent medical claims after investigating behaviors and health data from varied sources such as claims history, hospital-related information, and patient attributes.

5.2.4.3 Research and development: drug discovery and genomics analysis

The influences of DL models in unearthing the discovery of drugs have been increasing with new-fangled hi-tech improvements. DL algorithms can recognize possible combinations of drugs after quick processing of genomic, clinical, and population data. It is evident that DL models intensify interpretability and offer superior understanding, especially in biological data.

The collaboration of some of the more highly developed methodologies, higher computing power, and rising preference for creative methods of diagnosing, controlling, and cutting healthcare costs would undoubtedly help DL. With a large handful of excellent applications, a substantial increase from major manufacturers, and an increasing volume of information to facilitate cutting-edge analytics, DL will surely play an extremely important role in the effort to give the highest expected quality healthcare to clients for the future decades.

5.3 DL for medical image analysis

In medical image analysis, DL is the fastest-growing field today. The medical image field is vital for endoscopy, CT scan, mammography, MRI scan, ultrasound image, etc. In medical imaging, deep learning concepts are applied to find the disease. DL mainly supports feature extraction from images. These DL techniques easily detect the presence of disease. Implementing deep learning and artificial NNs is necessary to analyze medical images. This chapter explains the process of medical image detection, recognition, segmentation, registration, and disease diagnosis.

5.3.1 Medical image detection

Estimating and removing noise from the medical image is a vital image detection field. Denoising MRI, CT scan, and X-ray images is an impressive method. In detection, whether that medical image belongs to a particular class and whether it will appear in a region of interest. Another name for detection is localization. It will localize the position of an object in the image. In medical image analysis, detection aims to identify the abnormality of a patient using image analysis. The detection task includes the detection of lung nodules in chest images. Detection can be categorized into two approaches: anchor-based and nonanchor-based approaches. This anchor-based algorithm can be further divided into single-stage and multistage algorithms. Single-stage computation can provide efficient computation, and a multistage algorithm does good performance detection (Figure 5.2). The single-shot multibox detector is a single-stage detector with a simple detector.

The architecture mentioned above in Figure 5.2 is based on convolutional anticipation networks, which produce a fixed number of bounding boxes and their equivalent scores for object instances of certain classes in the boxes. A non-maximum suppression phase is applied to produce the final predictions. SSD uses a multiscale feature map; that is why it provides better performance. Two-stage structures generate a lot of ROI and classify it via the network. The Mask-RCNN generates objects via a regional proposal network after classifying these proposals. The significant difference between Mask-RCNN and Faster-RCNN is a segmentation branch. CornerNet is another type of network known as a single convolutional neural network mentioned in Figure 5.3. It will eliminate anchor boxes by using paired vitalpoints. In the top-let and bottom-left corners, the object bounding box is mentioned.

The detection method's performance has been measured using evaluation measures: mean average precision (MAP) and false-positive per image (FP/I). The aggregate of all mean precisions (APs) of all classes is computed using the MAP. The FP/I recall rate is the ratio of each false-positive (FP) image under a certain recall rate, which takes into account the proportion between false positives and missing rates.

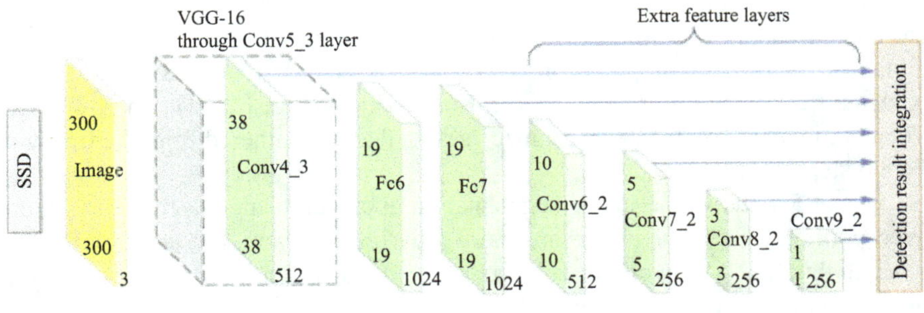

Figure 5.2 SSD for image detection

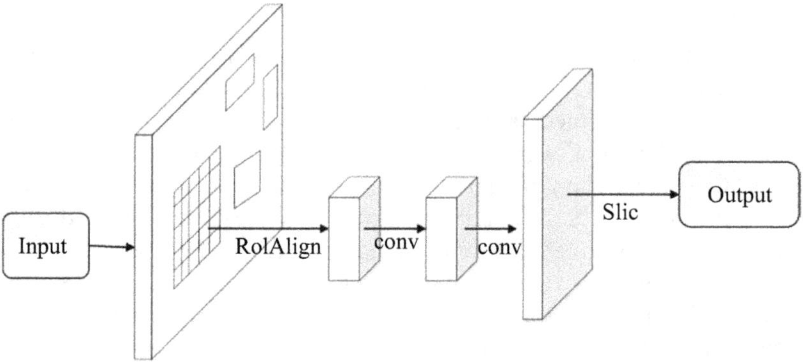

Figure 5.3 Mask-RCNN and CornerNet for image detection

5.3.2 Medical image recognition

Recognition or classification is the process of identifying and categorizing images into different classes. Traditional approaches depend entirely on form, color, and texture features, as well as their combinations, the majority of whom are the issue and supplementary in medical imaging, resulting in a high-level picture recognition challenge. Latest deep learning algorithms offer a quick and easy way to build an end-to-end model that can calculate classification process labels from raw medical image pixels. Deep learning methods, on the other hand, suffer from severe computational costs and limits in structural models and streams depending on the quality of medical pictures and the short size of the dataset. This problem can be solved by integrating a coding network and multilayer perceptron. High-level features and some selected features are extracted using deep CNN (DCNN). Overall, medical image classification can be categorized into two steps:

- The first step is to extract the most intriguing qualities of the image.
- The characteristics are then used to create models that categorize the image dataset in the second stage.

The DCNN would be first trained as a scripting network, similar to supervised learning. It can extract information from primary pixels in medical images that indicate a high level of precision. Next, a selected set of standard features is extracted depending on previous medical image knowledge. At last, an NN-based efficient model is designed to recognize various features gathered from the first and second steps.

Figure 5.4 shows the image segmentation approach using CNN. Convolution layer, pooling layers, one or perhaps more fully linked layers, and a softmax layer make up this CNN. Convolutional layers are coupled with pooling layers to reduce dimensionality. A softmax layer is used to classify the data. The following is a description of the deep learning model's concept design:

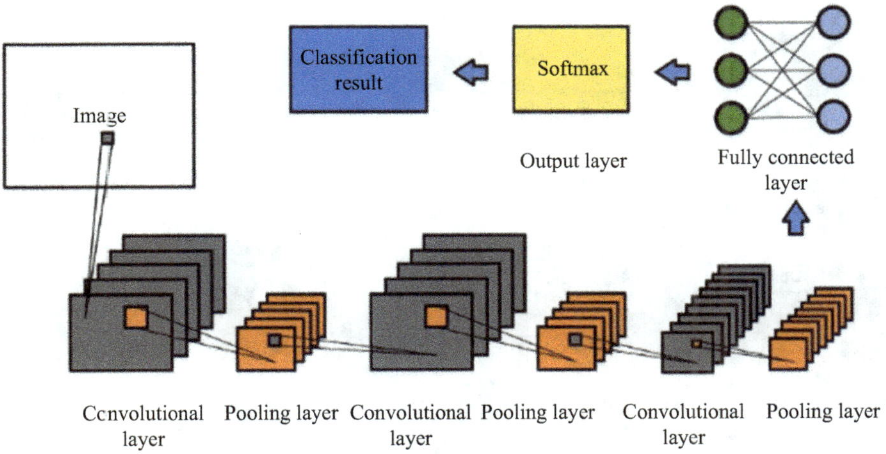

Figure 5.4 Image segmentation approach using CNN

- After subtracting the mean red, green, blue (RGB) value and zero component analysis (ZCA) whitening, the image is preprocessed by choosing the optimum activation function.
- Initial weights are a critical factor. The deep net would not train if the initial weights are too small, and if they are too large, the initial weights will differ significantly.
- The next set of critical image recognition tasks is data augmentation. This should retrieve random regions from the actual picture and diagonally rotate them in the frame.
- It will also use dropout to minimize overfitting and local response normalization to minimize the error rate.
- Select the proper learning rate. It is essential in each step.

According to the steps mentioned above, CNN is trained as a coding network in the following way.

5.3.3 Medical image segmentation

Segmentation is an image analysis technique that divides a selected image into distinct sections with similar features. It is the technique of recognizing organs or tumors using CT scans or MRI images in order to acquire data about their shapes and sizes. Formerly, image filters were used in the approach of automating this technique. Deep learning seems to have become the dominating approach since the introduction of AI, demonstrating impressive abilities in image recognition tasks. It is really a pixel labeling issue, whereby an image is divided into sections with similar characteristics. Segmentation is used in computer-aided diagnosis to detect the shape of a tissue or biological component in the images. Using multiple modalities of medical imaging, segmentation activities in clinical applications include

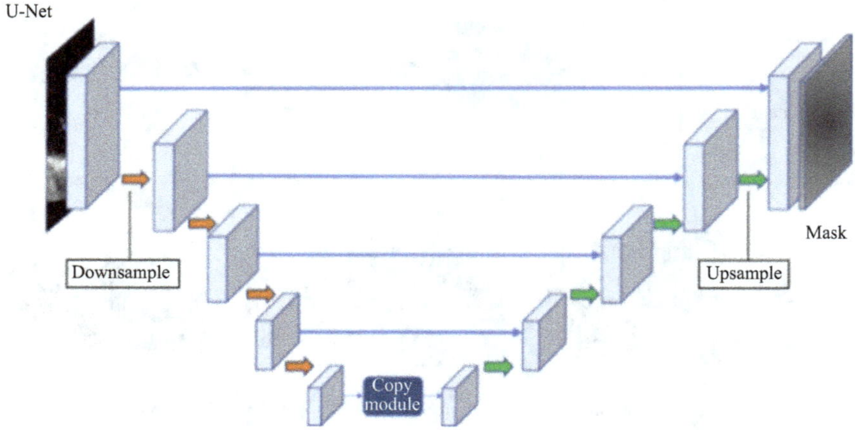

Figure 5.5 U-Net architecture for medical image segmentation

the segmentation of various organs such as the complete heart and pancreas, tumors, and lesions such as liver and liver cancer (Figure 5.5).

The main challenge of deep learning in medical imaging. Image segmentation, the holy grail of quantitative image analysis, is the practice of partitioning an image into multiple regions sharing similar attributes, allowing localization and quantification. For 50 years, these segmentation concepts have been used in the medical field. The multispectral tissue classification report is the most widely used ML segmentation technique. Statistical pattern recognition is a leading technology used in the medical field for segmentation.

5.3.4 Medical image registration

Image registration is also denoted as image fusion, image matching, and image warping. It is the process of arranging or aligning two or more images based on appearance. It was an essential preprocessing step applied in various medical applications. The goal is to establish optimal correspondence within the acquired images at various times by different patients, different image modalities, and different views. The purpose of medical image registration is to discover the appropriate spatial modification for aligning the fundamental anatomical characteristics. It is used in various therapeutic settings, including image guidance, motion detection, categorization, dosage accumulation, and picture restoration. The registration of medical images is a broad topic that can be broken down into numerous categories. Image registration can be classified as unimodal, multimodal, interpatient, or intrapatient depending on the input image. The registration methods can be separated into rigid, affine, and deformable approaches based on the deformation model. Recording methods may be separated into 3D, 3D into 2D, and 2D into 2D/3D depending on anatomical sites such as brain, lung recording, and so on. Image pair registration methods can also be split into 3D, 3D into 2D, and 2D into 2D/3D.

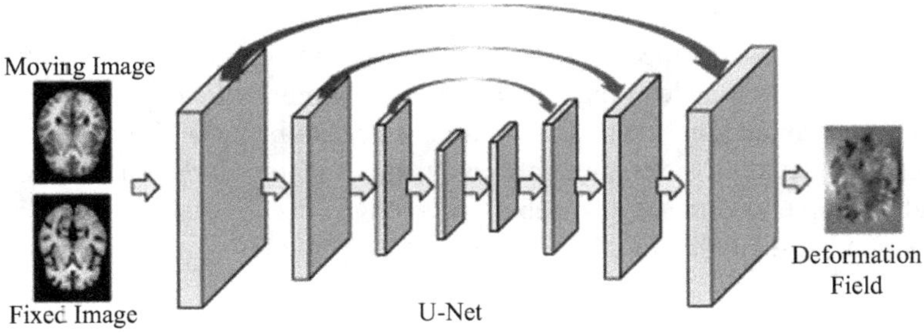

Figure 5.6 Image registration using U-Net

Figure 5.6 shows the image registration process. It combines moving and field images into a single image as a deformation field.

5.3.5 Disease diagnosis and quantification

An early diagnosis of a disease may prevent the death rate due to the disease. An efficient deep learning method can easily predict the disease by processing input images. The disease can be adequately quantified while applying medical image detection, recognition, segmentation, and registration. The accurate CNN segmentation method predicts the disease in the spine, kidney, brain, and lungs. The growth of advanced deep learning methods has facilitated abundant success in analyzing medical images with high precision, efficiency, stability, and scalability.

5.4 Conclusion

Deep learning is a powerful tool in the medical field. It has been used with great success for target identification, fragmentation, categorization, and identification. Deep learning's advancement in medicine depends on the development of massive data that has cross features and supplies a massive amount of valuable information for deep learning. Deep learning can indeed locate the lesion location and differentiate and categorize individual lesions in effect on disease therapy. Whereas deep learning offers its benefits, it also has some drawbacks. Large datasets play a significant role in deep learning algorithms.

Deep learning's rapid development in the medical area is inextricably linked to a wide variety of therapeutic procedures. Bettering the application of deep learning, including all phases of medical care, gets more complicated. It is dependent on two factors: one is the continual refinement of technology, and another is the development of medical background. In the disciplines of autonomous cars, computational linguistics, machine vision, and other areas, many good techniques have recently emerged. These techniques have garnered considerable attention in particular disciplines, and figuring out how to employ these powerful deep learning techniques is something academics should keep thinking about.

With the arrival of the 5G revolution, a significant and large arena for deep medical learning has opened up. The standard ML techniques are focused at the system level, while deep learning employs relatively minimal hardware innovation. The machine can acquire genuine cognition, a perfect blend of 5G networks, and deep learning technologies. Simultaneously, the ongoing adoption of intelligent medical instruments and surgical robots encourages deep learning at the physical level, which substantially simplifies treatment for patients.

References

[1] Muller H and Unay D. Retrieval from and understanding of large-scale multi-modal medical datasets: a review. *IEEE Trans Multimedia* 2017; 19(9):2093–2104.

[2] Greenspan H, van Ginneken B, and Summers RM. Guest editorial deep learning in medical imaging: overview and future promise of an exciting new technique. *IEEE Trans Med Imaging* 2016;35(5):1153–1159.

[3] Das S and Nene MJ. A survey on types of machine learning techniques in intrusion prevention systems. In: *Proc Int Conf Wirel Commun Signal Process Netw*. Presented at IEEE'17, Chennai, India, March 22–24, 2017, pp. 2296–2299.

[4] Aggarwal LM. Advances in medical technology and its impact on health care in developing countries. *Int J Radiol Radiat Ther* 2017;2(2):569–576.

[5] Sasubilli G and Kumar A. Machine learning and big data implementation on health care data. In: *Fourth Int Conf Intell Comput Control Syst (ICICCS)*, 2020, pp. 859–864, doi: 10.1109/ICICCS48265.2020.9120906.

[6] Rivas R, Montazeri N, Le NX, and Hristidis V. Automatic classification of online doctor reviews: evaluation of text classifier algorithms. *J Med Internet Res* 2018;20(11):e11141.

[7] Zhang K, Liu X, Liu F, *et al.* An interpretable and expandable deep learning diagnostic system for multiple ocular diseases: qualitative study. *J Med Internet Res* 2018;20(11):e11144.

[8] Yeganegi H, Fathi Y, and Erfanian A. Decoding hind limb kinematics from neuronal activity of the dorsal horn neurons using multiple level learning algorithm. *Sci Rep* 2018;8(1):577.

[9] Bing D, Ying J, Miao J, *et al.* Predicting the hearing outcome in sudden sensorineural hearing loss via machine learning models. *Clin Otolaryngol* 2018;43(3):868–874.

[10] Hu X and Yu Z. Diagnosis of mesothelioma with deep learning. *Oncol Lett* 2019;17(2):1483–1490.

[11] Mathews SM, Kambhamettu C, and Barner KE. A novel application of deep learning for single-lead ECG classification. *Comput Biol Med* 2018;99:53–62.

[12] Chu Y, Zhao X, Zou Y, Xu W, Han J, and Zhao Y. A decoding scheme for incomplete motor imagery EEG with deep belief network. *Front Neurosci* 2018;12:680.

[13] Sasubilli SM, Kumar A, and Dutt V. Machine learning implementation on medical domain to identify disease insights using TMS. In: *2020 Int Conf Adv Comput Commun Eng (ICACCE)*, 2020, pp. 1–4, doi: 10.1109/ICACCE 49060.2020.9154960.

[14] Jain SK and Agrawal A. De novo drug design: an overview. *Indian J Pharm Sci* 2004;66(6):721.

[15] Jarada TN, Rokne JG, and Alhajj R. A review of computational drug repositioning: strategies, approaches, opportunities, challenges, and directions. *J Cheminform* 2020;12(1):46.

[16] Masoudi-Nejad A, Mousavian Z, and Bozorgmehr JH. Drug-target and disease networks: polypharmacology in the post-genomic era. *Silico Pharmacol* 2013;1:17.

[17] Chen X, Yan CC, Zhang X, *et al.* Drug–target interaction prediction: databases, web servers and computational models. *Brief Bioinform* 2016; 17(4):696–712.

[18] Sasubilli SM, Kumar A, and Dutt V. Improving health care by help of Internet of Things and bigdata analytics and cloud computing. In: *2020 Int Conf Adv Comput Commun Eng (ICACCE)*, 2020, pp. 1–4, doi: 10.1109/ ICACCE49060.2020.9155042.

[19] Dai H, Umarov R, Kuwahara H, Li Y, Song L, and Gao X. Sequence2Vec: a novel embedding approach for modeling transcription factor binding affinity landscape. *Bioinformatics* 2017;33(22):3575–3583.

[20] Swarna SR, Kumar A, Dixit P, and Sairam TVM. Parkinson's disease prediction using adaptive quantum computing. In: *2021 Third International Conf Intell Commun Technol Virtual Mobile Netw (ICICV)*, 2021, pp. 1396–1401, doi:10.1109/ICICV50876.2021.9388628.

Chapter 6

Deep-stacked autoencoder for medical image classification

J. Anitha[1], S. Akila Agnes[1], S. Immanuel Alex Pandian[2] and Malin Bruntha[2]

Image classification is the most important process in the computer-aided diagnosis system. Feature learning is the major challenge in the classification task where the representation features are refined from the high-dimensional input data. Autoencoder (AE) is a self-supervised neural network that maps the input data with the target output data through the sequence of encoder and decoder layers. It automatically learns the abstract-level representation features with the bottleneck feature of encoder–decoder based neural networks. The encoder layers are trained on various levels in extracting the essential features to represent the image. At last, all levels of encoders are stacked together to form a stacked network. The decoder part is replaced by the final softmax layer for classification. The complete stacked network is fine-tuned for classification in a supervised fashion. In this chapter, a stacked AE model is developed to classify the medical images, particularly skin cancer images, into benign or malignant. From the experimental results, it is noted that the classification accuracy of a stacked AE with fine-tuning is higher than that of a stacked AE without fine-tuning. The experimental results also confirm that the stacked AE model with fine-tuning provides improved results as compared with other classification methods.

6.1 Introduction

Developing the computer-aided diagnosis (CAD) system provides a second opinion to the clinicians that guarantees early detection and provides an increase in the sensitivity rate in a cost-effective manner. The general framework of the CAD system follows four stages: (1) data collection, (2) feature extraction, (3) building classifier model, and (4) model evaluation phase.

[1]Department of Computer Science and Engineering, Karunya Institute of Technology and Sciences, India
[2]Department of Electronics and Computer Engineering, Karunya Institute of Technology and Sciences, India

The classifier used in the CAD system is trained with the feature extracted from the training images and the performance of the classifier is evaluated with the testing images.

The traditional classification methods use intensity, gradient, statistical, time-series, and morphological information as a feature to represent the input images for the classifiers. But these right set of features has to be extracted in a handcrafted manner and given as an input to the machine learning models, namely logistic regression [1–3], naïve Bayes [4–6], support vector machine (SVM) [7–9], etc. These methods are completely dependent on the handcrafted features that require a lot of computations, human intervention, and prior knowledge.

Nowadays, deep learning, the subfield of machine learning, provides a powerful solution to extract prominent features from the input with feature engineering that supports intelligent classification systems. This feature engineering works through multilayer architecture such as artificial neural networks [10,11], autoencoders (AEs) [12–16], restricted Boltzmann machine [17,18], and convolutional neural networks (CNN) [19–21], where it extracts the hierarchy of low-level features to high-level features from the raw images. As compared to traditional feature extraction methods, deep learning methods are automatic and eliminate the influences of handcrafted features. The hyperparameters used in machine learning models and the classification patterns are trained together. Hence, the deep learning approach is mainly adopted in a wide range of image classification problems.

CNN is a feed-forward deep learning architecture that has achieved promising results in solving segmentation and classification problems in the field of computer vision. CNN solves the classification problem by progressively extracting the hierarchical levels of features directly from the input image with various convolution filters. Since the feature learning and classification tasks are confined to each other, the model could accurately estimate the weightage of features toward the target application. In recent years, several works [22–26] have been reported in the field of image classification with CNN.

AE is one of the widely used deep learning techniques that attracted considerable attention in feature reduction tasks employed in the field of denoising [15,27], segmentation, and classification [28–30]. AE model examines the complex relationships in the dataset. Also, it extracts the high-level features with a reduced dimension that provides better classification performance. Since the AE focuses on image-to-image learning, many researchers have worked in stacked AE-based classification algorithms to simplify the process of early diagnosis in various medical image diagnosis systems.

A deep-stacked autoencoder (DSAE) is proposed by Ahmad *et al.* [31] for liver segmentation. The unsupervised features present in the input images have been learned and fine-tuned with the softmax layer using the labels given for the input images. They have also adopted patch-based learning instead of a pixel by pixel learning to reduce the complexity of the training algorithm with efficient classification accuracy.

A cervical cancer diagnosis system [13] has been developed to classify the various types of cancer such as Schiller, Cytology, Biopsy, and Hinselmann from the UCI database with the help of stacked AE and softmax classification. The performance of the softmax classifier is compared with the other machine learning methods where the proposed stacked AE with softmax classification achieves a classification rate of 97.8%.

Xu *et al.* [32] proposed a stacked sparse autoencoder (SSAE) to capture the high-level features from image patches and classify them as nuclear or non-nuclear on breast cancer using high-resolution histopathology images. Praveen *et al.* [33] also proposed an SSAE model to learn the features from the MRI brain images to accurately segment the stroke lesions. The sparse AE with an SVM classifier is utilized to classify the patches as lesions or normal with a mean accuracy of 90.4%.

A Type 2 diabetes classification system [12] has been developed using stacked AEs with a softmax classifier to experiment on UCI machine learning data. The network is further fine-tuned with the training dataset using backpropagation. This model achieves a classification accuracy of 86.26% as compared to other existing classification methods.

Skin disease is a widespread critical disease in the world that can be visually inspected and identified on the skin surface. There are several types of skin diseases present, such as melanoma, seborrheic keratosis, nevus, melanocytic, and basal cell carcinoma. Detection and diagnosis of suspicious lesions in the skin mainly depend on the visual examination by dermatologists. Melanoma is considered an important skin cancer type where the mortality rate is high. It is difficult for dermatologists to manually examine all melanoma cases visually in clinical-based diagnosis methods. Early detection helps in providing effective treatment. Several automated detection of skin cancer systems [34–39] has been developed using machine learning and deep network approaches to perform early diagnosis of skin cancer.

Lu *et al.* [40] have proposed a variational AE to detect anomalies in the skin disease images. The experiment is carried out on the ISIC2018 Challenge Disease Classification dataset where the model can able to detect the diseases with an area under the curve of 0.779. An automated basal cell carcinoma detection technique has been developed by Kharazmi *et al.* [41] where the visual features are extracted with the sparse AE (SAE). The compact features extracted from the dermoscopy images are used as a filter. The feature map is created by convolving the filter values with the input image. These features are further integrated with the patient profile information and given as an input to the softmax classifier to further improve the classification performance.

Segmentation of skin tones suffers from varying illumination conditions and different ethnicities. This problem is overcome by designing stacked AEs [42] to learn the discriminative information from the patches of the skin area. Rundo *et al.* [43] proposed a two-level stacked AE with the softmax classifier to perform skin cancer analysis that classifies the image as benign or malignant.

In this chapter, a DSAE model is proposed to classify the medical images, especially for skin cancer diagnosis. The performance of the proposed classification scheme is evaluated with the skin cancer images collected from the Kaggle*

*https://www.kaggle.com/fanconic/skin-cancer-malignant-vs-benign

dataset to classify the images as benign or malignant. The rest of the section is ordered as follows: Section 6.2 describes the general architecture of the AE and its variants. Section 6.3 deals with the proposed framework for skin cancer diagnosis using DSAE. The results and discussions are reported in Section 6.4. Finally, Section 6.5 presents the conclusion.

6.2 Autoencoder

AE is a self-supervised learning technique that takes the image as raw input data and learns to compress and decompress the image with minimum error. During the model learning phase, the network senses the best representation of the input image at the latent space. AE learns the unique aspects of an image by downsampling the image from high dimension to low dimension using non-linear transformation. While learning, the AE preserves the significant features while eliminating the insignificant ones. As compared to principal component analysis with AE, both provide dimensionality reduction. But the earlier one works in a linear manner and the latter one works in a non-linear manner. This property enables the AE to be used as a feature extractor in the classification task.

The AE process contains two main parts: encoder and decoder. The encoder is responsible for downsampling or compresses the input into a reduced number of bits. The decoder decompresses or upsamples the encoded data into the image. The decoder trains to reconstruct the image exactly provided that the encoder provides the best-encoded data. Figure 6.1 shows a simple AE for feature learning with three layers where the input and output layers have the same number of neurons and the hidden layer has less number of neurons. Consider an input image with 16 pixels which are compressed by the encoder into 4 pixels. The decoder progressively learns to reconstruct the original image from the 4 pixels obtained in the hidden layer.

The important characteristic of the AE [13] is it has an equal number of neurons in the input and output layer, and the series of hidden layers with a decreased

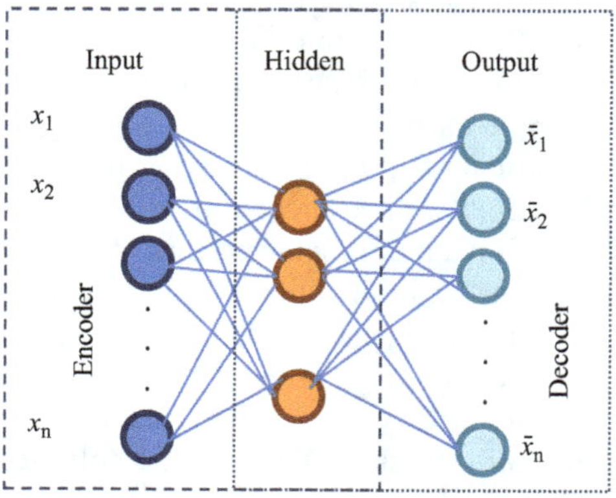

Figure 6.1 AE network for feature learning

number of neurons make this model perform well than other data reduction methods. The number of nodes in the encoder's successive layers is progressively reduced in size, which causes a reduction in dimensions. The decoder segment is the inverse structure of the encoder that comprises a series of hidden layers and an output layer. The transpose weights of the encoder are taken for decoder weights instead of random initialization. Throughout the learning process, model weights are set in order to learn abstract features using the backpropagation algorithm. The encoder derives the features from the input image, and the decoder restores the image with the lowest error between the actual and the replicated image.

AE learns the adaptive patterns from the input so that the recreated image $\overline{x}$ is identical to the image reference x, where the input image is composed of a pixel vector. The network weights are modified during the training to mitigate the loss of reconstruction [error $= x - \overline{x}$]. The AE is limited to data; the encoder model that is learned for one domain may not be suitable for another. It recreates the image that closely resembles the actual image. Encoder trains to compress input data in a lossy manner.

$$h(x) = s(x * w + b) \tag{6.1}$$

$$\overline{x} = s\left(h(x) * \overline{w} + \overline{b}\right) \tag{6.2}$$

where w and $\overline{w}$ represent the weight matrices, b and $\overline{b}$ represent the bias of the encoder and decoder, respectively. The loss of reconstruction should be minimized as follows:

$$\min_{\left(w,b,\overline{w},\overline{b}\right)} = \sum_{i=1}^{n} \|x_i - \overline{x}_i\|^2 \tag{6.3}$$

6.2.1 Stacked AE

The encoders present in the AE extract features from the input data. The stacked AE model is formed by stacking the encoders together from the different AEs. This model is utilized for feature extraction. Further, a softmax layer is added at the end of the model to construct a stacked AE network for classification. Figure 6.2 shows a stacked AE model with two levels of stacking. The first AE is trained with the input data and the features are learned. The features extracted from the first encoder are fed as an input to the second AE. The two encoders are trained independently with each other. After the training process is completed the hidden layers are stacked to formulate a stacked encoder with reduced feature representation.

6.2.2 Sparse AE

When converting high-dimensional input vectors into low-dimensional function vectors, overfitting is a major issue. Regularization is a technique that prevents the model from overfitting by ignoring some of the features, or by resetting weights to about zero. The sparse AE is an improved AE incorporating the sparse penalty concept, which restricts the hidden layer neuron's average activation value. It can

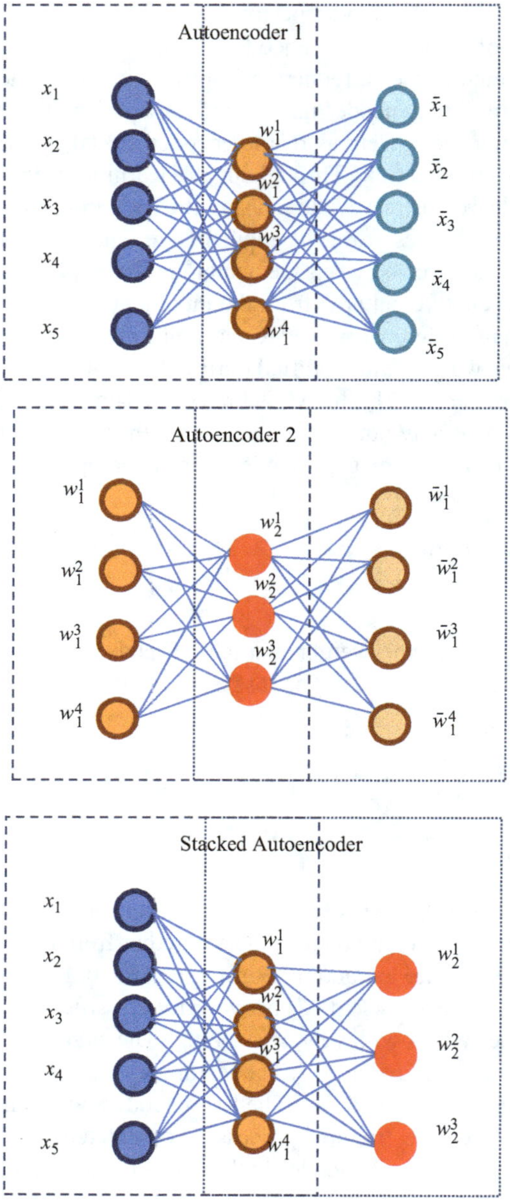

Figure 6.2 A general stacked AE network for feature learning

have fewer nodes on the input layer than the hidden layer. AE is a dense neural network; the network can have few null weight connections during training. The efficiency of the network improves by removing zero-weight links, which is called sparsity. The down sparsity applied between the first input layer and the hidden

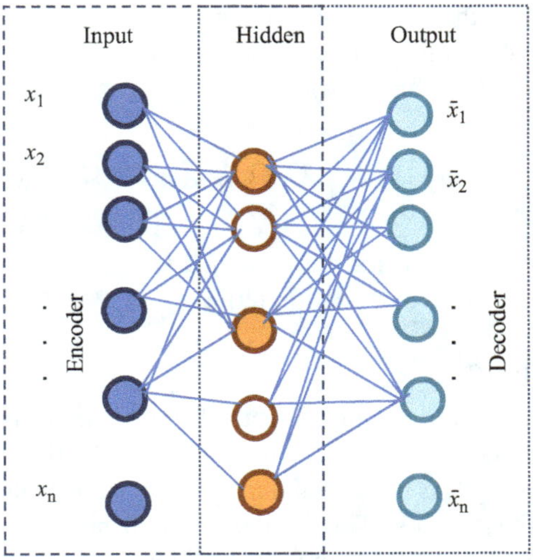

Figure 6.3 A general model of the sparse AE

layer sets the sparsity point, which specifies the active connections in order to map the data into functions [44]. Figure 6.3 shows a sample model of sparse AE where the active nodes in the hidden layer are represented in green color and the inactive nodes are represented with no color.

The level of the network sparsity may vary between 0.1 and 0.9. When the sparsity is set to 0.1, it is presumed that 10% of the connections will have a minimum weight that is very close to zero and may be skipped at the time of calculation. In order to monitor the average output activation of a neuron, sparsity is applied to the cost function, formulated as

$$\widehat{p}_i = \frac{1}{m}\sum_{j=1}^{m} z_i(x_j) \tag{6.4}$$

where i refers to the ith neuron, m is the total number of training samples, and j is the jth training sample. The sparsity regularization is characterized by the Kullback–Leibler divergence [45] as

$$\Omega_{\text{sparsity}} = \sum_{i=1}^{d} p\log\left(\frac{p}{\widehat{p}_i}\right)\log\left(\frac{1-p}{1-\widehat{p}_i}\right) \tag{6.5}$$

where p is the desired activation value called the sparsity ratio and d is the total neurons in a layer. Additionally, the L_2 standard regularization term is included in the cost function to control the weights.

$$\Omega_{\text{weights}} = \frac{1}{2}\sum_{l}^{L}\sum_{j}^{N}\sum_{i}^{K}\left(w_{ji}^{l}\right)^2 \tag{6.6}$$

where L, N, and K are the number of hidden layers, the number of observations, and the number of features, respectively. The updated cost function after adding the regularization terms in (6.5) and (6.6) is given by

$$E = \frac{1}{N}\sum_{j=1}^{N}\sum_{i=1}^{K} \|x_{ij} - \bar{x}_{ij}\|^2 + \lambda * \Omega_{\text{weights}} + \beta * \Omega_{\text{sparsity}} \tag{6.7}$$

where λ defines the coefficient for the L_2 regularization term and β represents the coefficient for the sparsity regularization term.

6.2.3 Convolutional AE

Convolutional AEs discover features directly from the input images by using the self-supervised learning method. In a convolutional AE, the location-based relationship between the pixels is utilized and makes it more acceptable than other one-dimensional vector-based AEs. This model learns a number of dynamic spatial filters that could extract abstract features from all hierarchical levels. Convolutional AE has encoder and decoder blocks, where the context-specific features are extracted by the filters of the encoder block. The decoder block is composed of deconvolutional layers that reconstruct the image from the extracted features. The sizes of the input and the output image remain the same as in the simple AE, but the encoder and decoder parts need not be symmetrical in the convolutional AE network. Figure 6.4 shows a general convolution AE with convolution layers, max-pooling layers, upsampling layers, and deconvolution layers.

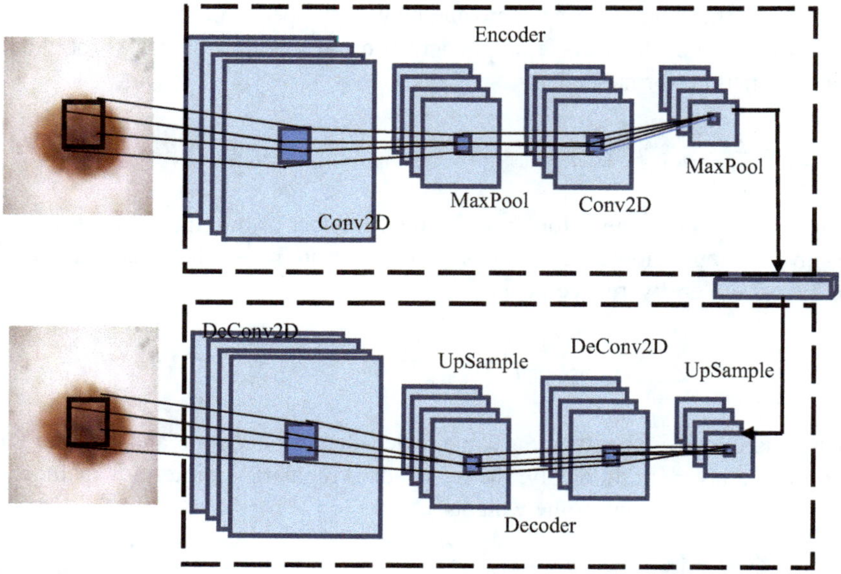

Figure 6.4 A general model of convolutional AE

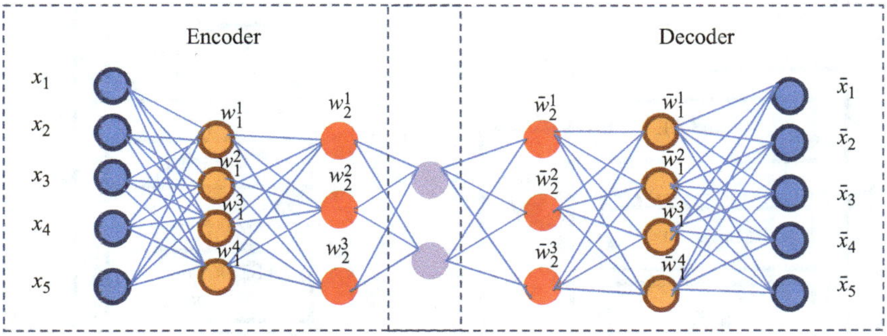

Figure 6.5 A general model of deep AE

6.2.4 Deep AE

A deep AE contains a symmetrical network, where the image is encoded with a sequence of encoding layers in the encoder section and decoded with an equivalent number of decoding layers in the decoder section. The encoder section contains a decreasing number of neurons in the subsequent encoding layer that enforces dimensionality reduction. The decoding part does the transition from the reduced features into the original image with the increasing number of neurons in the respective decoding layers. The principal objective of this process is to recreate the original image from the reduced latent space with minimum error. Figure 6.5 shows a deep AE with three levels of encoding and decoding layers.

6.3 Proposed method

The proposed method uses a DSAE for identifying the suitable features to classify the skin cancer images into benign or malignant. The feature extracted through the DSAE is fed into various classifier algorithms such as Softmax, SVM, and K-nearest neighbor (KNN) for skin cancer classification. The block diagram of the proposed system for classifying the skin cancer images is shown in Figure 6.6. The system contains both feature learning and classification phases. In the representation learning phase, the representation features are learned from the input skin images using a DSAE model. In this DSAE, two-level AEs are stacked and ended with a classification layer. In the classification phase, the features extracted from the AE are used as input and the classifier is trained with the extracted features by supervised learning.

The features extracted from the first-level encoder–decoder network are fed as an input to the second-level encoder–decoder network. After the models are trained, the encoder part of the encoder–decoder networks is stacked to form a stacked AE for feature extraction. The classifier uses the features extracted by the stacked AE as input for classifying the input image as benign or malignant. To improve the classification further the DSAE is fine-tuned. During fine-tuning, the model is tuned with supervised learning to minimize the classification error and to improve the performance of the model. Further, the model is evaluated by adding sparsity to the AE that prevents model overfitting by discarding some of the features.

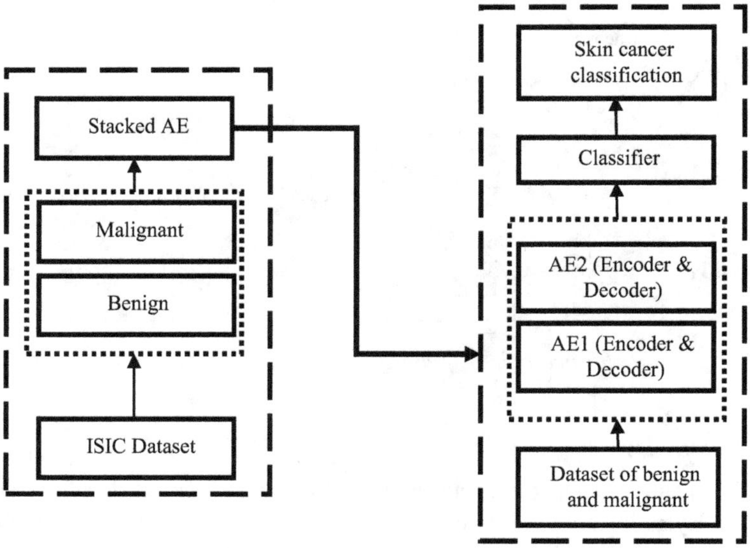

Figure 6.6 The block diagram of the DSAE for skin cancer classification

6.3.1 Representation learning using AE

A deep AE contains an AE that condenses high-dimensional spatial information into a low-dimensional set of features without the lack of critical information. For this experiment, the input images of skin cancer belong to two classes, known as benign and malignant, where representational characteristics are studied by self-supervised learning techniques. The bottleneck property of AE enables the model to be utilized in noise removal and dimensionality reduction applications. Bottleneck property determines the degree of compactness and the level of retainability. A bottleneck is a middle layer placed in between an encoder and a decoder. The representation features of an image are achieved by AE using the back-propagation algorithm. The model could retain the relevant information and also reject the irrelevant and duplicate information through representation learning. The dimensional reduction scale is defined by the size of the bottleneck layer. In this proposed model four different configuration of the two-level AE is stacked for feature extraction and the model details are presented in Table 6.1. As there are no harden rules for fixing the values of weight normalization parameters, the suitable values for this application are taken by trial and error method.

6.3.2 Softmax layer

In this proposed model the neurons at every hidden layer use the logistic sigmoid activation function ($(1 + e^{-x})^{-1}$) and the final classification layer uses the softmax activation function. A softmax function is a type of linear function which normalize the data in terms of probability in the range of 0–1. A softmax layer is a final layer

Table 6.1 Specification of various stacked AE model

Model name	Size of bottleneck		Weight normalization parameters
	AE1	AE2	
Model 1: Stacked AE (200–100)	200	100	$\Omega_{\text{weights}} = 0.001$ Sparsity proportion $= 0.05$ $\Omega_{\text{sparsity}} = 1$
Model 2: Stacked AE (200–50)	200	50	$\Omega_{\text{weights}} = 0.001$ Sparsity proportion $= 0.05$ $\Omega_{\text{sparsity}} = 1$
Model 3: Stacked SAE (200–100)	200	100	$\Omega_{\text{weights}} = 0.004$ Sparsity proportion $= 0.15$ $\Omega_{\text{sparsity}} = 4$
Model 4: Stacked SAE (200–50)	200	50	$\Omega_{\text{weights}} = 0.004$ Sparsity proportion $= 0.15$ $\Omega_{\text{sparsity}} = 4$

in any neural network that uses logistic regression for classification. The number of neurons at the softmax layer is determined by the number of classes considered for classification. The softmax function takes an N-dimensional vector of arbitrary real values and computes the equivalent N-dimensional probability vector where the sum of the probability of all the classes will be one.

$$\text{softmax}(X) : \begin{bmatrix} x_1 \\ \cdots \\ x_N \end{bmatrix} \rightarrow \begin{bmatrix} s_1 \\ \cdots \\ s_N \end{bmatrix} \tag{6.8}$$

$$s_i = \frac{e^{x_i}}{\sum_{j=1\ldots N} e^{x_j}} \tag{6.9}$$

6.3.3 Support vector machine

SVM is a supervised learning approach that is used for classification problems. SVM classifier categorizes a set of features by determining the optimal hyperplane that divides the feature set into various groups. The objective of the SVM classifier is to find an optimal hyperplane that gives the largest margin between classes. The margin of the classifier is computed as the distance between the hyperplane and the support vectors. Here the support vectors are the data points that are closer to the hyperplane. There are various kernel functions utilized in SVM to find the distance between the hyperplane and support vectors such as the linear kernel, polynomial kernel, and radial basis function kernel/Gaussian kernel. In this experiment, the Gaussian kernel function is used to compute the distance in the SVM classifier.

6.3.4 K-nearest neighbor

KNN is a non-parametric supervised learning approach used for classification problems. The distance metric provides a similarity between features in the dataset. The features set is ordered by the increasing distance. Then the category of the feature is estimated from the KNNs in the dataset. In this experiment, a Euclidean distance function is used as a distance metric to find (K=5) the nearest features, and the feature is classified by a plurality voting among its neighbors.

6.3.5 Fine-tuning

Fine-tuning is an idea of transfer learning. Transfer learning helps to utilize the knowledge gained by solving one problem and using it in the related problem. In the proposed model, the AE networks are trained to learn the essential features from the image, and then the features gained from the AEs are used for classifying the images. The fine-tuning process corrects features weights for the classification problem in a supervised manner. The objective of this fine-tuning process is to minimize the classification error with backpropagation.

6.3.6 Sparsity and regularization in AE

AE uses a backpropagation neural network to convert input pixel data to a set of features with minimal reconstruction error. The dense AE has approximately 5 million relations between the input layer neurons and the output hidden neurons. Almost all neurons in the input layer are bound to one neuron in the next hidden layer. Of all these connections, only a few are powerful, while some are weak or useless in the secret layer to trigger the neuron. In order to highlight the latent structure of input data, these sparsely active connections should be strengthened. Small variation in the regularization parameters of the sparse autoencoder model makes a considerable difference in the performance of the model. When designing a sparse AE, it is a difficult task to decide the optimal values for these hyperparameters. Furthermore, the hyperparameter values of the AE are not standardized, which means that the model has been trained for one domain and cannot match another domain.

6.4 Results and discussions

The simulations are performed in MATLAB 2018a on an Intel(R) Core(TM) i7-8550U CPU processor with 16 GB RAM and 3.79 GHz on the Microsoft Windows 10 platform. The features extracted from the images with the help of the DSAE are classified through various classifiers such as Softmax, SVM, and KNN. In this experiment, stacked AE models with various configurations are developed to investigate the effect of feature extraction with respect to the number of neurons in encoder layers, number of encoder layers, and regularization parameters. In stacked AE every encoder layer in the encoder section is independently trained and stacked. The DSAE network is developed by stacking trained encoder layers with its learned weights. Also, the performance of DSAE models in classifying skin cancer images is analyzed with and without fine-tuning.

6.4.1 Datasets

The dataset used in this experiment is obtained from the Kaggle website. This contains benign and malignant samples of nevus and melanoma skin cancer collected from the ISIC-Archive. The images in the ISIC-Archive are categorized based on various attributes such as diagnostic, clinical, technological, and database information. Each image contains metadata about the clinical information,

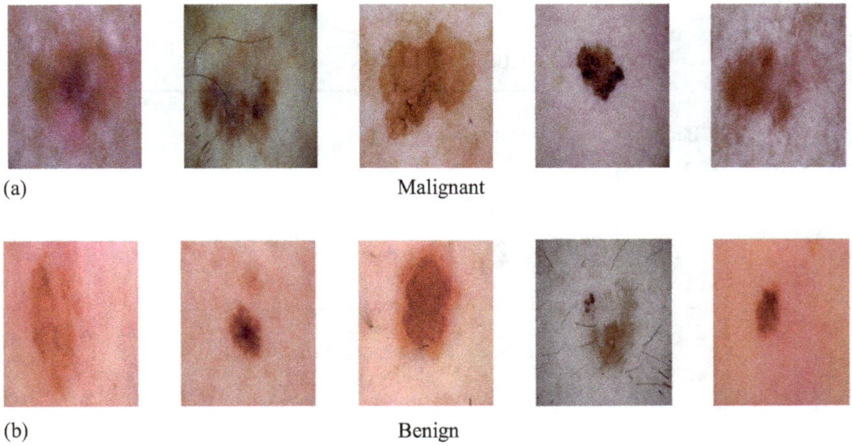

Figure 6.7 Sample set of images from the dataset

dimension, tag, info, and segmentation results with expert marked cancer area. The clinical information includes age, sex, diagnosis type, benign/malignant, and melanocytic. The collected dataset contains a separate set of training and testing images of size 224 × 224 × 3 including malignant and benign cases. The training set contains 1,440 images of benign type and 1,197 images of malignant type. Similarly, the testing set contains 360 and 300 images of benign and malignant types, respectively. In this experiment, the images are resized by 64 × 64 × 3 to speed up the process. The proposed system develops a model to learn features through DSAE and classify the skin cancer images into benign or malignant. The model utilizes the stacked AE as a feature extractor with dimensionality reduction capability. Figure 6.7 shows a sample set of benign and malignant types of skin cancer images.

6.4.2 Evaluation metrics

The performance of the proposed model is evaluated in terms of accuracy, specificity, precision, recall, and F-score that are calculated from the confusion matrix as shown in Figure 6.8.

$$\text{Accuracy} = \frac{\text{TP} + \text{TN}}{\text{TP} + \text{FP} + \text{TN} + \text{FN}} \tag{6.10}$$

$$\text{Specificity} = \frac{\text{TN}}{\text{TN} + \text{FP}} \tag{6.11}$$

$$\text{Precision} = \frac{\text{TP}}{\text{TP} + \text{FP}} \tag{6.12}$$

$$\text{Recall} = \frac{\text{TP}}{\text{TP} + \text{FN}} \tag{6.13}$$

$$\text{F-score} = \frac{2 * \text{Precision} * \text{Recall}}{\text{Precision} + \text{Recall}} \tag{6.14}$$

		Actual values	
		Benign	Malignant
Predicted values	Benign	True Positive (TP)	False Positive (FP)
	Malignant	False Negative (FN)	True Negative (TN)

Figure 6.8 Confusion matrix

6.4.3 Analysis of the simple AE

The simple encoder–decoder network consists of one encoder block and a decoder block. The encoder block will have one hidden layer with 200 neurons. The decoder block is symmetric to the encoder as shown in Figure 6.9. The network is initialized with random weights and the sigmoid activation function is used in the hidden layer neurons.

The weight normalization parameters such as L_2 weight regularization, sparsity proportion, and sparsity regularization are fixed as per the specifications given in Table 6.1. The encoder–decoder network is trained for 500 epochs. Mean squared error (MSE) loss function is used for fitting the model. Figure 6.10 shows the learning performance curve of the encoder–decoder network. It is observed that the MSE of the model is unchanged after 450 epochs, which convinced that the model is trained adequately for learning the features.

Then the encoder part of this trained model is used as a feature extractor for the skin cancer classification task. The stacked AE is constructed by cascading the trained encoder part with an output softmax layer which is shown in Figure 6.11. The stacked AE and other classifiers such as SVM and KNN are implemented for classifying the skin cancer images.

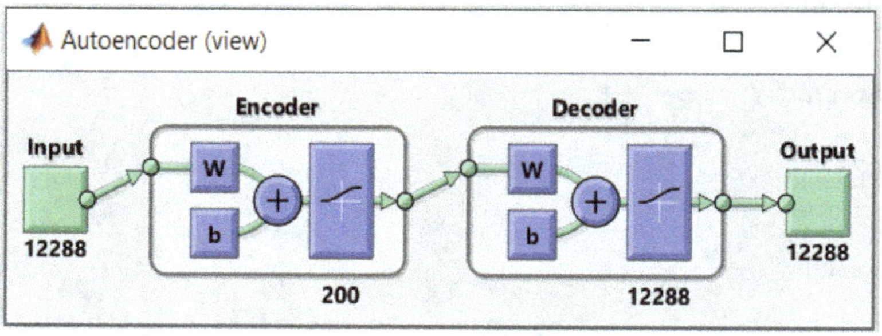

Figure 6.9 A simple encoder–decoder network for representation learning

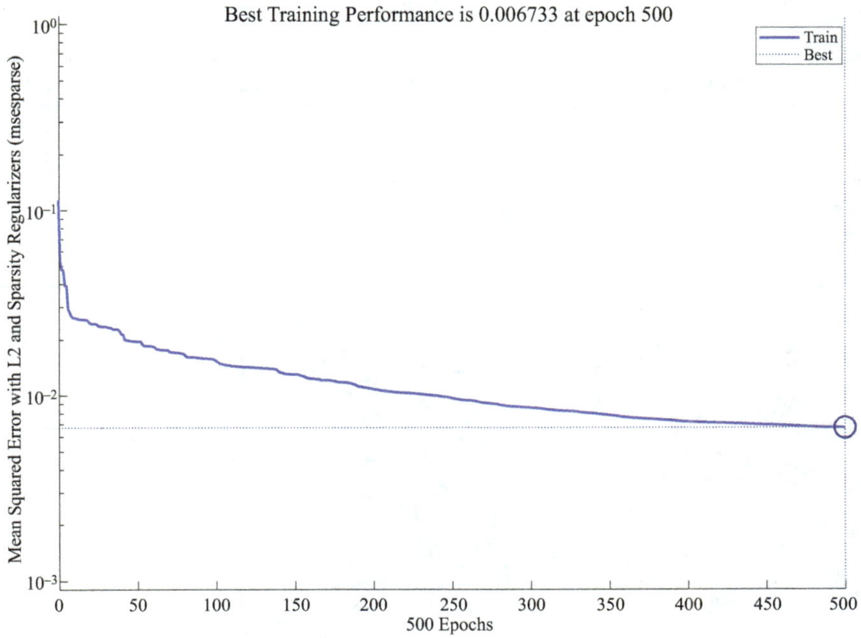

Figure 6.10 Learning performance curve for feature extraction using AE

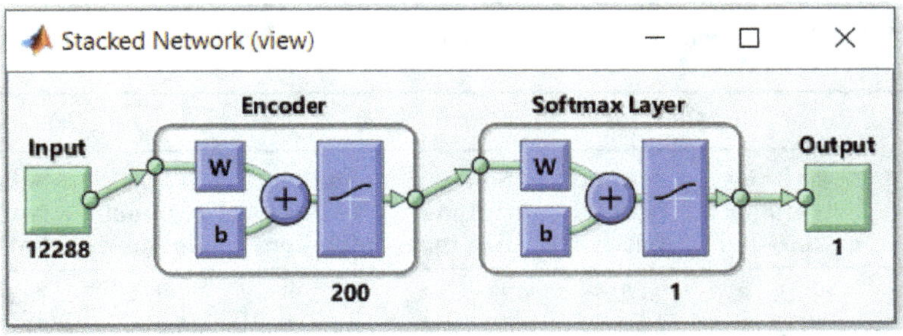

Figure 6.11 Stacked network designed with an AE and a softmax layer

The efficiency of the extracted features is examined with various classifiers and the experimental results are presented in Table 6.2. From the experimental results, it is noticed that the obtained features contain sufficient information for classifying the images. As compared with SVM and KNN classifiers, the AE classifier produces marginally improved results in terms of accuracy, specificity, precision, recall, and F-score. However, the area under the curve (AUC) of the AE classifier with fine-tuning is 1, which is 7% higher than the SVM classifier.

Figure 6.12 presents the visualization of learned features of the trained AE with 200 hidden neurons after 100, 200, and 500 iterations, respectively; each mini square

Table 6.2 Performance of different classifiers on the features extracted from AE

	Softmax (without fine tune)	Softmax (with fine tune)	SVM	KNN
Accuracy	79.8	80.2	80.0	78.9
Specificity	82.2	84.7	80.4	84.2
Precision	78.3	80.3	75.0	79.3
Recall	77.0	74.7	79.6	72.7
F-score	70.5	77.4	77.2	75.8
AUC	0.9510	1	0.9320	0.8622

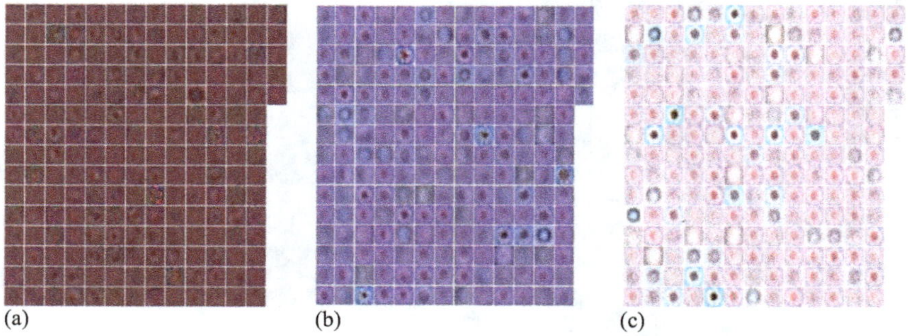

Figure 6.12 Visualization of learned features set from AE1 (a) 500 iterations, (b) 200 iterations, and (c) 100 iterations

Table 6.3 Performance of the softmax classifier in different iterations

	500		200		100	
	Softmax (without fine-tune)	Softmax (with fine-tune)	Softmax (without fine-tune)	Softmax (with fine-tune)	Softmax (without fine-tune)	Softmax (with fine-tune)
Accuracy	79.8	80.2	78.5	79.7	77.7	78.8
Specificity	82.2	84.7	79.2	84.7	80.9	77.4
Precision	78.3	80.3	75.6	80.1	78.0	69.7
Recall	77.0	74.7	77.7	73.7	74.3	81.0
F-score	70.5	77.4	76.6	78.8	76.1	74.9
AUC	0.9510	1	0.9238	1	0.9089	0.9999
MSE	0.0015	0.0015	0.0040	0.0040	0.0050	0.0050

in the figure shows the activated pixels of the input image at one of 200 hidden neurons. It is obvious that the different hidden neurons have learned different shapes, textures, and edge features from various positions and orientations in the image.

The performance of the softmax classifier with and without fine-tuning using the 200 features extracted by a trained AE is examined after 100, 200, and 500 iterations, and the classification results are presented in Table 6.3. It is noticed that the model with fine-tuning influenced the results of the classifier positively and the performance of the model with fine-tuning outperforms the model without fine-tuning in terms of all performance metrics. This exhibits that the fine-tuning process enables the model to update the weightage of the features with respect to the classification task. The accuracy of the softmax classifier is gradually increased as the number of iteration is increased. The MSE of the encoder–decoder network is gradually minimized as the number of iterations increases.

From the graph, it is observed that the receiver operating characteristic (ROC) curve of AE with a softmax classifier is closer to the upper left corner and the AUC is higher than SVM and KNN classifier. The performance of the classifier is evaluated with a false positive rate and true positive rate and the obtained results confirm that the AE classifier has a low false positive rate and the high true positive rate. The ROC curves shown in Figure 6.13 reveal that the performance of the AE classifier separates the classes with a maximum AUC of 0.95 for 500 iterations.

6.4.4 Effect of sparsity in AE

Sparsity is one of the regularization criteria that helps to restrict the activations in a fully connected neural network by supporting the most relevant connections and suppressing the least contributing connections. The sparse AE classifier is obtained from the trained AE with 200 neurons. The hyperparameters such as L_2 weight

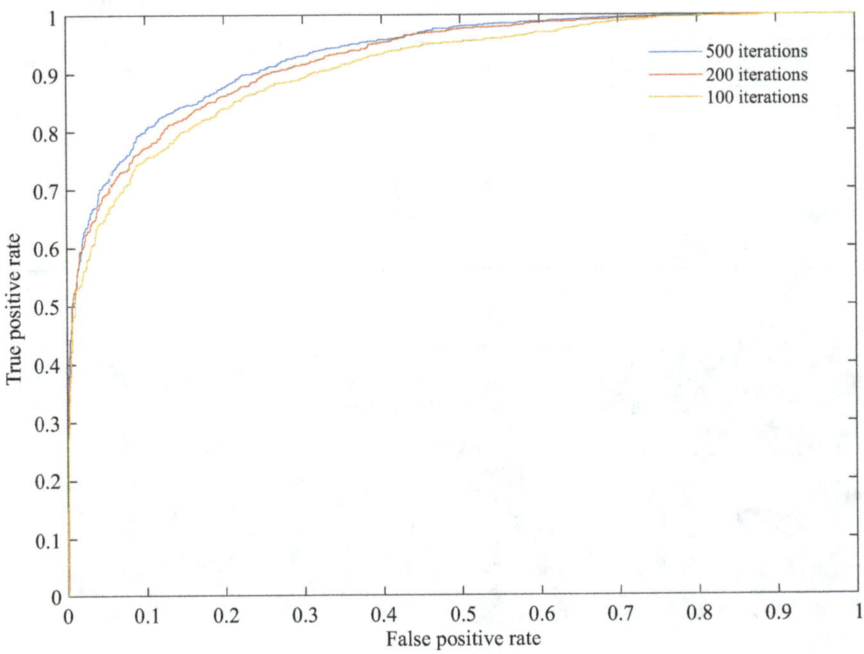

Figure 6.13 ROC curve of AE softmax classifier without fine-tuning for different iterations

regularization, sparsity proportion, and sparsity regularization are set as 0.004, 0.15, and 4, respectively. In this experiment, the performance of the various classifiers on the set of features extracted from the SAE is analyzed. Table 6.4 shows the performance of different classifiers on skin cancer classification with reduced features set which are extracted by SAE after 500 iterations. The performance of SAE with a softmax classifier is marginally better than SVM and KNN.

6.4.5 Effect of squeezing bottleneck in AE

The bottleneck layer determines the amount of dimensionality reduction in AE. The bottleneck layer restricts the amount of information that passes from encoder to decoder and ensures good feature learning. The effect of reducing the nodes of the bottleneck layer in feature extraction is examined with 200, 100, and 50 nodes. The influence of squeezing the bottleneck in AE is qualitatively analyzed by visualizing the activated images at the encoded part. From the visualization shown in Figure 6.14, it is observed that the minimum weighted pixels are suppressed while squeezing the bottleneck.

Table 6.5 shows the performance of different classifiers on skin cancer classification with the various number of nodes in the bottleneck layer of AE such as 200, 100, and 50. The efficiency of AE with the softmax classifier is not affected by

Table 6.4 Performance of different classifiers on the features extracted from SAE

	Softmax (without fine tune)	**Softmax (with fine tune)**	**SVM**	**KNN**
Accuracy	80.0	83.0	78.6	66.4
Specificity	81.7	82.8	77.5	62.3
Precision	78.0	78.3	70.3	29.7
Recall	78.0	83.3	80.2	89.0
F-score	78.0	80.7	74.9	44.5
AUC	0.9194	1	0.8514	0.7321

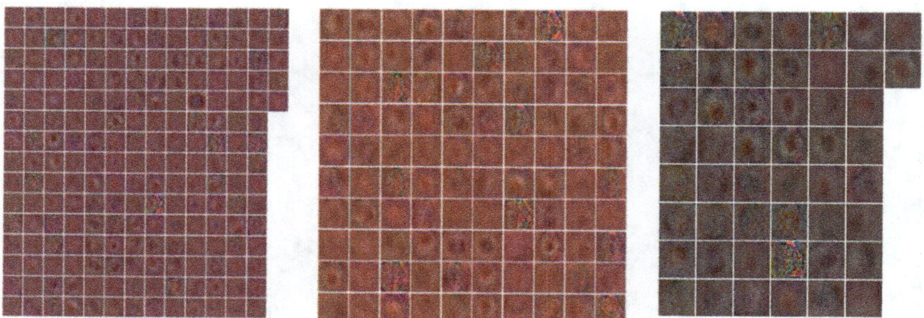

(a) (b) (c)

Figure 6.14 Visualization of activated features learned from AE (a) 200 nodes, (b) 100 nodes, and (c) 50 nodes

Table 6.5 Performance of different classifiers with various bottleneck size in AE

No. of nodes in bottleneck layer	200		100		50	
	Softmax (without fine-tune)	Softmax (with fine-tune)	Softmax (without fine-tune)	Softmax (with fine-tune)	Softmax (without fine-tune)	Softmax (with fine-tune)
Accuracy	79.8	80.2	78.5	82.3	78.6	83.2
Specificity	82.2	84.7	80.8	81.7	82.6	82.7
Precision	78.3	80.3	77.3	76.7	80.7	78.0
Recall	77.0	74.7	75.8	83.0	74.5	83.9
F-score	70.5	77.4	76.5	79.7	77.4	80.8
AUC	0.9510	1	0.9291	1	0.9060	1
MSE	0.0015	0.0015	0.0018	0.0018	0.0021	0.0021

squeezing the number of nodes in the bottleneck layer. This confirms that the AE removes the redundant details from the feature set. There is no standard procedure to choose the number of neurons in the hidden layer. Based on the target application, the minimum number of features could be decided.

6.4.6 *Performance of deep stacked encoder*

The proposed DSAE classifies the images in two stages. At first, the input image of size 12,288 pixels is mapped into 100 or 50 features with a backpropagation encoder–decoder network by minimizing the reconstruction error. Figure 6.15

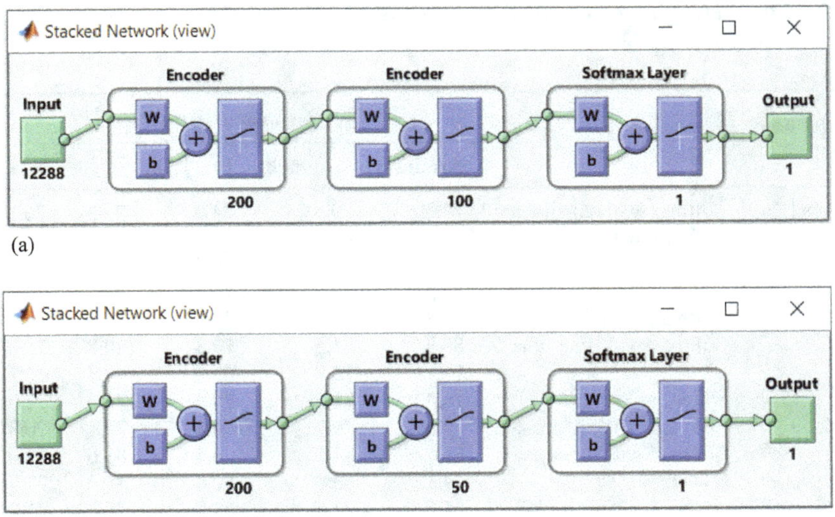

(a)

(b)

Figure 6.15 Structure of DSAE network with softmax layer (a) 12,288-100 and (b) 12,288-50 shows the structure of DSAE with a softmax layer. Later, the mapped features are classified into two classes such as benign and malignant using various classifiers. The various stacked AE models are implemented and presented in Table 6.1.

In Model 1, the first encoder–decoder (ED1) network has 200 neurons at the bottleneck layer which are used to reconstruct the input image size of 12,288 pixels. Then the second encoder–decoder (ED2) network has 100 neurons at the bottleneck layer that are used to reconstruct the feature set with a size of 200. The ED1 and ED2 models are fitted regularly using a backpropagation algorithm for 500 epochs. Once these models have been trained, the encoder part of ED1 and ED2 are stacked as a single segment called stacked AE. This stacked AE acts as a feature extractor for the skin cancer classification problem.

Model 2 is designed as same as Model 1, except for the ED2 network, where the second encoder–decoder (ED2) network has 50 neurons at the bottleneck layer that are used to reconstruct the feature set with a size of 200.

Model 3 is designed as Model 1, but the sparsity parameters are also added in Model 3 to normalize the weights of the encoder–decoder network. The hyperparameters such as level of sparsity, L2R coefficient, and the sparse regularizing coefficient for Model 3 are fixed, as specified in Table 6.1. Model 4 is designed the same as Model 3 except for the ED2 network, where the second encoder–decoder (ED2) network has 50 neurons at the bottleneck layer that are used to reconstruct the feature set with a size of 200.

In this experiment, various classifiers such as softmax classifier, softmax with fine-tuning, SVM, and KNN are implemented for classifying skin cancer images using the features extracted by the various proposed stacked AE models. Table 6.6 presents the performance of different classifiers on skin cancer classification with the

Table 6.6 Performance of different classifiers with various proposed stacked AE models

Model name	**Classifier**	**Performance metrics**				
		Accuracy	**Specificity**	**Precision**	**Recall**	***F*-score**
Model 1	Softmax without fine-tune	79.3	79.5	74.0	79.3	76.6
	Softmax with fine-tune	**80.2**	82.4	79.3	77.5	78.4
	SVM	78.5	79.5	74.7	77.2	75.9
	KNN	78.2	79.7	75.3	76.4	75.8
Model 2	Softmax without fine-tune	77.8	79.9	77.0	75.5	76.2
	Softmax with fine-tune	**80.9**	81.6	77.3	80.8	78.6
	SVM	75.9	75.9	69.0	75.8	72.2
	KNN	76.4	78.3	74.0	74.0	74.0
Model 3	Softmax without fine-tune	79.5	79.7	74.3	79.4	76.8
	Softmax with fine-tune	**81.5**	81.5	76.7	81.6	79.0
	SVM	79.1	83.0	81.0	75.0	77.9
	KNN	75.6	75.6	68.3	75.6	71.8
Model 4	Softmax without fine-tune	78.2	77.4	70.3	79.3	74.6
	Softmax with fine-tune	**81.9**	81.6	76.7	82.4	79.4
	SVM	79.8	81.1	77.0	78.3	77.6
	KNN	75.9	76.1	69.3	75.6	72.3

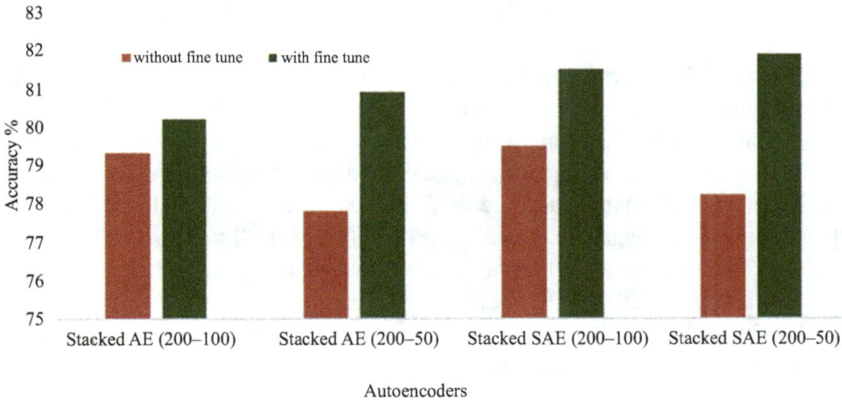

Figure 6.16 Comparative analysis of various stack AE models

features set extracted by various DSAE models after 500 iterations. The performance of the softmax classifier with fine-tuning classifies the images better than SVM and KNN classifiers for all the models. Softmax classifier with fine-tuning works with stacked AE and allows the classifier to update the weight of features extraction phase concerning the target skin cancer classification task. The comparative result analysis is shown in Figure 6.16. However, in softmax, SVM and KNN classifiers, the feature learning, and classification phases are not connected and the images are classified based on the static set of features extracted from stacked AE.

6.5 Conclusion

Representation learning is a major challenge for machine learning algorithms. In this work, a stacked AE is implemented for automatic feature extraction and the extracted features are refined further for the classification task using transfer learning. The transfer learning technique helps to transfer the knowledge from one model to another model. The proposed DSAE model recreates the image from the reduced feature set with a minimum MSE of 0.0015. Since stacked AE model follows self-supervised learning, it can learn essential information from a limited set of data. Therefore, this feature extraction approach is suggested for medical image classification tasks where the dataset is scarce. In this work, the stacked AE acted as a feature extractor in classifying skin cancer images. The experimental results confirm that the DSAE model with a softmax classifier provides improved results as compared with other classification methods. For this reason, the stacked AE can be used as a feature extractor for computer-aided medical image classification systems.

References

[1] Y. K. Singh, and N. D. Singh, "Binary face image recognition using logistic regression and neural network," In: *2017 International Conference on*

Energy, Communication, Data Analytics and Soft Computing (ICECDS), 2017, pp. 3883–3888.

[2] J. Yan and J. Lee, "Degradation assessment and fault modes classification using logistic regression," *J. Manuf. Sci. Eng.*, 2005, 127(4), 912–914.

[3] F. Cao, Z. Yang, J. Ren, W.-K. Ling, H. Zhao, and S. Marshall, "Extreme sparse multinomial logistic regression: a fast and robust framework for hyperspectral image classification," *Remote Sens.*, 2017, 9(12), 1255.

[4] A. Sitthi, M. Nagai, M. Dailey, and S. Ninsawat, "Exploring land use and land cover of geotagged social-sensing images using naive Bayes classifier," *Sustainability*, 2016, 8(9), 921.

[5] Y. Tan and P. P. Shenoy, "A bias-variance based heuristic for constructing a hybrid logistic regression-naive Bayes model for classification," *Int. J. Approx. Reason.*, 2020, 117, 15–28.

[6] V. Muralidharan and V. Sugumaran, "A comparative study of naive Bayes classifier and Bayes net classifier for fault diagnosis of monoblock centrifugal pump using wavelet analysis," *Appl. Soft Comput.*, 2012, 12(8), 2023–2029.

[7] Z. Li, B. Niu, F. Peng, G. Li, Z. Yang, and J. Wu, "Classification of peanut images based on multi-features and SVM," *IFAC-PapersOnLine*, 2018, 51(17), 726–731.

[8] Y. Zhang, Z. Dong, A. Liu, *et al.*, "Magnetic resonance brain image classification via stationary wavelet transform and generalized eigenvalue proximal support vector machine," *J. Med. Imaging Heal. Inform.*, 2015, 5(7), 1395–1403.

[9] A. Widodo and B.-S. Yang, "Support vector machine in machine condition monitoring and fault diagnosis," *Mech. Syst. Signal Process.*, 2007, 21(6), 2560–2574.

[10] M. A. Mohammed, T. A. H. Naji, and H. M. Abduljabbar, "The effect of the activation functions on the classification accuracy of satellite image by artificial neural network," *Energy Procedia*, 2019, 157, 164–170.

[11] S. Heo and J. H. Lee, "Fault detection and classification using artificial neural networks," *IFAC-PapersOnLine*, 2018, 51(18), 470–475.

[12] K. Kannadasan, D. R. Edla, and V. Kuppili, "Type 2 diabetes data classification using stacked autoencoders in deep neural networks," *Clin. Epidemiol. Glob. Heal.*, 2019, 7(4), 530–535.

[13] K. Adem, S. Kiliçarslan, and O. Cömert, "Classification and diagnosis of cervical cancer with stacked autoencoder and softmax classification," *Expert Syst. Appl.*, 2019, 115, 557–564.

[14] Z. Mehmood, F. Abbas, T. Mahmood, M. A. Javid, and A. Rehman, and T. Nawaz, "Content-based image retrieval based on visual words fusion versus features fusion of local and global features," *Arab. J. Sci. Eng.*, 2018, 1–20.

[15] P. Görgel and A. Simsek, "Face recognition via deep stacked denoising sparse autoencoders (DSDSA)," *Appl. Math. Comput.*, 2019, 355, 325–342.

[16] G. Liu, H. Bao, and B. Han, "A stacked autoencoder-based deep neural network for achieving gearbox fault diagnosis," *Math. Probl. Eng.*, 2018, 1–10.

[17] A.-A. Nahid, A. Mikaelian, and Y. Kong, "Histopathological breast-image classification with restricted Boltzmann machine along with back-propagation," *Biomed. Res.*, 2018, 29(10), 2068–2077.

[18] X. Zeng, F. Chen, and M. Wang, "Shape group Boltzmann machine for simultaneous object segmentation and action classification," *Pattern Recognit. Lett.*, 2018, 111, 43–50.

[19] E. Rehn, A. Rehn, and A. Possemiers, "Fossil charcoal particle identification and classification by two convolutional neural networks," *Quat. Sci. Rev.*, 2019, 226, 106038.

[20] G. Ciocca, P. Napoletano, and R. Schettini, "CNN-based features for retrieval and classification of food images," *Comput. Vis. Image Underst.*, 2018, 176, 70–77.

[21] W. Zhang, R. Li, H. Deng, *et al.*, "Deep convolutional neural networks for multi-modality isointense infant brain image segmentation," *NeuroImage*, 2015, 108, 214–224.

[22] A. Krizhevsky, I. Sutskever, and G. E. Hinton, "Imagenet classification with deep convolutional neural networks," In: *Advances in Neural Information Processing Systems*, 2012, pp. 1097–1105.

[23] S. A. Agnes, J. Anitha, S. I. A. Pandian, and J. D. Peter, "Classification of mammogram images using multiscale all convolutional neural network (MA-CNN)," *J. Med. Syst.*, 2020, 44(1), 30.

[24] T. Shanthi and R. S. Sabeenian, "Modified Alexnet architecture for classification of diabetic retinopathy images," *Comput. Electr. Eng.*, 2019, 76, 56–64.

[25] S. Wan, Y. Liang, and Y. Zhang, "Deep convolutional neural networks for diabetic retinopathy detection by image classification," *Comput. Electr. Eng.*, 2018, 72, 274–282.

[26] Y. Guo and A. S. Ashour, "Neutrosophic multiple deep convolutional neural network for skin dermoscopic image classification," In: *Neutrosophic Set in Medical Image Analysis*, Elsevier, 2019, pp. 269–285.

[27] P. Vincent, H. Larochelle, I. Lajoie, Y. Bengio, and P.-A. Manzagol, "Stacked denoising autoencoders: Learning useful representations in a deep network with a local denoising criterion," *J. Mach. Learn. Res.*, 2010, 11 (Dec), 3371–3408.

[28] D. Oszutowska-Mazurek, P. Mazurek, and O. Knap, "Stacked autoencoder for segmentation of bone marrow histological images," In: *Computer Science On-line Conference*, 2018, pp. 425–435.

[29] M. Zia ur Rehman, S. O. Gilani, A. Waris, *et al.*, "Stacked sparse autoencoders for EMG-based classification of hand motions: a comparative multi day analyses between surface and intramuscular EMG," *Appl. Sci.*, 2018, 8(7), p. 1126.

[30] Y. Ju, J. Guo, and S. Liu, "A deep learning method combined sparse autoencoder with SVM," In: *2015 International Conference on Cyber-Enabled Distributed Computing and Knowledge Discovery*, 2015, pp. 257–260.

[31] M. Ahmad, J. Yang, D. Ai, S. F. Qadri, and Y. Wang, "Deep-stacked auto encoder for liver segmentation," In: *Chinese Conference on Image and Graphics Technologies*, 2017, pp. 243–251.

[32] J. Xu, L. Xiang, Q. Liu,*et al.*, "Stacked sparse autoencoder (SSAE) for nuclei detection on breast cancer histopathology images," *IEEE Trans. Med. Imaging*, 2015, 35(1), 119–130.

[33] G. B. Praveen, A. Agrawal, P. Sundaram, and S. Sardesai, "Ischemic stroke lesion segmentation using stacked sparse autoencoder," *Comput. Biol. Med.*, 2018, 99, 38–52.

[34] S. Serte and H. Demirel, "Gabor wavelet-based deep learning for skin lesion classification," *Comput. Biol. Med.*, 2019, 113, 103423.

[35] B. Harangi, "Skin lesion classification with ensembles of deep convolutional neural networks," *J. Biomed. Inform.*, 2018, 86, 25–32.

[36] Y. Li and L. Shen, "Skin lesion analysis towards melanoma detection using deep learning network," *Sensors*, 2018, 18(2), 556.

[37] S. M. Sasubilli, A. Kumar, and V. Dutt, "Improving health care by help of internet of things and bigdata analytics and cloud computing," In: *2020 International Conference on Advances in Computing and Communication Engineering (ICACCE)*, 2020, pp. 1–4, doi: 10.1109/ICACCE49060.2020.9155042.

[38] P. K. Kotturu and A. Kumar, "Comparative study on machine learning models for early diagnose of Alzheimer's disease: multi correlation method," In: *2020 5th International Conference on Communication and Electronics Systems (ICCES)*, 2020, pp. 778–783, doi: 10.1109/ICCES48766.2020.9137872.

[39] A. Esteva, B. Kuprel, R.A. Novoa, *et al.*, "Dermatologist-level classification of skin cancer with deep neural networks," *Nature*, 2017, 542(7639), 115–118.

[40] Y. Lu and P. Xu, "Anomaly detection for skin disease images using variational autoencoder," arXiv Prepr. arXiv1807.01349, 2018.

[41] P. Kharazmi, S. Kalia, H. Lui, Z. J. Wang, and T. K. Lee, "A feature fusion system for basal cell carcinoma detection through data-driven feature learning and patient profile," *Ski. Res. Technol.*, 2018, 24(2), 256–264.

[42] S. R. Swarna, A. Kumar, P. Dixit, and T. V. M. Sairam, "Parkinson's disease prediction using adaptive quantum computing," In: *2021 Third International Conference on Intelligent Communication Technologies and Virtual Mobile Networks (ICICV)*, 2021, pp. 1396–1401, doi:10.1109/ICICV50876.2021.9388628.

[43] G. Sasubilli and A. Kumar, "Machine learning and big data implementation on health care data," In: *4th International Conference on Intelligent Computing and Control Systems (ICICCS)*, 2020, pp. 859–864, doi: 10.1109/ICICCS48265.2020.9120906.

[44] J. Anitha, S. A. Agnes, and S. I. A. Pandian, "Self-supervised representation learning framework for remote crop monitoring using sparse autoencoder," In: *Intelligence in Big Data Technologies-Beyond the Hype*, Springer, 2021, pp. 219–227.

[45] B. A. Olshausen and D. J. Field, "Sparse coding with an overcomplete basis set: a strategy employed by V1?," *Vis. Res.*, 1997, 37(23), 3311–3325.

Chapter 7

Comparison of machine learning and deep learning algorithms for prediction of coronary heart disease

Sajeev Ram Arumugam[1], E. Anna Devi[2], T. Abimala[3] and Oswalt Manoj[4]

The aim of the study is to introduce deep learning (DL) systems in the medical domain. Earlier, machine learning (ML) systems were used in the medical domain for the prediction of clinical data, so we aim to use DL systems and compare the performance of the DL system with ML systems.

For this comparative study, we need a clinical dataset. We are using a database of the Framingham Heart Study, which could be used for predicting coronary heart disease. The dataset consists of 4,240 observations and 16 features. These features have both numeric and categorical data. We have also narrated data visualization, data normalization methods, and data splitting. In this study, we have used six different ML methods, namely, naïve Bayes, decision tree, random forest, support vector machine, K-nearest neighbours, and logistic regression. After implementing ML algorithms, we have used neural networks (NNs) for classification and the results of all the models were compared. To implement the models, we have used the Google Colab platform (Python Notebooks).

We have implemented both ML and DL models, and to evaluate the models few parameters have been used, namely, accuracy, sensitivity, specificity, positive predictive value, negative predictive value, and area under the curve. Analysing the parameters of all the models, it is observed that the random forest model performs well with 87.25% accuracy, which was the same as the NN model.

[1]Department of Artificial Intelligence and Data Science, Sri Krishna College of Engineering and Technology, India
[2]Department of Electronics and Communication Engineering, Sathyabama Institute of Science and Technology, India
[3]Department of Instrumentation and Control Engineering, St Joseph's College of Engineering, India
[4]Department of Computed Science and Business Systems, Sri Krishna College of Engineering and Technology, India

On observing the results, it is noted that the random forest and NN model perform the best than all other methods for the dataset used. With these observations, we could not conclude that one of the machines or DL algorithms is the best, as the dataset had only 4,240 observations and every algorithm has different assumptions and working nature.

7.1 Introduction

7.1.1 Coronary heart disease (CHD)

In 2017, World Health Organization (WHO) declared cardiovascular diseases (CVDs) as the leading cause of death globally. CVDs are a collection of various disorders, including CHD, cerebrovascular disease, peripheral arterial disease, rheumatic heart disease, congenital heart disease, and deep vein thrombosis and pulmonary embolism ('Cardiovascular Diseases (CVDs)' n.d.). CHD is a disorder that occurs when deposits of cholesterol are accumulated inside the arteries, which narrow down the space of it, decrease the flow of oxygen-rich blood to the heart and lead to sudden heart damage and heart attack [1]. CHD does not occur suddenly; it happens as a slow process over a long period of time and will be known to the patients when it causes any serious effects.

The major risk factors for CHD include smoking, alcohol consumption, unhealthy diets and less physical activity. Symptoms of CHD include discomfort in the chest region, arms and shoulders and shortness of breath ('Coronary Artery Disease | Cdc.Gov' n.d.). Even though CHD is of the same type, symptoms are not common in all patients. It varies from patient to patient. Many people do not have any symptoms until they feel discomfort in their chest or get cardiac arrest ('Coronary Heart Disease | NHLBI, NIH' n.d.). Early detection of CHD will help to treat the patients at the right time and avoid serious issues. It is not too easy to detect the CHD at the early stages but it can be predicted based on health conditions of a person such as cholesterol level, diabetics, age, etc. So, computer-based prediction techniques could be used to predict the risk of CHD, which would help to start with the right treatment at the right time.

In recent times, researchers have been introducing data-mining-based classification techniques in almost every field. Similarly, we could use machine learning (ML) or deep-learning-based (DL) classification techniques also to predict patients having the risk of CHD. Classification algorithms analyse the database consisting of several features and identify the high risk of CHD in patients [2].

7.1.2 ML and DL techniques

Artificial intelligence (AI) systems are a resemblance to human behaviour which is performed in a smarter manner. These AI systems not only have inbuilt algorithms but are also updated regularly. Some simple AI systems include robots, self-driving cars, automized computer applications and many more ('What Is the Difference between AI, Machine Learning and Deep Learning? – Geospatial World' n.d.).

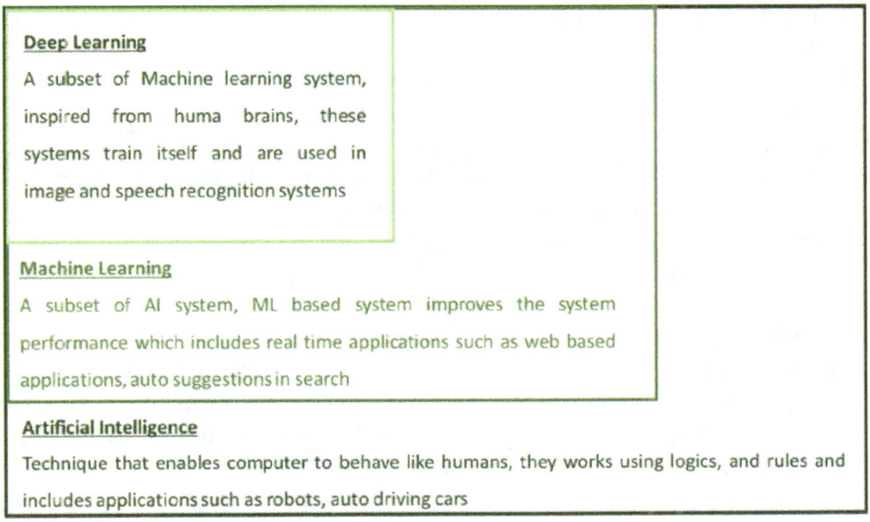

Figure 7.1 Comparison between AI, ML and DL systems

ML is a sub-branch of AI that focuses on developing self-learning applications that could learn from the provided data and improve the performance of the system in terms of accuracy. ML algorithms are developed in such a way to identify patterns in the data and make predictions and classifications. Some of the real-time examples of ML examples include digital assistants, autofill options, web-based recommendations, robotic vacuum cleaner, medical imaging analysis systems and much more ('What Is Machine Learning? - India | IBM' n.d.).

DL algorithms are ML techniques that are inspired by human brains. These algorithms work more similar to humans. DL algorithms perform classification tasks directly from the data such as image, video, audio and text. It does not require any pre-processing or feature extraction steps before passing through the DL systems. Real-time examples include image captioning, audio to text conversions, emotion detections, voice-controlled devices, etc. The relation between AI, ML and DL-based systems can be understood clearly from Figure 7.1.

In this research work, we would try to predict the risk of CHD using both ML and DL classification systems, compare the results of all the methods and try to identify the right algorithm which works for this particular dataset. Also, this research will also help researchers to identify the right algorithm for their problem and how to implement it.

7.2 Related works

Atkov *et al.* [3] developed a classification system based on an artificial neural network (ANN) for the classification of CHD. The authors have used a dataset from Central

Clinical Hospital—Russia, comprising details of 487, of which 327 with CHD and the rest without CHD. The authors have also tried to improve the performance of the system by modifying the input features. They finally managed to obtain an accuracy level ranging between 64 and 94% for a different combination of features.

El-Bialy *et al.* [4] have applied ML algorithms on four different datasets, namely the Cleveland dataset, Hungarian dataset, V.A dataset and Statlog project dataset for coronary artery disease. The authors tried to compare the results between different ML techniques and projected the results of those algorithms on various datasets. The average accuracy of individual classification of the dataset was identified to be 75.48%, and finally, they combined all the datasets and performed classification using a fast decision tree classifier, which gave an accuracy of 78.06%.

Kim *et al.* [5] used fuzzy logic and decision tree for developing a classification system for classifying the risk of CHD. They used the KNHANES-VI dataset which holds the data of 302 men, 186 women with risk and 135 men, 107 women with no risk. The major features that the authors focused on were age, gender, cholesterol level, smoking habit, and diabetes. The accuracy of the developed system was 69.51%.

Kim and Kang [6] developed a system based on a neural network (NN) for CHD risk classification. They have implemented NN-based feature correlation in two stages. In the first stage, the features are analysed and they are reduced on the basis of the importance and feature correlation analysis in the second stage. They have used the KNHANES-VI dataset which holds 4,146 individuals of which 3,031 had a lower risk of CHD and 1,115 had higher risks. The classification system provided an area under the curve (AUC) of 0.749.

Lim *et al.* [7] have proposed an optimized deep belief network as an alternate for the Framingham study as it causes overestimates of risk. Altering the basic layer levels of the deep belief network, the authors have tried developing an optimized algorithm of DBN to enhance the results. They have used two layers holding 5 and 7 nodes in each layer. Using the optimized algorithms, they got an accuracy of 89.24%, which is much better than the compared Framingham algorithm which holds 50.15%.

Wiharto *et al.* [8] carried over the feature selection process using tiered multivariate analysis. They had a multilayer perceptron NN classifier by which the accuracy, sensitivity and specificity of the system were identified to be 86.3, 84.80 and 88.20%.

Latha and Jeeva [9] have tried to improve the accuracy of the classification system by combining multiple classification algorithms. The authors have used a dataset consisting of 303 instances with 14 attributes from the UCI ML repository. They have identified that ensemble techniques, namely bagging and boosting, could help in improving the performance of weak classifiers. They managed to bring a 7% increase in the accuracy of the classifier and also further used future reduction techniques to enhance the performance of the classifier.

Alaa Id *et al.* [10] developed an ML-based classification system for classifying risk in CVD. To develop the classification system, they have used 423,604 participants' data. They have also compared their parameters with different algorithms and managed to get an AUC score of 0.774.

7.3 Materials and methods

We have used a comparative study to analyse the performance of the most used ML and DL methods, including naïve Bayes, decision tree, random forest, support vector machine (SVM), K-nearest neighbour (KNN), logistic regression and NN algorithms. We have considered different performance parameters, including accuracy, sensitivity, specificity, positive predictive value (PPV), negative predictive value (NPV) and AUC, and the results were discussed in the forthcoming sections.

The dataset used in the study work was available in the Kaggle portal, and the complete work is performed in the Google Colab platform (Python).

7.3.1 Data preparation

To develop the classification system, we have used a publicly available dataset available in the Kaggle portal ('Heart Attack Analysis & Prediction Dataset | Kaggle' n.d.) which was used for an ongoing study belonging to residents of the town of Framingham, Massachusetts. The dataset consists of 4,240 records with 15 attributes, as shown in Table 7.1.

The first step of the work will be to import the dataset into our environment for which we use the pandas data frame through which we import the comma-separated value (CSV) file into the environment. Once the data are being imported, we look into all the variables and we could find a variable education which definitely could not play a major role in CHD, so our first step would be removing the variable 'education' from the data frame. Once it is done, we try to describe the dataset to find the min, max and standard deviation values of every variable, and the plot for the distribution of all data is given in Figure 7.2.

Table 7.1 Descriptions of the variables in the dataset

Attribute name	**Description**
male	Gender of the person: 1 – Male and 0 – Female
age	Age of the person; mentioned as whole numbers
education	Level of education: 1 – Lower school level, 2 – Higher school level, 3 – UG level, 4 – PG level
currentSmoker	Is the person a smoker: 0 – Non-smoker, 1 – Smoker
cigsPerDay	No. of cigarettes the person smokes on an average per day
BPMeds	Whether the patient is under meditation for blood pressure: 0 – Not taking medicines for BP, 1 – taking medicines for BP
prevalentStroke	Had the patient previously had a stroke: 0 – No, 1 – Yes
prevalentHyp	Whether the person was hypertensive: 0 – No, 1 – Yes
diabetes	Whether the person had diabetes: 0 – No, 1 – Yes
totChol	Total cholesterol level of the person
sysBP	Systolic blood pressure level of the person
diaBP	Diastolic blood pressure level of the person
BMI	Body mass index of the person
heartRate	Heart rate of the person
glucose	Glucose level of the person
TenYearCHD	Ten years risk for CHD: 0 – No risks, 1 – Risk for CHD

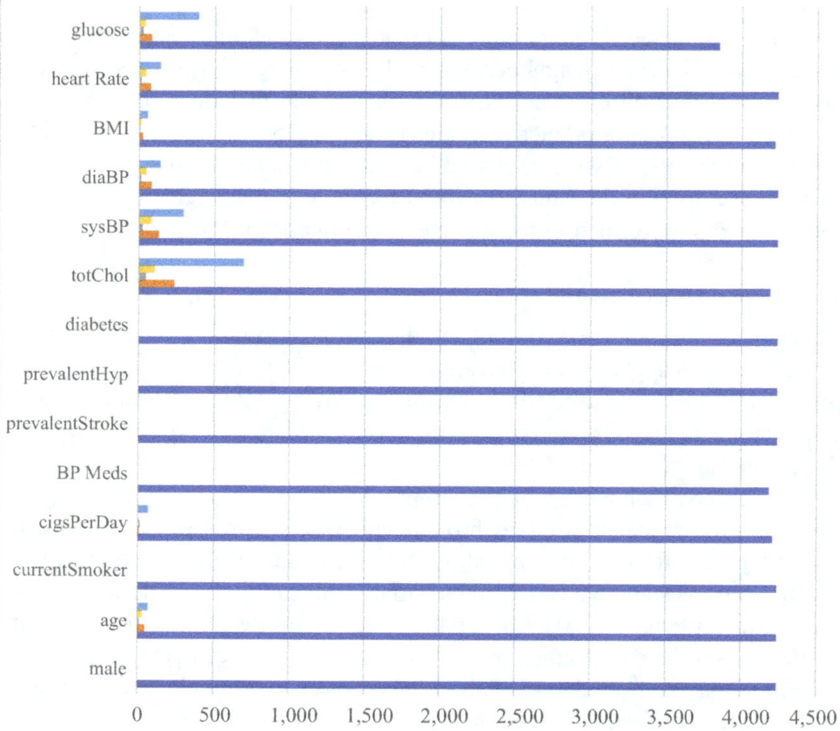

	male	age	current Smoker	cigsPer Day	BPMeds	prevalent Stroke	prevalent Hyp	diabetes	totChol	sysBP	diaBP	BMI	heart Rate	glucose
Max	1	70	1	70	1	1	1	1	696	295	142.5	56.8	143	394
Min	0	32	0	0	0	0	0	0	107	83.5	48	15.54	44	40
Std	0.495	8.573	0.500	11.922	0.170	0.077	0.463	0.158	44.591	22.033	11.910	4.080	12.025	23.954
Mean	0.429	49.580	0.494	9.006	0.030	0.006	0.311	0.026	236.70	132.35	82.898	25.801	75.879	81.964
Count	4240	4240	4240	4211	4187	4240	4240	4240	4190	4240	4240	4221	4239	3852

Max Min Std Mean Count

Figure 7.2 Description of the data including the count, minimum value, maximum value, mean and standard deviation

7.3.2 Fixing the missing data issue

Analysing the data description, we came across a few details on the variables. Of the 14 variables, 6 variables are categorical data type and the rest are numeric data type. The very first step is to identify the missing values among the dataset as the missing value could not be placed in a data frame when we train the system. From the data analyses, we could identify missing values in cigsPerDay, BPMed, totChol, BMI, heartRate and glucose variables. Figure 7.3 describes the missing value in each variable.

We had six variables with missing values among which glucose has 388 missing values. Similarly, BPMed – 53, totChol – 50, cigsPerDay – 29, BMI – 19 and heartRate have one missing data in the dataset. Collectively, we have 12% of the data in the actual dataset and hence we have to drop the missing data from the dataset. After dropping the missing value from the dataset, we will be having 3,751 records. We have to look into the distribution of the data which makes it easier for us to proceed further; hence we plot the histogram of the data as shown in Figure 7.4.

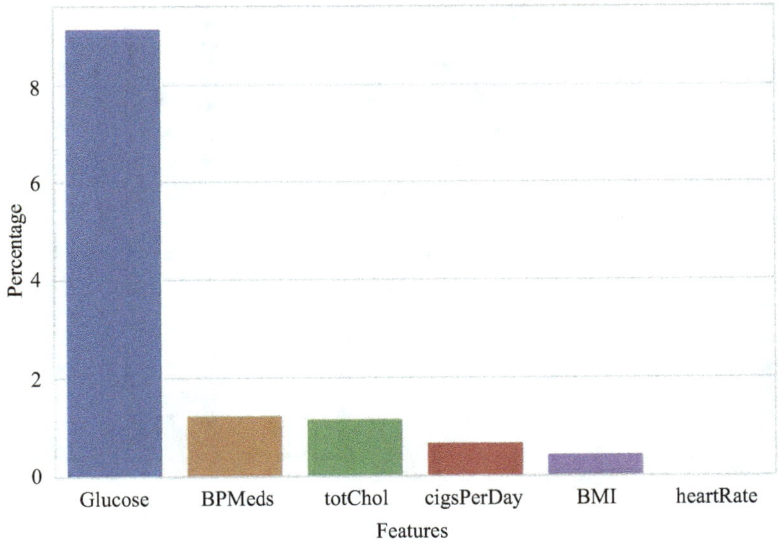

Figure 7.3 Percentage of missing values in different features

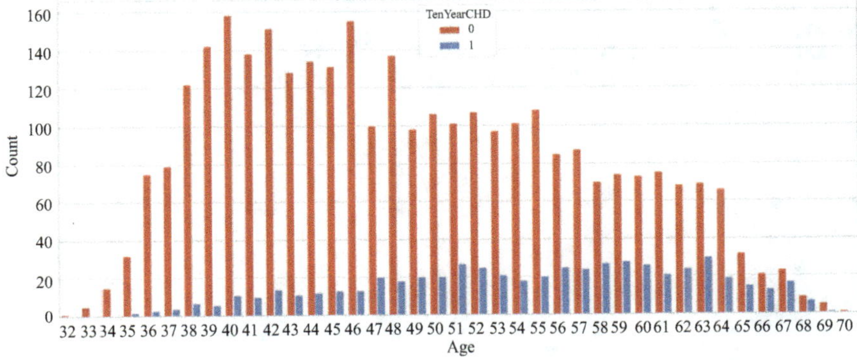

Figure 7.4 Distribution of dataset after removing the missing data

7.3.3 Data analysis

We shall now perform data analysis so that we could identify the relationship between the variables. In relation to it, we try to bring the relations between the variables and the classes of the dataset such as age, gender, currentSmoker, diabetes, BPMed and prevalentHyp. These relations are shown in Figure 7.5.

We could clearly observe that more risky people are from the age group of 51–63 and hence the age could be considered an important feature for

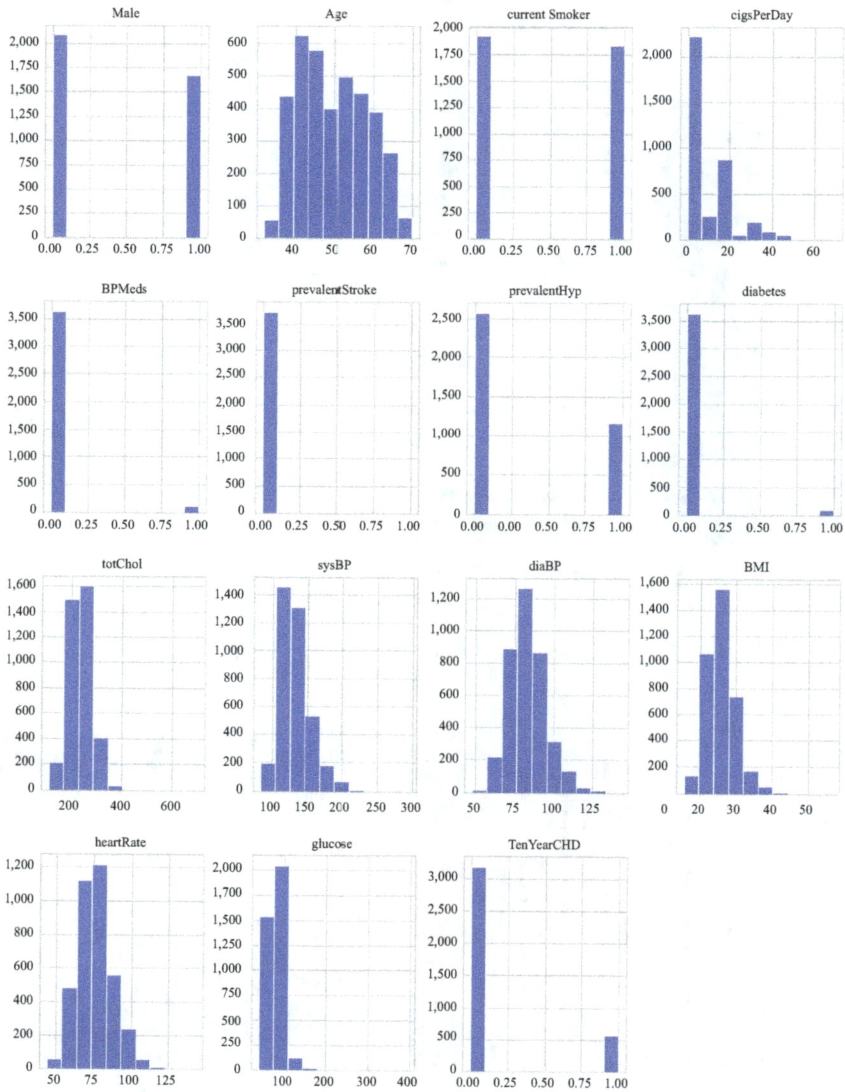

Figure 7.5 Age vs CHD class

classification. Similarly plotting the data between the classes gives us good clarity of which to consider and could be observed in Figure 7.6.

Observing Figure 7.6 we could understand that male is more affected compared to females and similarly smokers, and people with diabetes, blood pressure and hypertension are riskier of CHD. So, all these features could be considered for classification. But before confirming the class distribution also should be considered.

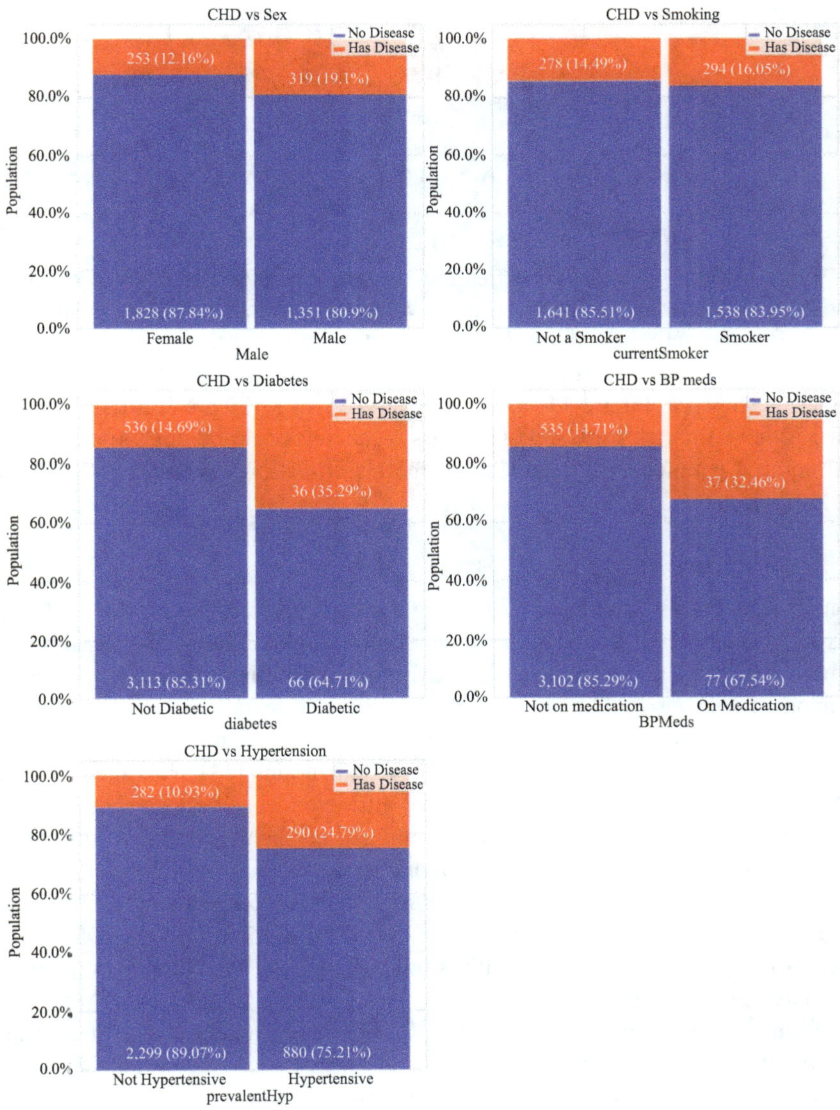

Figure 7.6 Class distribution on variables

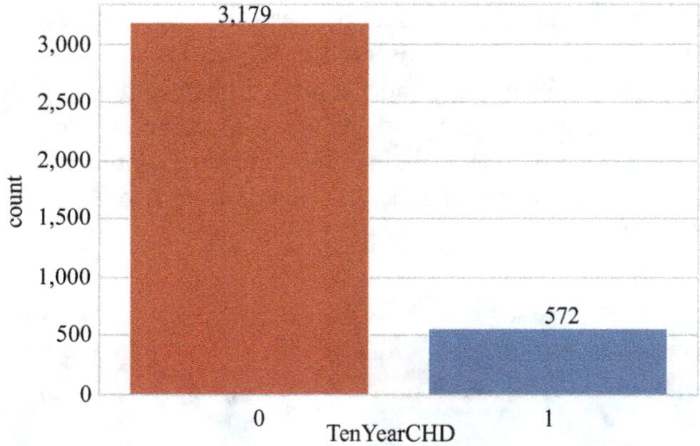

Figure 7.7 Class count: 0 – No risk for CHD, 1 – Risk for CHD

For developing a good classification system, it is expected to have a uniform distribution of classes in the dataset. Developing a classification over an imbalanced class dataset will result in a poor predictive score on the minimum class and hence class balancing becomes more important in this stage [11]. The class difference is shown in Figure 7.7, and it is observed that 3,179 are non-risks for CHD and 572 have the risk for CHD. The class balancing could be done after selecting the required parameters for developing the classification system, class balancing could be done.

7.3.4 Feature selection

Feature selection is a method of choosing the right features needed for building a classification system. It also has some benefits including reduced training time, less complexity, good accuracy score and avoiding overfitting issues [12]. Principal component analysis (PCA) is one of the widely used feature reduction techniques due to its advantages, including removing correlative features, improving performance, avoiding overfitting and improved visualizations [13]. This method transforms the available attributes into new attributes considering enhancing the classification system, hence maintaining and adding the data in the future. In this study, we have avoided using PCA to avoid serious changes in the results.

Liu and Rudy (2010) proposed a feature reduction technique by excluding the features one after the other and retraining the NN again. This type of reducing the features is a good option but it requires more time as all the combinations of features have to be retrained repeatedly.

Bourta is a feature selection method. Unlike other models, it does not find the possible combination for good prediction; instead, it tries to find all the features which carry the necessary information for prediction [14]. While implementing feature reduction using Bourta, the features are not being evaluated among themselves but instead with random versions of the features.

7.3.4.1 Bourta methodology

Step 1: Create duplicate features, randomize the features and shuffle the values of each column (shadow features).

Step 2: Train a classifying model using the random forest algorithm and calculate the Z-score.

Step 3: Identify the maximum Z-score for the shadow attributes and the attributes which hold a higher Z-score than the maximum Z-score.

Step 4: Filtering the attributes that are not yet considered important, and a two-sided quality test is performed with the maximum Z-score.

Step 5: If the importance level of the attribute is identified to be lower than the maximum Z-score among shadow attributes, tag the variable as 'unimportant' and if higher it is tagged as 'important'.

Step 6: Removes all the shadow attributes

Repeat Steps 1–6 until the importance is computed for all attributes

We tried to implement feature reduction using Bourta, and initially, we had 14 features and implemented the feature technique for 100 iterations. At the end of 100 iterations it is identified that only three features have been identified as selected variables, and they are 'age', 'sysBP' and 'BMI'. As training the system is more complicated with two variables and hence we are considering the top eight features shortlisted by Bourta which are 'age', 'cigsPerDay ','totChol', 'sysBP', 'diaBP', 'BMI', 'heartRate' and 'glucose'.

7.3.5 Balancing the dataset

After the feature reduction now, we have eight features in hand, as observed in Figure 7.7. It is clear that positive and negative classes are imbalanced and hence class balancing has to be performed for obtaining an efficient classification system. The popular way of balancing the classes in a dataset is changing the composition of the training dataset. As the existing data samples are sampled, they are referred to as resampling methods or sampling methods.

Sampling is generally performed in the training datasets. In order to maintain the class bias in the model, class balancing is avoided in the testing dataset. Although there are hundreds of sampling techniques, a particular method could not be believed as the best sampling method. Every method behaves differently based on the choice of ML algorithm used. Some of the most used sampling methods include Random Oversampling, Synthetic Minority Oversampling Technique (SMOTE), Borderline-SMOTE and Borderline Oversampling with SVM.

Random Oversampling is the simplest oversampling, the minority values are just duplicated in order to increase the minority class [15]. SMOTE is one of the most successful sampling techniques, which works by selecting the samples which is near the dividend line and creating a new sample nearby the classification line [16].

Borderline-SMOTE method selects the misclassified classes with the help of KNN classification algorithm and generates samples that are difficult to classify [17]. Borderline Oversampling is another extension to SMOTE, using the decision tree classification technique, that tries to fit the dataset to an SVM classifier [18].

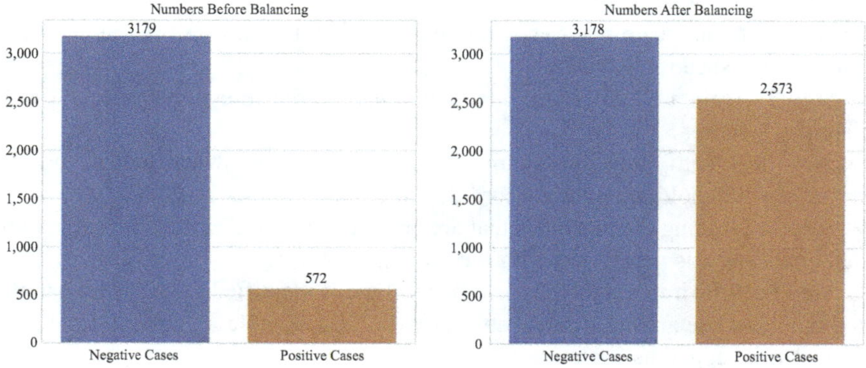

Figure 7.8 Class distribution before and after data balancing using SMOTE

In this work, we are using SMOTE technique for solving the class imbalance issue. Figure 7.8 shows the class distribution in the dataset before and after applying SMOTE algorithm. Before SMOTE the class distribution was 3,179:572 for negative and positive cases, and it was 3,178:2,573 after balancing using the SMOTE algorithm.

7.3.6 Feature scaling

Feature scaling is performed on train and test data before passing it for training and standardizing the independent features of the dataset. This process brings all the values in the independent variables to the same magnitude. The two most important scaling techniques are min–max normalization and standardization [19].

Min–max normalization is one of the most common and easiest normalization methods, in which the minimum value of the variable is transferred to 0 and the maximum value is transferred to 1 and the other values are transferred into a decimal value between 0 and 1 [20]. The value is calculated using (7.1)

$$x_{\text{new}} = \frac{x_i - \min(x)}{\max(x) - \min(x)} \tag{7.1}$$

where x_{new} is the new value; x_i is the current value; min(x) is the minimum value of the variable; and max(x) is the maximum value of the variable.

Standardization is the process of transformation of variable value between 0 and 1 such that the mean of the variable is 0 and the standard deviation of the variable is 1 [21]. The values are transformed using (7.2)

$$x_{\text{new}} = \frac{x_i - x_{\text{mean}}}{\text{standard deviation}} \tag{7.2}$$

where x_{new} is the new value; and x_{mean} is the mean for variable x.

Table 7.2 Min, max and standard deviation of the variables before and after scaling

	Before scaling			**After scaling**		
	Min	**Max**	**Standard deviation**	**Min**	**Max**	**Standard deviation**
Age	32	70	8.387718	−2.286495	2.244434	1.000109
cigsPerDay	0	70	11.910544	−0.790519	5.087268	1.000109
totChol	113	600	44.374775	−2.852165	8.123738	1.000109
sysBP	83.5	265.141083	23.541311	−2.230541	5.486146	1.000109
diaBP	51	142.5	12.313639	−2.710231	4.721366	1.000109
BMI	15.96	56.8	3.959873	−2.557827	7.756761	1.000109
heartRate	44	143	11.70683	−2.726505	5.731021	1.000109
glucose	40	394	29.680767	−1.484229	10.443989	1.000109

In this work we have used a standard scalar for normalizing the data. Table 7.2 describes the min, max and the standard deviation of the data before and after scaling which gives us a clear view of how the data is being normalized.

7.3.7 Methodology

7.3.7.1 Logistic regression

Logistic regression is one of the classification algorithms which comes under supervised learning which helps to classify the target variable. The target might be two or more classes [22]. There are two types of logistic regression binary or binomial, and multinomial binary or binomial. Binomial: Dependent variable will be of binary type (either 0 or 1). Multinomial: Dependent variable will be of more than two classes (Type A, Type B and Type C).

In logistic regression, a linear function is fed as an input to the other function which is shown in (7.3)

$$h_\theta(x) = g\left(\theta^T x\right), \text{where } 0 \leq h_\theta \leq 1 \tag{7.3}$$

where g is the sigmoid function as shown in Figure 7.9. The classes of it could be either positive or negative, which could be observed in Figure 7.9, holding y value between 0 and 1

The loss function which helps us in measuring the performance of logistic regression is defined in (7.4):

$$J(\theta) = \frac{1}{m}\left(-y^T \log(h) - (1-y)^T \log(1-h)\right) \tag{7.4}$$

Once the loss function is calculated the primary objective is to minimize it. To reduce it, the possible way is to adjust the weights of the model. To monitor the change in the loss function gradient descent is obtained using (7.5)

$$\frac{\delta J(\theta)}{\delta \theta_j} = \frac{1}{m} X^T (g(X\theta) - y) \tag{7.5}$$

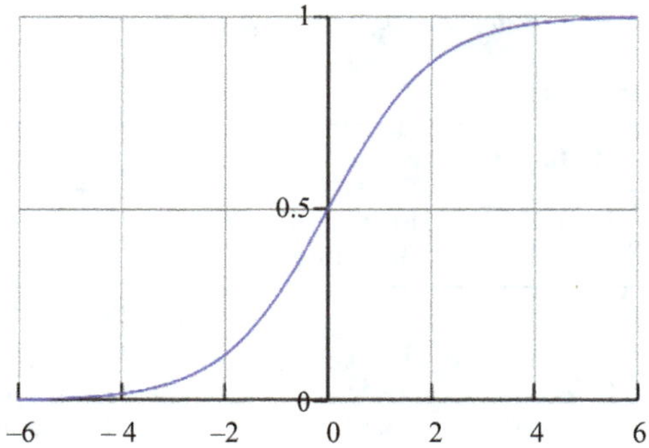

Figure 7.9 Sigmoid function for logistic regression

7.3.7.2 Decision tree

Decision tree splits the dataset in different ways on the basis of different conditions. Under the supervised algorithms category, decision trees are considered to be one of the most powerful algorithms, which could be used for both classification and regression tasks. The entities of decision tree are nodes and leaves. Nodes are used for splitting the data and leaves are used to get outcomes [23]. The structure of decision tree is shown in Figure 7.10.

Building the tree

Once the root node has been created for a decision tree, the next process is executed in two steps:

Step 1: Terminal node creation

Creating the terminal node includes the decision of when to stop creating the terminal node and when to create a new terminal node. The decision is made by calculating the maximum tree depth and minimum node records. Terminal nodes are only used in taking the final decision.

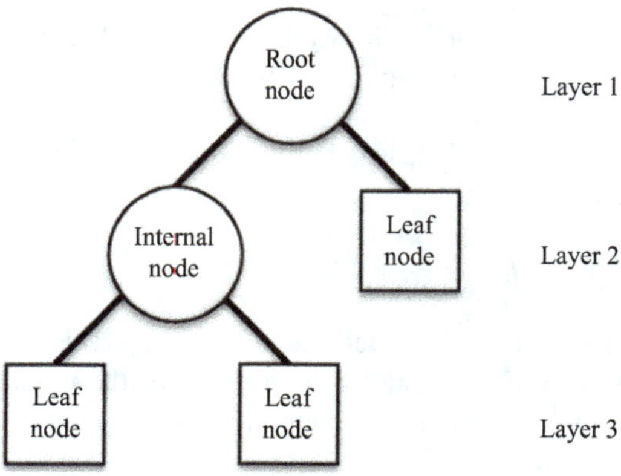

Figure 7.10 Architecture of decision tree

Step 2: Recursive splitting

Recursive splitting is used for building the tree. Once the nodes are created the child nodes are recursively created by splitting up the dataset. This could be achieved by calling the same function again and again.

7.3.7.3 KNN algorithm

KNN, an efficient lazy learning algorithm, is also classified under a supervised ML algorithm. It is capable of handling both the classification and regression problems but is mostly used for classification problems [24].

Implementation of KNN

Step 1: For any classification/regression system the initial step will be loading the dataset to the environment.

Step 2: The next step will be choosing the value of K, which is the number of nearest points. K value must be an integer.

Step 3: For every test data,

3.1: Calculate the distance between the test data and every training data, which could be done using any one of the following algorithms: Hamming distance, Manhattan and Euclidean algorithm. Out of which Euclidean algorithm is used mostly.

3.2: Arrange the distance values 1 in ascending order, which was calculated using step 3.1.

3.3: Next the top K rows will be selected from the array

3.4: Based on the maximum class in the K rows it will assign the class to every test variable

The process of the KNN algorithm is well shown in Figure 7.11.

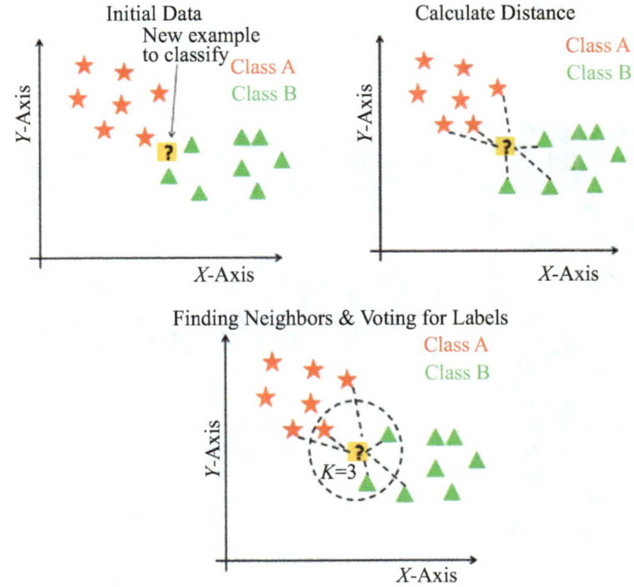

Figure 7.11 Step by step process of KNN algorithm (Source: www.datacamp.com)

7.3.7.4 Random forest

Random forest is a supervised ML algorithm that can be used for both regression and classification problems. Random forest algorithm develops decision trees based on classification systems on the dataset and makes predictions from each of them, and will be selecting the optimal solution based on the voting [25]. This method is considered better than an individual decision tree algorithm as it avoids overfitting the algorithm by averaging the results of all the decision trees.

Implementation of random forest

Step 1: Selecting random samples from the dataset.
Step 2: Constructing a decision tree for every sample.
Step 3: Obtain the prediction result from all the decision trees.
Step 4: Adding votes to all the predicted results.
Step 5: The most voted prediction result is fixed as the final predicted result.

Figure 7.12 describes the working model of random forest.

7.3.7.5 Naïve Bayes algorithm

Naïve Bayes is a probability-based ML algorithm used to perform classification tasks. Naïve Bayes algorithm is developed by applying the Bayes theorem having the assumption that all the prediction variables are not dependent on each other. Bayes theorem is a mathematical function used for computing conditional probabilities [26]. The main task in naïve Bayes is the probability of a task assuming another is already occurred (i.e. assuming the event has already occurred). Equation (7.6) defines the naïve Bayes function

$$P(i|j) = \frac{P(j|i) * p(i)}{p(j)} \tag{7.6}$$

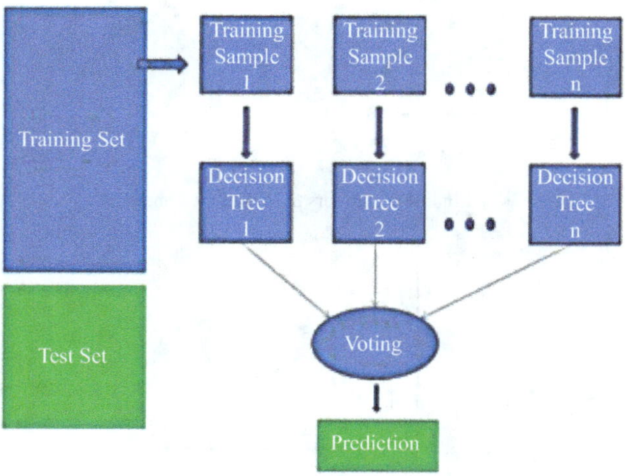

Figure 7.12 Architecture of random forest algorithm

where $P(i|j)$ is the probability of i assuming j has already occurred; $P(j|i)$ is the probability of j assuming i has already occurred; $P(i)$ is the probability of i occurring and $P(j)$ probability of j occurring.

7.3.7.6 Support vector machine

SVM is a supervised ML algorithm, which could be used for both regression and classification tasks. This is one of the oldest ML algorithms introduced in late 1960 and has been refined year after year. It is capable of handling continuous and categorical variables [27]. The data items are placed in an n-dimensional space (n being the features available in the dataset). The hyperplane that differentiates the classes is located and the classification is performed, which is shown in Figure 7.13.

7.3.7.7 Artificial neural network

ANNs are computational logarithms in network structure inspired by a human brain. The network holds nodes that are similar to neurons in a human brain. Nodes from all the layers are interconnected to each other which helps to make predictions [26]. ANNs primarily consist of three layers, as shown in Figure 7.14.

Input layer: This will be the entry point of the NN which accepts different data types as desired by the programmer. *Hidden layer:* This layer is responsible for identifying the features and performing calculations. *Output layer:* The output of the network is informed through this layer.

Training data plays a major role in making the NN more efficient. Once the network is trained and fine-tuned for higher accuracy, the network will be more efficient than any other prediction system.

Implementation

Step 1: Feeding the data to the input layer from which it is transferred to the hidden layer.

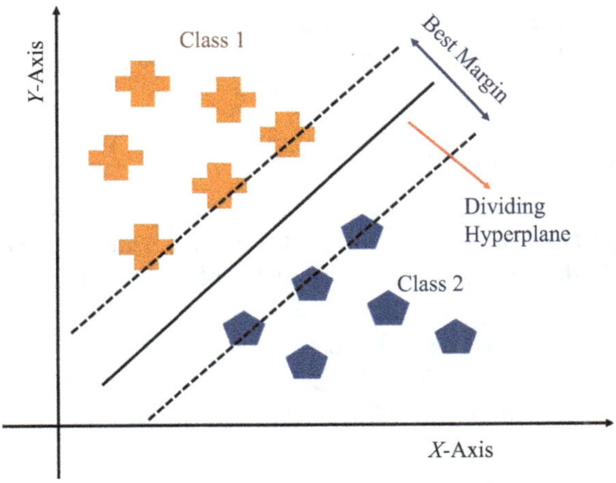

Figure 7.13 SVM model (Source: https://towardsdatascience.com/support-vector-machines-for-classification-fc7c1565e3)

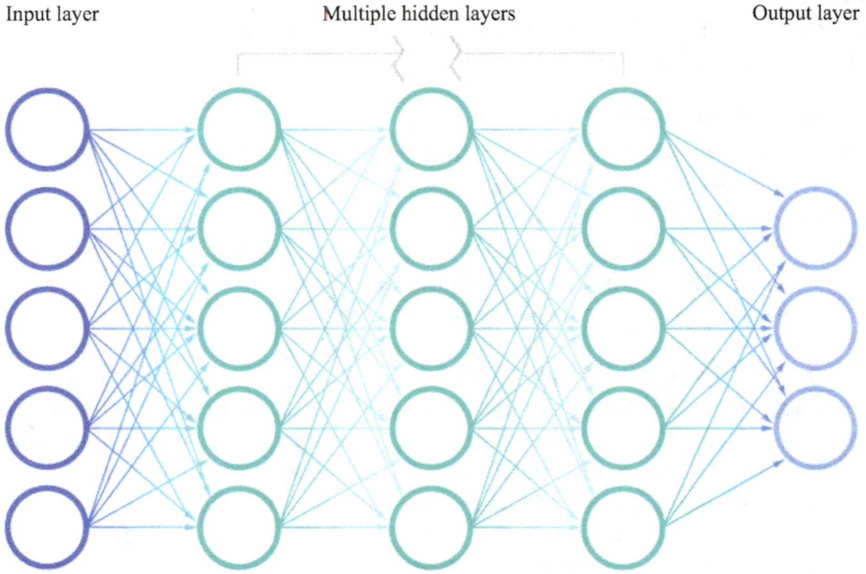

Figure 7.14 Architecture of ANN (Source: https://www.ibm.com/cloud/learn/neural-networks)

Step 2: Random weights are assigned to each input.

Step 3: Bias is added to every input; the combination is calculated as follows:

Combination = Bias + Weights * Inputs

and the combination is transferred to the activation function.

Step 4: The activation function decides which node has to be activated for performing the functions.

Step 5: The application function is applied to the output layer to deliver the output

Step 6: To enhance the accuracy and reduce the loss, backpropagation is performed.

7.3.8 Performance metrics

After successfully implementing the ML and DL algorithms, the next step will be identifying how effective the model is performing. For this, performance metrics are being used. In this work, we have used different performance measures, namely confusion matrix, accuracy (ACC), sensitivity (SE), specificity (SP), PPV, NPV, area under the curve (AUC) and F1-score.

The confusion matrix is a table that holds four different combinations of actual and predicted values, which include true positive (TP), false positive (FP), true negative (TN) and false negative, as shown in Figure 7.15. These values help us in the calculation of accuracy, sensitivity, specificity, precision, PPV and NPV.

F1-score considers both false positive and false negative, and it is the weighted average of recall and precision, as shown in (7.7). Classification models are

		Predicted Class: Positive	Predicted Class: Negative	
Actual Class	Positive	True Positive (TP)	False Negative (FN) Type II Error	Sensitivity $\frac{TP}{(TP+FN)}$
	Negative	False Positive (FP) Type I Error	True Negative (TN)	Specificity $\frac{TN}{(TN+FP)}$
		Precision $\frac{TP}{(TP+FP)}$	Negative Predictive Value $\frac{TN}{(TN+FN)}$	Accuracy $\frac{TP+TN}{(TP+TN+FP+FN)}$

Figure 7.15 Confusion matrix (Source: https://manisha-sirsat.blogspot.com/2019/04/confusion-matrix.html)

considered to be more efficient when the score is 1 and have to be improved when the score is less than 1 and are considered to be the worst if the score is 0.

$$\text{F1} - \text{score} = 2 * \frac{RC * PC}{RC + PC} \tag{7.7}$$

where RC is the recall; RC = (TP+TN)/(TP+FP+FN+TN)

PC is the precision; PC = TP/(TP+FP)

AUC represents how efficiently the model differentiates the classes; the greater the AUC, the better the classification.

7.4 Results and discussion

The first model built was logistic regression. The optimum parameters were identified using grid search, and the model was trained and validated with the testing data. The logistic regression model is managed to give an accuracy of 66.55%. The F1-score for the model was identified to be 0.64. The AUC value was identified to be 0.719. The traditional logistic regression model managed to give a decent performance.

While implementing the decision tree model, we could notice the accuracy was identified to be 71.97% with an F1-score of 66.87, and the AUC score was identified to be 0.755. The time taken by the algorithm was observed to be a little more while compared with the traditional logistic regression. KNN algorithm execution time was found to be lower than the previous models. The nearest neighbour value was set to 5 and executed. The model managed to give an accuracy of 84.72% and an AUC score of 0.859. F1-score was identified to be 84.36%.

The random forest model is built with 50 estimators and the model was performing well with 87.25% accuracy, 85.19% F1-score and 0.934 AUC curve. The model was performing better than the KNN model and the execution time was similar to KNN. The naïve Bayes model could not produce a more accurate value and managed to get an accuracy of 66.38%, F1-score of 53.78% and an AUC score of 0.721. The last model in the ML model is SVM. The model got an accuracy of 86.03%, an AUC score of 0.926, and F1-score to be 84.25%.

Finally, as the last model, ANN was built with one input layer, two hidden layers and one output layer. The description of layers is shown in Figure 7.16. The input layer could get the input variables and nine nodules were placed in it, followed by two hidden layers with nine nodules each, and finally the output layer with one nodule. For the input and hidden layers, the Relu activation function was used. For the output layer softmax activation was used for binary classification. For the compilation process, three parameters are required, as its binary classification cross-entropy is used. Adam optimiser and accuracy metrics were also used. The model was trained with a batch size of 10 and 60 epochs. The model managed to give an accuracy of 87.5%, F1-score was found to be 85.19 and the AUC value is 0.934.

The confusion matrix for the models is considered to be more important as many of the performance parameters are derived from the confusion matrix and Figure 7.17 provides the confusion matrix of all the models.

True negative is higher in random forest and ANN with 579 predictions. False positive is low with 68 predictions for random forest and ANN. False negative is low with 26 predictions in ANN, and finally, true positive is high in KNN with 472 predictions. The ROC–AUC curve for all the models is shown in Figure 7.18. The AUC value was found to be a maximum of 0.934 in random forest and ANN.

```
Model: "sequential"
_________________________________________________________________
Layer (type)                 Output Shape              Param #
=================================================================
dense (Dense)                (None, 9)                 72
_________________________________________________________________
dense_1 (Dense)              (None, 9)                 90
_________________________________________________________________
dense_2 (Dense)              (None, 9)                 90
_________________________________________________________________
dense_3 (Dense)              (None, 1)                 10
=================================================================
Total params: 262
Trainable params: 262
Non-trainable params: 0
_________________________________________________________________
```

Figure 7.16 Description of layers in ANN

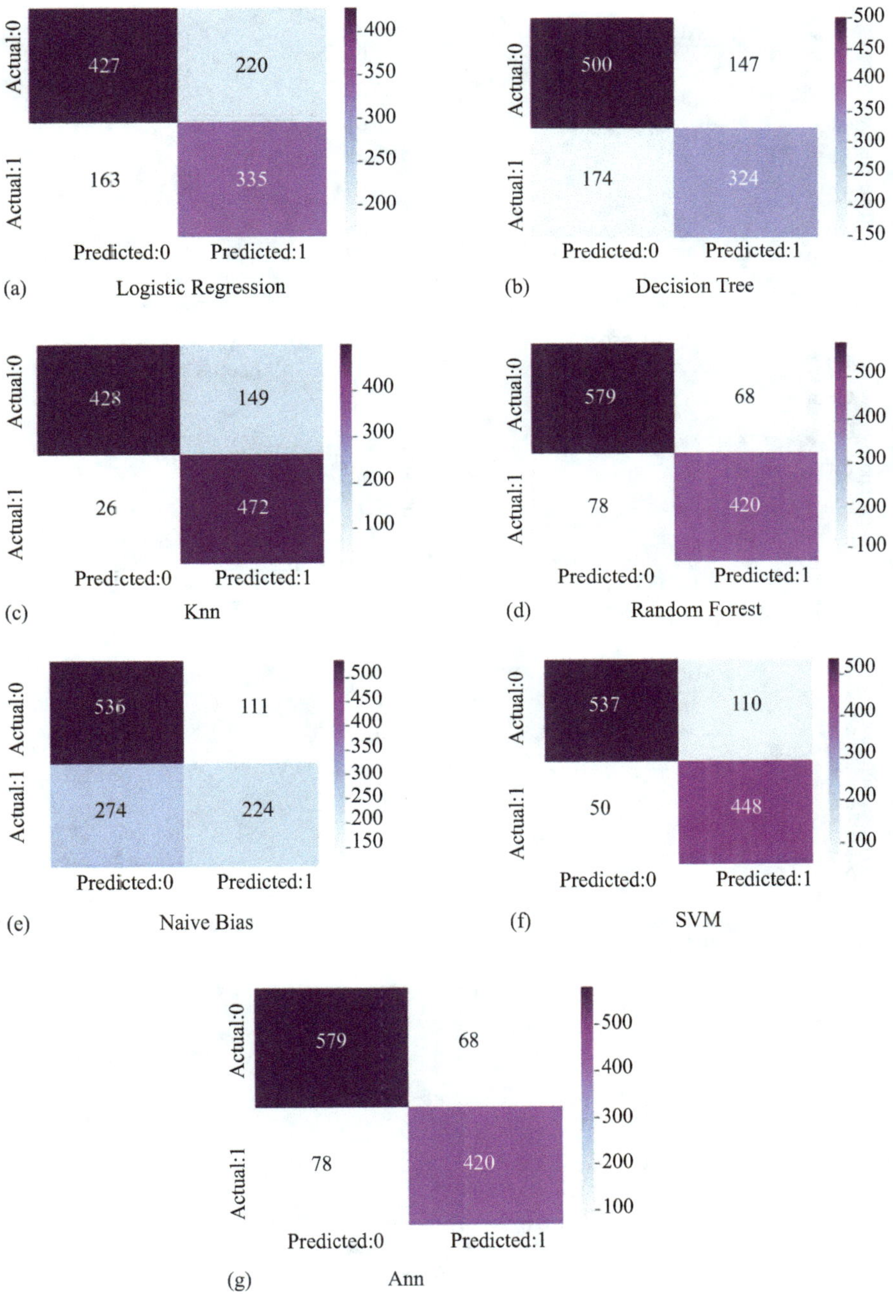

Figure 7.17 Confusion matrix of ML and DL algorithms

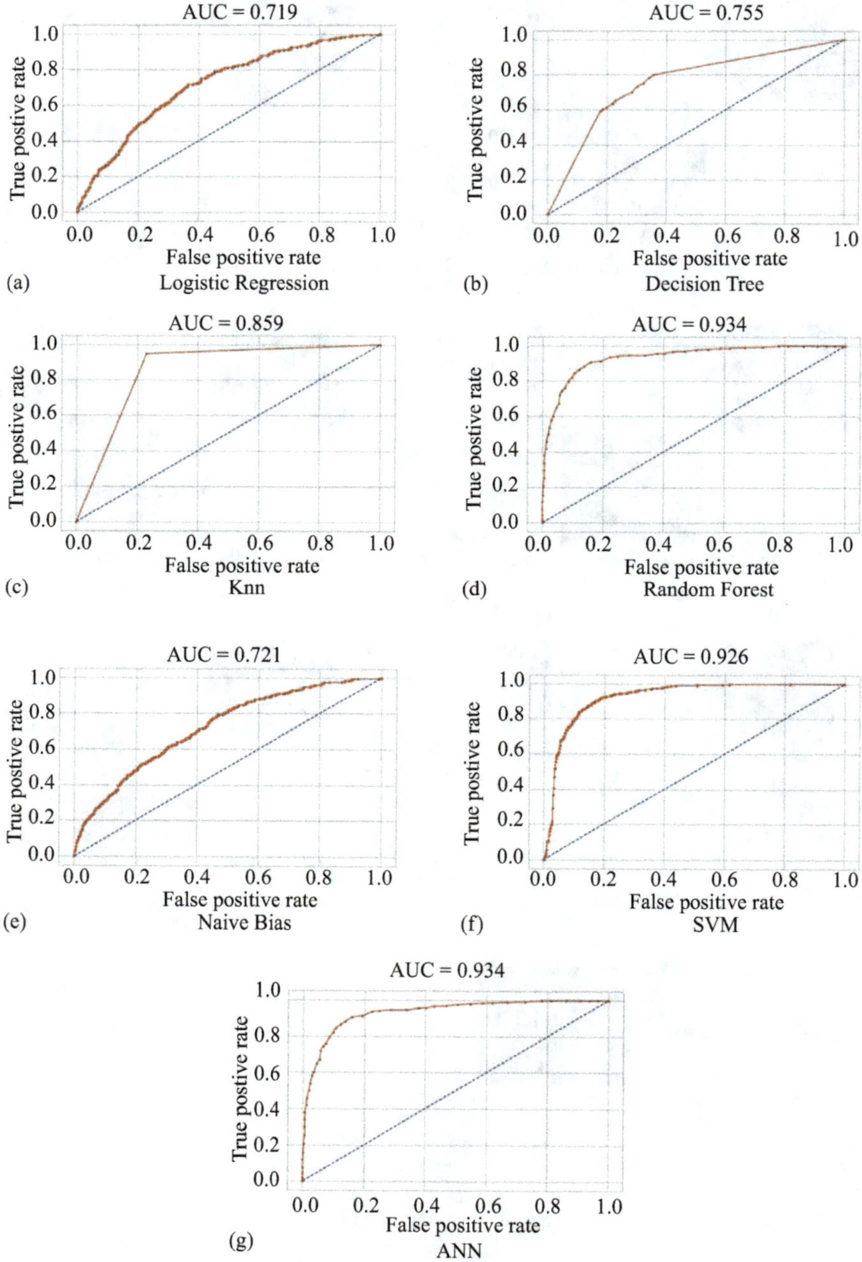

Figure 7.18 ROC–AUC curve for all the ML and DL algorithms

The performance metrics such as accuracy, sensitivity, specificity, PPV, NPV, AUC and F1-score of all the ML and DL algorithms are tabulated for comparison in Table 7.3.

Accuracy, AUC and F1-score are plotted separately in Figure 7.19, and it is observed that accuracy is higher in random forest and ANN models. F1-score is nearly 1 in random forest and ANN with a value of 0.85, and finally, the AUC value is also nearly 1 in random forest and ANN with a value of 0.934, and hence we could observe that random forest and ANN performs the best when compared with other ML methods.

Table 7.3 Performance metrics of ML and DL systems

Algorithm	Accuracy	Sensitivity	Specificity	PPV	NPV	AUC	F1-score
Logistic regression	66.55	67.27	66	60.36	72.37	0.719	0.64
Decision tree	71.97	65.06	77.28	68.79	74.18	0.755	0.67
K-nearest neighbours	84.72	94.78	76.97	76.01	95.04	0.859	0.84
Random forest	87.25	84.34	89.49	86.07	88.13	0.934	0.85
Naïve Bayes	66.38	44.98	82.84	66.87	66.17	0.721	0.54
Support vector machine	86.03	89.96	83	80.29	91.48	0.926	0.85
ANN	87.25	84.34	89.49	86.07	88.13	0.934	0.85

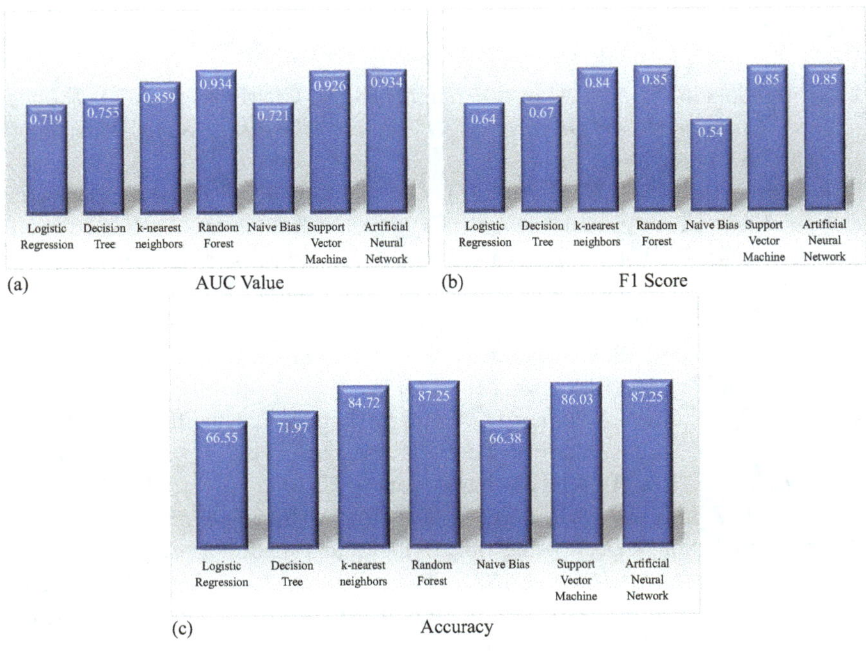

Figure 7.19 AUC score, F1-score and accuracy of ML and DL methods

7.5 Conclusion

The primary aim of this work was to compare the performance of the ML and DL algorithms for the CHD dataset. The dataset holds 4,240 with 16 features. We have almost done the entire process step by step, starting from visualizing the dataset, removing the missing values, followed by selecting the best performing features using the Bourta algorithm, which identified AGE, BP and BMI as the best. Since the number of features was too low we took the top eight features. The next step was to balance the class in the dataset as the class was identified to be imbalanced and hence applied SMOTE method to balance the class, and finally, standard scalar was applied to normalize the dataset. Once all these data preparation and pre-processing methods were done, ML and DL methods, namely, logistic regression, decision tree, k-nearest neighbours, random forest, naïve Bayes, support vector machine and feed-forward neural network was deployed, to evaluate the performance of the models. Several metrics were calculated and compared, from which it was identified that the random forest and neural network methods gave the same results. Although the working methodology and the assumptions made by the algorithms also differ, with these results it could not be concluded that a particular classification method is the best. One of the major drawbacks is that the dataset available does not hold all the features which are mandatory to confirm the disease. The dataset was not approved by any doctors in that field and hence the model developed could not be used for real-time implementation.

References

[1] Puig-Cotado, F., E. Tursan d'Espaignet, S. St. Claire, *et al.* 2020. *Tobacco and Coronary Heart Disease: WHO Tobacco Knowledge Summaries*. World Health Organization, pp. 1–10.

[2] Amarbayasgalan, T., K. H. Park, J. Y. Lee, and K. H. Ryu. 2019. "Reconstruction error based deep neural networks for coronary heart disease risk prediction." *PLoS ONE*, 14(12): 1–17. https://doi.org/10.1371/journal.pone.0225991.

[3] Atkov, O. Y., S. G. Gorokhova, A. G. Sboev, *et al.* 2012. "Coronary heart disease diagnosis by artificial neural networks including genetic polymorphisms and clinical parameters." *Journal of Cardiology*, 59(2): 190–4. https://doi.org/10.1016/j.jjcc.2011.11.005.

[4] El-Bialy, R., M. A. Salamay, O. H. Karam, and M. E. Khalifa. 2015. "Feature analysis of coronary artery heart disease data sets." *Procedia Computer Science*, 65 (Iccmit): 459–68. https://doi.org/10.1016/j.procs.2015.09.132.

[5] Kim, J., J. Lee, and Y. Lee. 2015. "Data-mining-based coronary heart disease risk prediction model using fuzzy logic and decision tree." *Healthcare Informatics Research*, 21(3): 167–74. https://doi.org/10.4258/hir.2015.21.3.167.

[6] Kim, J. K., and S. Kang. 2017. "Neural network-based coronary heart disease risk prediction using feature correlation analysis." *Journal of Healthcare Engineering*, 1–13. https://doi.org/10.1155/2017/2780501.

[7] Lim, K., B. M. Lee, U. Kang, and Y. Lee. 2018. "An optimized DBN-based coronary heart disease risk prediction." *International Journal of Computers, Communications and Control*, 13(4): 492–502. https://doi.org/10.15837/ijccc.2018.4.3269.

[8] Wiharto, H. K., and H. Herianto. 2017. "Hybrid system of tiered multivariate analysis and artificial neural network for coronary heart disease diagnosis." *International Journal of Electrical and Computer Engineering*, 7(2): 1023–31. https://doi.org/10.11591/ijece.v7i2.pp1023-1031.

[9] Latha, C. B. C., and S. C. Jeeva. 2019. "Improving the accuracy of prediction of heart disease risk based on ensemble classification techniques." *Informatics in Medicine Unlocked*, 16: 100203. https://doi.org/10.1016/j.imu.2019.100203.

[10] Alaa Id, A. M., T. Bolton, E. Di Angelantonio, J. H. F. Ruddid, and M. Van Der Schaar. 2019. "Cardiovascular disease risk prediction using automated machine learning: a prospective study of 423,604 UK Biobank participants." *PLoS ONE*, 14(5): e0213653. https://doi.org/10.1371/journal.pone.0213653.

[11] Zheng, W., and M. Jin. 2020. "The effects of class imbalance and training data size on classifier learning: an empirical study." *SN Computer Science*, 1 (2): 1–13. https://doi.org/10.1007/s42979-020-0074-0.

[12] Kwak, N., and C. H. Choi. 2002. "Input feature selection for classification problems." *IEEE Transactions on Neural Networks*, 13(1): 143–59. https://doi.org/10.1109/72.977291.

[13] Karamizadeh, S., S. M. Abdullah, A. A. Manaf, M. Zamani, and A. Hooman. 2013. "An overview of principal component analysis." *Journal of Signal and Information Processing*, 4(3): 173–75. https://doi.org/10.4236/jsip.2013.43b031.

[14] N. Bhalaji, K. B. Sundhara Kumar, and S. Chithra. 2018. "Empirical study of feature selection methods over classification algorithms." *Int*ernational Journal of *Intelligent Systems Technologies and Applications*, 17: 98–108.

[15] Zhang, J., and L. Chen. 2019. "Clustering-based undersampling with random over sampling examples and support vector machine for imbalanced classification of breast cancer diagnosis." *Computer Assisted Surgery*, 24(2): 62–72. https://doi.org/10.1080/24699322.2019.1649074.

[16] Zhu, T., Y. Lin, and Y. Liu. 2017. "Synthetic minority oversampling technique for multiclass imbalance problems." *Pattern Recognition*, 72: 327–40. https://doi.org/10.1016/j.patcog.2017.07.024.

[17] Xu, X., W. Chen, and Y. Sun. 2019. "Over-sampling algorithm for imbalanced data classification." *Journal of Systems Engineering and Electronics*, 30(6): 1182–91. https://doi.org/10.21629/JSEE.2019.06.12.

[18] Gosain, A., and S. Sardana. 2017. "Handling class imbalance problem using oversampling techniques: a review." In *2017 International Conference on Advances in Computing, Communications and Informatics, ICACCI*, 2017, pp. 79–85. https://doi.org/10.1109/ICACCI.2017.8125820.

[19] Singh, D., and B. Singh. 2020. "Investigating the impact of data normalization on classification performance." *Applied Soft Computing*, 97: 1–24. https://doi.org/10.1016/j.asoc.2019.105524.

[20] Jain, S., S. Shukla, and R. Wadhvani. 2018. "Dynamic selection of normalization techniques using data complexity measures." *Expert Systems with Applications*, 106: 252–62. https://doi.org/10.1016/j.eswa.2018.04.008.
[21] Ioffe, S. 2017. "Batch renormalization: towards reducing minibatch dependence in batch-normalized models." *Advances in Neural Information Processing Systems*, 1702.03275: 1946–54.
[22] Subasi, A., and E. Erçelebi. 2005. "Classification of EEG signals using neural network and logistic regression." *Computer Methods and Programs in Biomedicine*, 78(2): 87–99. https://doi.org/10.1016/j.cmpb.2004.10.009.
[23] Yan Song, Y., and Y. Lu. 2015. "Decision tree methods: applications for classification and prediction." *Shanghai Archives of Psychiatry*, 27(2): 130–5. https://doi.org/10.11919/j.issn.1002-0829.215044.
[24] Deng, Z., X. Zhu, D. Cheng, M. Zong, and S. Zhang. 2016. "Efficient KNN classification algorithm for big data." *Neurocomputing*, 195: 143–48. https://doi.org/10.1016/j.neucom.2015.08.112.
[25] Liaw, A., and M. Wiener. 2002. "Classification and regression by random forest." *R News*, 2(3): 18–22.
[26] Paquin, F., J. Rivnay, A. Salleo, N. Stingelin, and C. Silva. 2015. "Multiphase semicrystalline microstructures drive exciton dissociation in neat plastic semiconductors." *Journal of Material Chemistry C*, 3: 10715–22. https://doi.org/10.1039/b000000x.
[27] Mathur, A., and G. M. Foody. 2008. "Multiclass and binary SVM classification: implications for training and classification users." *IEEE Geoscience and Remote Sensing Letters*, 5(2): 241–5. https://doi.org/10.1109/LGRS.2008.915597.
[28] "Cardiovascular Diseases (CVDs)." n.d. Accessed May 15, 2021. https://www.who.int/news-room/fact-sheets/detail/cardiovascular-diseases-(cvds).
[29] "Coronary Artery Disease | Cdc.Gov." n.d. Accessed May 15, 2021. https://www.cdc.gov/heartdisease/coronary_ad.htm.
[30] "Coronary Heart Disease | NHLBI, NIH." n.d. Accessed May 15, 2021. https://www.nhlbi.nih.gov/health-topics/coronary-heart-disease.
[31] "Heart Attack Analysis & Prediction Dataset | Kaggle." n.d. Accessed May 17, 2021. https://www.kaggle.com/rashikrahmanpritom/heart-attack-analysis-prediction-dataset?select=heart.csv.
[32] Liu, H. M., R. Setiono, and Z. Zhao. 2010. "Feature selection: an ever evolving frontier in data mining." *Journal of Machine Learning Research: Workshop and Conference Proceedings 10: The Fourth Workshop on Feature Selection in Data Mining*, 10: 4–13.
[33] "What Is Machine Learning? – India | IBM." n.d. Accessed May 15, 2021. https://www.ibm.com/in-en/cloud/learn/machine-learning.
[34] "What Is the Difference between AI, Machine Learning and Deep Learning? – Geospatial World." n.d. Accessed May 15, 2021. https://www.geospatialworld.net/blogs/difference-between-ai-machine-learning-and-deep-learning/.

Chapter 8

Revolution in technology-enabled healthcare: Internet of Things

S. Sangeetha[1], Deepa Shanmugham[2], S. Balamurugan[3] and K. Maharaja[4]

The Internet of Things (IoT) is a humongous network, designed through extended internet connectivity to a physical device. Recently, internet technology has revolutionized the medical sector. This renowned new technology offers new prospects for medical professionals to monitor their patients and collect health matrices for timely intervention and proper management. IoT device collects patient data and feeds the data to a software application where the healthcare professional can view it. IoT with sensors can also interpret the data based on the algorithm and generate an alert so that healthcare professionals can intervene. The implementation of information technology in the field of medicine is a growing niche in the healthcare sector and has vast applications in different subspecialties of medicine. It has huge potential in diagnostics, monitoring of treatment, and follow-up of patient care. Internet of Medical Things (IoMT) applications are constantly evolving as technology becomes more powerful and sophisticated.

Medical informatics reduces the costs involved in treatment, improves its efficiencies, and saves the lives of the people. Although there are many benefits to using Internet of Things (IoT) technology in healthcare settings, there are also challenges in its use, such as the security of patient data, adoption of new technology, and difficulties in implementation. Compared to cost and challenges there are more benefits in the usage of the Internet and its facilities by the patients. This revolution along with IoT is just beginning and many innovative IoMT solutions are on the horizon for the future.

[1]Department of Electrical and Electronics Engineering, Sri Ramakrishna Institute of Technology, India
[2]Department of Obstetrics and Gynecology, AarupadaiVeedu Medical College, India
[3]Department of Uveitis and Ocular Inflammation, Aravind Eye Hospital, India
[4]Engineering Department, University of Technology and Applied Sciences, Oman

8.1 IoT and healthcare information systems

The digital technology-enabled self-management is the need of the hour to tackle key challenges of delivering care to the aging population with limited healthcare resources [1]. Internet facility and increasing the availability of smartphones and other portable devices help in sharing bio parameters and make remote monitoring of the emergency condition of the patient possible. IoT is a combination of physical objects, devices, and systems that are interconnected, communicates with each other electronically, processes the received data, and transfers the same with less input from human [2]. Technology-enabled smartwatches have the capability to sense the physical quantities, like temperature, blood pressure, breathing ability, bodily movement, heart rate, and oxygen saturation of blood [3]. It also can measure stress and sleep quality. Technologies are considered smart, if they communicate in both ways by monitoring, collecting, and sending data wherever it is required.

The implementation of the IoT in the medical field is called the Internet of Healthcare Things (IoHT). The integration of electronic devices used by the medical industry and information technology serves the healthcare industry better in providing efficient patient care with the remote monitoring facility. It consists of two broad components, one is the mobile application (Apps) that is designed to view and log the data and the other component is a sensor used to acquire and transmit data continuously from the patient's body, which could be analyzed further for necessary action [3]. Some devices with IoT are smart glucometers, smart inhalers, connected thermometers, and blood pressure meters which are frequently used in the healthcare industry.

In general, the main objective of IoHT is to facilitate easy access to the healthcare database, provide proficiency in treatment, and make the patient have comfort experience. Its role in the different stakeholders of the medical sector includes:

- **Patients:** IoT in the medical industry offers an opportunity for the patients to observe their health conditions and control the intensity of various diseases. For example, diabetic patients can be assisted to intake the insulin based on the data obtained by a smart glucometer with the help of automated insulin delivery systems.
- **Health professionals:** Internet-connected devices used by the patients help them to afford quality and instant medical care from doctors. Monitoring the state of health with the help of IoT allows medical professionals to diagnose the case more precisely and lend treatment to exact observance.
- **Hospitals:** Implementing IoHT in hospitals helps to track the status and location of the equipment on their premises, invigilates sanitation, and delivers required medicine to the inpatient wards. It also keeps records of the patients to update automatically.
- **Health insurance companies**: Data received with the help of IoHT can improve and fasten the claim management processes of health insurance

companies. They also have more details like the patient's behavior and their health condition based on the obtained data in order to avoid duplicate claims.

Even in large environment scenarios like multispeciality hospitals, there will be a large number of connected devices to deliver an uninterruptable service and support. From this, we can infer that these devices are self-configured, more adaptive to any environment, and even with a large network of connected devices. It renders support to medical officers through smart devices with unique identification to monitor health from a remote place. This technology also provides scope for data mining, analysis of data, and future trend in the medical field [4].

The main characteristics features of an IoT system consist of the following:

- Connected devices
- Connecting all the physical devices to the Internet
- Unique identification of each device
- Versatile in nature
- Identifying
- Ability to embed artificial intelligence, machine learning, etc.
- Reliable communication facility
- Self-adaptability
- Programmable

IoHT empowers individuals to monitor their health parameters of common diseases like pulse, blood pressure, glucose level and make them aware of their health condition. This technologically developed medical care facility assists the physicians to deliver timely care to the patients and analyze their case history and causes for the abnormal condition in patient health.

8.2 Remote health monitoring and telehealth

The medical IoT (MIoT) is a recently developed digital transformation of the healthcare system. Wearable outfits and mobile applications help to maintain physical fitness, provide health education, and track the symptoms in case of a pandemic situation. These technologies also support collaborative disease management and integrated care. Medical informatics, which is the convergence of medicine and information technologies, will transform healthcare into more efficient and cost-effective, saving many lives.

The participation of digital technology in the medical field leads to the emergence of Digital Health Advisors, who take care of the overall health of the population. They have the ability to interpret and diagnose the health issues from the collected health parameters from the patients. These health professionals also play a vital role in the different sectors of medical departments, like chronic disease care, dietary and nutrition, clinical psychology, and psychiatry. The global population consists of more aged people which raises year by year leading to the increasing importance of the above-mentioned services.

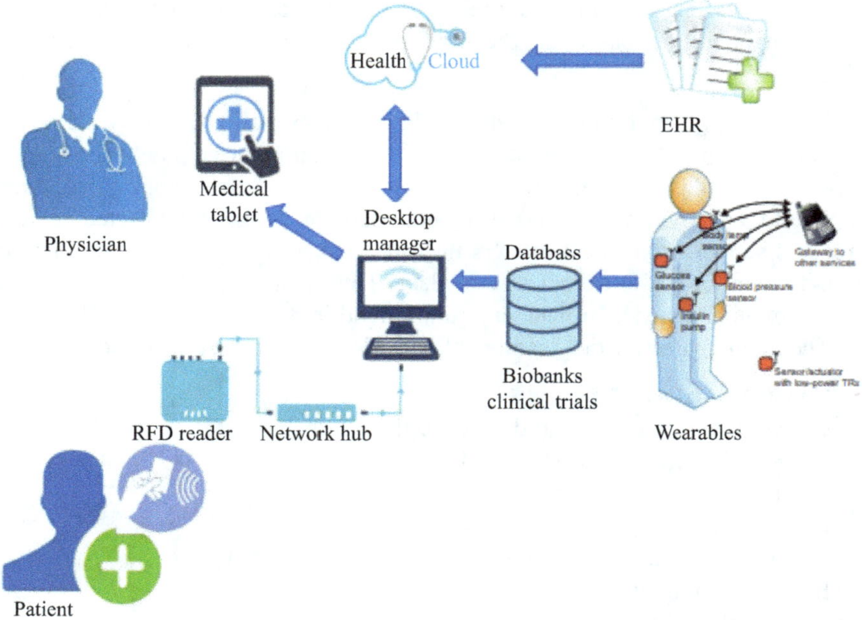

Figure 8.1 Implementation of the IoT in hospitals

Figure 8.1 illustrates the implementation of digital technology in the medical field. A patient with some common disease will have an identity card issued by the hospital that can be scanned and linked to the allotted secured cloud. Patients' health records created by the hospitals, their clinical lab results, medical histories, and previous prescription are stored and accessed whenever required from the cloud. Doctors can have access to this record on their laptops or desktop computers [10]. Instead of maintaining the patient information in one place as filed papers, these electronic health records will improve the efficiency of maintenance of data, easily sharable, and accessible at any place.

The major challenge in the implementation of IoT is communication between different devices as they communicate with the cloud and server in their own language. Different manufacturer has their own exclusive protocols that again create the problem that they cannot communicate between themselves. Different software environment, privacy constraints, and the administrative difficulty of handling all collected information make restricted use of IoT in the field of medical science.

8.2.1 PharmaIoT

The concept of PharmaIoT involves the implementation of information technology services in the drug development process and clinical trials of the developed drugs with various levels of the public based on age, gender, location, etc. In recent days, the best solution for patients suffering from diseases like Parkinson's and multiple sclerosis is to use wearable sensors that facilitate medical management, improving their performance and quality of life [11]. The commonly used devices such as insulin pens can be embedded with the wearable devices which will be connected to the cloud through a communication network to gather data for analysis. These data can be used to forecast the intensity of the disease and also further research on the occurrence and control of the disease. Patients can transfer their health data across different service providers to access them for the care they need. It leads to new service platforms for the companies to have new business models along with quality healthcare.

8.2.2 Mobile applications for healthcare

In this information technology era, knowledge and data play a key role where the output of smart electronic devices is monitored with the help of a mobile application. The video game player uses an app called "The Myo" to control movements. It is now used by orthopedics patients to aid them in doing exercises correctly after a fracture or any alignment problem. With the help of the above application, both doctors and patients can monitor the angle of limb movement and their progress, respectively. There is another application to measure heart rate and electrocardiogram (ECG) called "The Zio Patch"[12]. The pharmaceutical companies have also taken an initiative to use the technology to focus on healthcare as medicine-centric. To promote micro-stimulation of nerves Glaxo Pharma works on electroceuticals and bioelectrical drugs which are found to be good for many neurological diseases. Similarly, companies like J&J and Google joined hands to develop new technology that executes the surgery with the help of robots. This initiative will bring revolution in surgical procedures in all the medical departments. They also started their research in collaboration with Philips to design blood pressure monitors that can be used by an individual as a wearable device. The blood glucose levels also can be monitored by a portable device smart lens developed by a combined research of Novartis and Google.

8.2.3 Big data in healthcare

The main challenge in the MIoT and wearable sensors is the volume of data gathered from wearable devices in the pharmacy sector. The software companies are trying to streamline these data for the purpose of research and commercial use. There are "three Vs" in "big data": first is the volume which denotes a vast amount of data to be handled; second is the variety of data where there are different types of data available in the set; and the last one is velocity which is accessibility speed [10]. Now compared with other sectors, the healthcare sector has become a prominent user of big data analytics. Compact and easy-to-wear devices like Fitbit can provide access to vast biometric data to researchers in the medical and pharmaceutical sectors. It can also be used to carry out research in the area of nutrition, physical fitness, and disease progression.

8.2.4 Challenges in MIoT

IoT in any field should provide easy access to its device applications and data. It facilitates the users for analytics and also to view the same in the dashboard. It also should offer easy device management with a reduced security risk. Any technology will have its own challenges to be addressed when a large mass of the population is using the same for different applications. The challenges in MIoT can be classified into two main categories: one is fiscal/policy and the other one is technological issues [10]. Fiscal and policy issues deal with the rate that is charged for the service provided to healthcare practitioners for teleconsultation with patients. It is necessary to streamline this fee based on the service, patient type, and disease type. Generally, payments are made to service organizations like hospitals and polyclinics to reduce unnecessary in-office encounters. This makes the medical system feel that direct visits to hospitals are actually a cost, thus this technology can reward positive health outcomes for most of the population.

8.3 Wearables and medical devices

Nowadays, the general population spent most of the time involving in various tasks ignoring health and fitness. General health check-up in a hospital takes significant time since this requires multiple diagnostic tests, validating the results, diagnosing the problem, and finally starting the treatment. This hassle causes the patients to delay their hospital visits and rush mostly when they are going through serious symptoms after attaining a severe stage. So it becomes mandatory to look for an alternative solution. A small device as wearable can be worn by the patients. These devices being close to the patients will monitor them continuously and publish their health metrics instantaneously. This will provide awareness of health status to both the patients and their doctor in a timely manner.

Elderly people suffering from aging-related diseases will be benefited from such instant health monitors and then preventive medicine at the right time. Wearable device technologies can monitor health indicators with reduced overall costs. Hence this will be a reliable tool for long-term health monitoring. Medical electronics include wearable devices that can be worn or embedded in human skin to monitor an individual's activities continuously without interrupting the user's motions. These devices are connected with a specifically designed mobile application called the MIoT [5].

8.3.1 Activity trackers

Activity trackers monitor and track human movement used in medical and sports fields. This monitoring includes walking step counting, elevation gain, breathing rate, heart rate, skin temperature, oxygen saturation, fall risk assessment; recognizing sports exercise, measuring sleep quality, and observing the daily routine. There are two main reasons for these activity trackers to be popular. These trackers motivate their users to perform exercise regularly and meet their health goals by showing their walking distance, calories burned, step count, badges, and trophies as gamification. It

Figure 8.2 Motion tracker

is provided directly as digital values with its appropriate units through a smartphone without manual calculation [6]. The high precision observation of human body movements is critical, which could be made possible by sensors like pedometers, accelerometers, altimeters, optical heart rate sensors, and bioimpedance sensors. These devices can be used to be aware of human movement in the modern edge computing world leveraging smartphones and network gateways.

Three types of data can be obtained from these types of trackers as shown in Figure 8.2. The device can be connected to a personal computer (PC) to view the values obtained from sensors. In the next type, the collected data are stored in a facility provided for future access. In the last data model, multiple units of devices are deployed on the body and connected through wires to a gateway unit with local memory. It is also designed as a hand-held device with a wireless connection to a PC [7]. The locally stored data from the device can be sent to a cloud from there to the hospital to store permanently.

8.3.2 Vital sign measurement

Wearable devices are also designed and implemented for monitoring and measuring critical health indicators. Some of them are parameters from electroencephalogram (EEG) and ECG measurements. The wearable device can be designed to measure multitask vital signs measurement from one device. Unfortunately, these technologies are difficult to develop in wearable form and are inefficient from the perspective of their performance and energy usage. We also have a challenge of converging the signals across various systems that makes it impossible to be used in daily patient care.

8.3.3 Smart jacket

Smart jacket are designed to obtain physiological indicators as signals from an individual for comprehensive health monitoring. Remote health monitoring methods which we have discussed previously collect fewer numbers of physiological data, which is insufficient for monitoring health in full range, especially for chronic diseases. In this jacket shown in Figure 8.3, all the sensors are integrated into textile

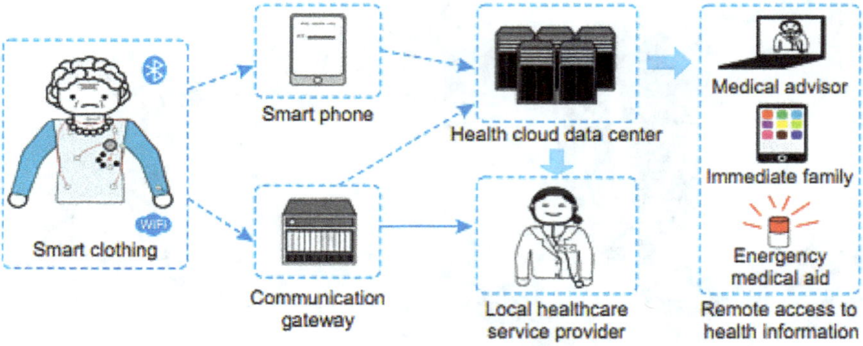

Figure 8.3 Implementation of smart clothing

clothing appropriately which measures health indicators. Arrangement of the sensor is important in smart clothing, proper care has to be taken while integrating it into the clothes for accurate performance. For efficient performance of the designed coat, the sensors quality, its placement, the ability to acquire even weak signals, low-power requirement, and comfort are the important factors [8].

All the sensors should be covered by a flexible layer and positioned appropriately to ensure comfort for the user while wearing the clothes. These small devices touch the user's skin to sense the physical parameter and transmit the data to the storage. It is possible to wash these clothes by removing some of these devices and leaving others in the cloth which is waterproof. The challenges in the design of a smart jacket are the design of communications protocol. While designing the cloth some factors to consider are user body shape, their daily habits, and parameters to be monitored. In addition to the above constraints, technical possibility, ease to wear, and cost are also to be considered during the manufacturing process [5].

8.3.4 Wire-based wearable devices

The MIoT uses wire-based wearable devices with Arduino and Raspberry Pi as controllers. This system allows medical practitioners to monitor a patient's health using different sensors. Sensitive data are transmitted for analysis that will be used for further medical diagnosis [9].

Study on the application of wearable devices discussed above plays the important role in upgrading the medical field with technology. The outcome of this study is to analyze the possibility and challenges of using these devices in real time, with reduced power consumption and improved performance.

8.4 IoT in chronic diseases

The prevalence of chronic diseases is on the rising trend in the world, being a threat to the development of the nation, in view of health. World Health Organization

(WHO) has predicted that the incidence of chronic diseases in low and middle-income countries tends to rise over time, till 2030 [13]. This can result in a healthcare crisis if left unattended. People with chronic diseases are needed to monitor their vital signs periodically so that appropriate intervention is possible. The costs incurred by chronic diseases and the associated morbidities are very high. Hence necessary steps have to be taken to avoid the economic burden. IoT is an emerging technology to tackle these chronic diseases, preventing complications and decreasing fatality. IoT contributes to building a healthcare practice that can meet the needs of people with chronic illnesses. Chronic diseases include cardiac disease, diabetes mellitus, malignant neoplasms, respiratory illness, and neuropsychiatry illness.

IoT-enabled devices include sensors and wearable devices used by patients to measure various health parameters in specific chronic diseases. Sinnapolu and Alawneh (2018) created an innovative model of Internet of service application meant for heart rate monitoring in cardiac patients, using a wearable sensor. This monitoring is of help in identifying heart rate changes during exertion, sleep, and stress [14]. This application can detect various conduction defects in the heart also. The data can be received on a real-time basis or stored on the cloud so that the physician can monitor it on his own, and also it is portable. Fog computing was developed by Cisco as an extension of the cloud, which is helpful when real-time data analysis is required by the physician. Fog computing provides real-time data as it acts as an intermediate between the user and the cloud. A mobile device for monitoring diabetes mellitus with added security features was developed by Saravanan *et al.* In their device, the security of data was ensured by the blockchain with trusted parties to collect data [15]. Analysis of the use of IoT in chronic diseases by Dadkah *et al.* concluded that IoT was used with the highest priority in cardiovascular diseases followed by diabetes mellitus in developing countries [16]. The reason might be the clear knowledge of the prevention and treatment of these diseases rather than malignant neoplasms and respiratory disorders. Since modification of lifestyle is one of the important management aspects of cardiovascular diseases and diabetes mellitus, IoT device works well for monitoring the patient's lifestyle. With advanced medical care facilities, new parameters may be identified which can monitor the other chronic diseases better. This may necessitate the need for newer IoT devices.

The present smart devices can be upgraded in the future for optimal healthcare solutions. Chinnasamy *et al.* evaluated the use of IoT for healthcare providers in treating elderly patients with non-communicable diseases (NCD) in Malaysia and stated that in a technologically developed country like Malaysia, the use of IoT is possible with greater accursed in treating early NCD patients [17]. IoT sensors can help to monitor and detect the signs of respiratory decompensation in patients with bronchitis and chronic obstructive pulmonary disease.

Smartphones can be used to monitor patients with chronic diseases such as diabetes, cardiac disease, respiratory illness, and neurological illness for their vital signs, and other parameters showing prognosis, through the inbuilt sensors in them. This can be then transmitted to a central server through the mobile wireless network for further processing, finally to the concerned physician. In patients with diabetes mellitus, smart devices can prompt the patients to check their glucose level, feet, renal parameters, and fundus examination for optimal glycemic control and also prevent complications like diabetic foot, kidney failure, and loss of vision due to diabetic retinopathy (DR). These results can be clouded or in real time so that the physician can communicate and treat the patients effectively. The main application of this technology is to remind patients about taking their medication. The smart device alerts the patient to take medication, whenever a dose is missed. Digestible sensors which will notify if the medication is not taken in time can be of great help in the management of chronic diseases.

Healthcare providers are expected to be at the forefront in introducing IoT and implementing the technology in their management of patients with IoT for optimal benefits. The entire community needs to be made aware of the utility of IoT devices such as smartphones and smartwatches, to make their use highly effective. This can be achieved through campaigns, social media, etc.

Cancer has become a global challenge in recent years. According to reports by the WHO, it is the second leading cause of death worldwide. The impact of cancer as a chronic illness is quite devastating for the patients in terms of quality of life. It poses an overwhelming challenge to healthcare setups, causing a threat to the economy of the country. Most cancer patients require continuous monitoring, frequent treatment, and support from specialist medical staff for a better prognosis. IoT-based healthcare systems can provide ways to detect and monitor cancer patients in real time for long periods, using smart devices and sensors.

Breast cancer is the most commonly occurring cancer in women worldwide. Early detection of breast cancer can save lives with a good prognosis. But, a simple and feasible method of screening or early detection of breast cancer is still in research. A machine learning-based diagnosis system proposed by Memon *et al.* can effectively classify benign and malignant tumors of the breast [18]. This IoT can be adopted by clinicians for early detection.

The early detection of breast cancer significantly poses a great challenge for clinicians and researchers. For early-stage detection of breast cancer, Memon *et al.* proposed a machine learning-based diagnostic system which effectively classifies the malignant and benign tumors of the breast in the environment of IoT [18]. The performance of the proposed method in terms of accuracy is good as compared with other methods. Similarly, in the diagnosis of ovarian cancer, many ultrasound-based prediction models have been reported in the literature and implemented in clinical practice after approval by the International Ovarian Tumor Analysis (IOTA) Group. Van Calster and his co-workers (2014) developed a web-based risk prediction model based on clinical and ultrasound parameters to distinguish benign, early-stage, and advanced stage and malignant ovarian tumors before surgery. He also temporally validated the score on 5,909 patients and published it [19]. This IOTA ADNEX model is available as a web-based and mobile application for clinicians and is of immense help for them in the planning of surgery.

Smart wearable devices named patient side units are aimed at identifying a particular type of cancer, with thermal, immunosensor, and nanosensors. Some of the sensors work in the detection of cancer by exploiting the ligand properties of DNA strands, proteins, and antibodies. Nanosensors are the latest innovations for the detection of leukemia, a form of blood cancer, in research, soon to be implemented in possible. Nanosensors are breathable and can detect molecules in our breath, correlating it with cancer and some metabolic diseases. Further innovations are needed in this regard for the development of devices in detecting these cancers at the earliest.

8.5 IoT in emergency medical care

Emergency medicine refers to providing life-saving measures during life-threatening situations of health. The ultimate goal of emergency medical services is to provide timely treatment for patients who need urgent medical care to save a life. The timely transfer of patients to a tertiary care center is also facilitated by emergency medical services. The emergency operation begins with the first call at the emergency coordination station, for which the response needs to be prompt, but often lacking. An IoT-based smart technology can overcome the challenges facing the emergency call by automatically providing additional data which will enhance the emergency response and save time as well as reduce the mortality rate. The success of the treatment given to a sick patient at an emergency care unit depends on the information collected in the golden hour. However, the information about the patient's health conditions received during emergency medical care is from the patient's relatives verbally, which is quite inaccurate. Hence, it will be better to have data of health records of the public which can be accessed by physicians during emergency hours. Electronic medical records have the potential to save time and tremendous support for the treating physician during emergencies. The wireless sensor networks are being used by the healthcare industries to measure a patient's vital parameters through medical sensors that are attached to his body. The monitors with sensors attached to the sick patients during transportation can be remotely accursed by the physician so that he can monitor the temperature, heart rate, blood pressure, oxygen saturation, and ECG and manage accordingly. In addition, remote medical centers use end-point devices like video and audio devices to perform advanced patient monitoring. Thierry Edoh proposed IoT-based information processing systems for the patients to improve emergency medical care services globally. His device can cause a revolution in the healthcare management systems, especially in developing countries [20].

Road traffic accidents are the leading cause of death worldwide in recent years. The number of deaths due to road traffic injuries is around 1.2 million as per the reports from WHO [21]. Timely intervention can save lives in a road traffic accident. IoT helps to prevent accidents by IoT-enabled crowd sensing-based road traffic congestion detection units to detect road traffic congestion and help to prevent accidents. This technology can help to reach the emergency center closer to the accident place by suggesting an alternative route. Also, an IoT-based calling system is proposed which forwards the data about the patient to the emergency doctor and the emergency transportation unit. This system is incorporated with electronic medical records (EMRs) with the health data of the injured person, place of occurrence of the accident, and geographical data, which can aid the physician in management. The same will be forwarded to the emergency doctor and emergency transportation unit. Hence the golden hour of emergency management is utilized well in saving the lives of these patients.

However, further innovations are needed in the future, on the use of the IoT in delivering emergency medical care and patient-centric medical care delivery.

8.6 IoT and pregnancy care

Quality of care during pregnancy is essential to detect the complications of pregnancy earlier and ensure the health and well-being of the mother and the fetus. Various quality improvement strategies have been studied and implemented in clinical practice to improve maternal and neonatal outcomes. Moreover, continuous monitoring of various health parameters is mandatory in patients with high-risk pregnancies. IoT-based systems can provide cost-effective health monitoring services for pregnant women with high-risk pregnancies, reducing their hospital stay. Literature suggests that remote monitoring systems of maternal and fetal parameters like blood pressure, heart rate, uterine contractions, and fetal heart rate can improve the obstetric outcome drastically [22]. Such monitoring also promotes a healthy lifestyle in pregnancy, alleviates the anxiety of patients, and reduces the risk factors for complications during pregnancy. The complications which affect pregnancy include anemia, hypertension in pregnancy, gestational diabetes mellitus, and preterm labor. Marko *et al.* conducted a study on the usage of a mobile application in monitoring eight pregnant women with low risk. The mobile application designed by him was connected to a blood pressure monitoring device and weighing scale, which can be monitored remotely by the obstetrician [23]. Grym *et al.* conducted a case study on 20 pregnant women for a period of 7 months and evaluated the feasibility of smart wrist bands in monitoring health parameters [24].

Sarhaddi and Fatemeh *et al.* (2020) presented a technology-based system designed for long-term monitoring of the health of the mother during pregnancy and postpartum [25]. This system enables the monitoring of pregnant women continuously and interpretation is made by the physician based on the subjective and objective data collected from the mothers. In his study, the Samsung gear sport smartwatch was used for monitoring the blood pressure and heart rate using photoplethysmography signals in pregnant women, and variability in the parameters was notified. This technology can be used in hypertension complicating pregnancy for appropriate intervention and prevention of complications.

The smart wristband with IoT technology is a feasible system to collect and store representative data on continuous variables of health parameters during pregnancy. Saarikko in his research illustrated that continuous monitoring can provide real-time information about patient conditions between scheduled

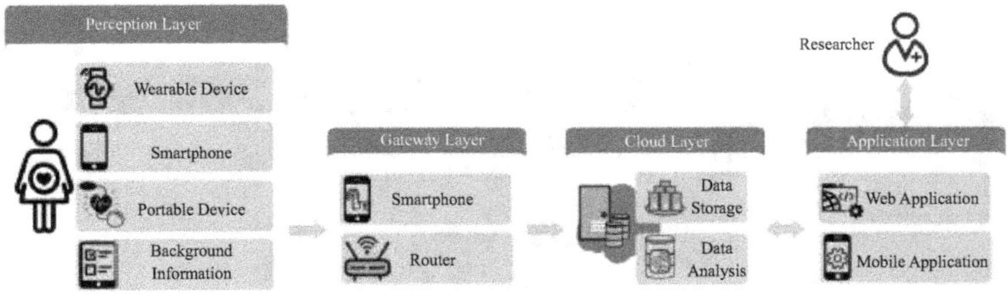

Figure 8.4 IoT-based maternal health monitoring system

appointments during the antenatal period. This may help the obstetrician target and tailor the pregnancy follow-up [26].

In practice, pregnant women are provided with wearable devices with sensors that help in monitoring maternal and fetal parameters. This will enable her and the healthcare professional to track the data remotely at any time using a web-based interface system as shown in Figure 8.4. This maternal health monitoring system consists of a perception layer with any portable device, a gateway layer for transmission, a cloud layer for data storage, and an application layer that includes the mobile application. Monitoring of maternal uterine activity and fetal condition can also be incorporated into that web application. Remote monitoring of pregnant women provides vast opportunities to identify pathological signs at the earliest, hence timely intervention can be done to improve obstetric outcomes. Future research on predicting the complications related to pregnancy and identifying the high risks is needed to decrease maternal morbidity and mortality.

8.7 IoT in eyecare

Ophthalmology is a wonderful data-driven specialty with tremendous potential to apply IoT. The data used in eye care range from simple numerical values (intraocular pressure, spherical equivalent of glasses) to complex images which could be two-dimensional (fundus photographs) or three-dimensional scans [optical coherence tomogram (OCT), corneal topography images]. In addition, the vast array of clinical and surgical records has huge potential for the application of IoT. In particular, IoT can facilitate eye care delivery by enhancing accessibility, making it cost-effective and personalized as well.

8.7.1 Visual acuity tester

An online mobile application-based visual acuity checker was developed compatible with Android and Mac devices as shown in Figure 8.5. In this test, the patient is instructed to hold the mobile at approximately 14 inches and opts to read among the six options or read as "not sure." The developers had evaluated the letters to correspond to the standard "Snellen'svisual acuity chart" [27]. It is validated and approved by Food and Drug Administration as "Paxoscheckup."

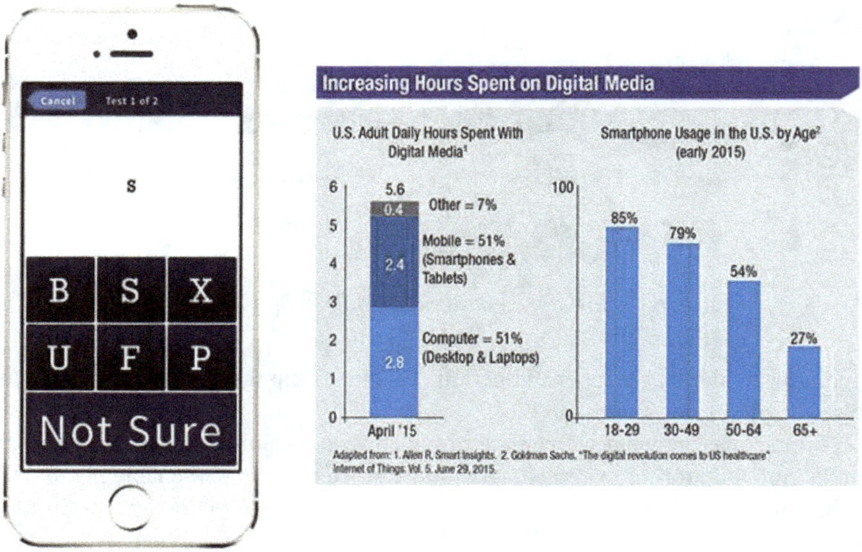

Figure 8.5 Visual acuity test

Paxoscheckup has recently been enhanced to use for contrast acuity, low illumination contrast acuity, contrast sensitivity, inverse acuity, Amslergrid (Digi grid), and color discrimination test [28]. To initiate its usage, the patient should initially register with the physician and perform the test as suggested by them. The test results are stored in accordance with the standard Health Insurance Portability and Accountability Act guidelines for protecting healthcare information. The data can be transmitted bi-directionally from the patient to the clinician based on pre-set trigger alerts or viewed from the physician's end on the serial performance of the test results as a graphical plot.

8.7.2 Mobile imaging

The Paxos analytics was developed by two Stanford ophthalmologists, namely David Myung and Robert T. Chang. It is useful for transmitting images between ophthalmologists in an elegant manner. The system standardized the external source of light using an external light of variable intensity. This standard source helped to avoid the dependence on the smartphone's internal flash which varies among different brands of smartphones and averted its drainage of power as well. This robust system adapts to any size and brand phone model using a mounting and lens alignment system [29,30].

The Paxos analytics system is also improvised to provide real-time data on the study outcome variables for the pharmaceutical companies when they perform clinical studies. With this platform of Digi Sight, the developer has a huge revenue-generating potential. The workflow pattern of a patient when they walk into a clinic can be streamlined and optimized to reduce the waiting time at various stations. When a patient needs to perform multiple tests, the patient can be directed to a particular test of lesser wait time, depending on their corresponding wait time. It will be icing on the cake if the newer imaging models like OCT are integrated into the EMR system by uploading the complex image system. Automation with trigger alerts set for the captured variables shall help the clinician to make a better and newer diagnosis. When the data on drug history of allergies changes dramatically, the drug warnings could be issued on them including the interactions. The dynamically changed newer prescriptions could be dispensed locally or delivered to avert waiting in a queue.

On a large scale, when healthcare data among different providers are integrated, an updated clinical summary of the most recent healthcare interactions will act as a game-changer in the field. Individual lifestyle tracker data like activity level data and DR data shall plan better eye care delivery models. For cataract surgeon who operates at different surgical centers, their individual personalized phacoemulsification settings for the specific dynamics/fluidics could be integrated to their comfort level to deliver the optimized outcome.

The IoT has a huge potential to improvise workflow efficiency in eye clinics and operating rooms. It can avert errors in the calculation of intraocular lens power calculation for cataract surgery by automation. In operating rooms, it can automatically update the lens stock, which helps in reducing the work burden and improves efficiency by avoiding the cancellation of surgeries due to the dearth of premium intraocular lenses/powers. A similar process of automation can avoid varied errors in human training, policies, and checklists. The process of interlinking a patient's mobile device with the operating room is a revolutionary change to aid in the automated delivery of chosen intraocular lenses and minimize human interactions. It is a revolutionary change to supersede the checklists designed by WHO.

In a conventional ophthalmic setup, the correlation of clinical assessment with the image-capturing investigatory devices is costly and a time-consuming process. This opens a plethora of opportunities for applying deep learning (DL) algorithms in the real world. Such DL techniques using visual fields and digital fundus photographs are validated for their accuracy in varied ophthalmic diseases like DR, glaucoma, retinopathy of prematurity, age-related macular degeneration (AMD), and uveitic diseases. It is a valuable adjunct to existing ophthalmic resources, especially in resource-constraint settings.

The existing detection of diseases like DR, AMD, and diabetic macular edema can be enhanced with increasing sensitivity with specific DL algorithms adopted for use in OCT [31]. It shall translate into better eye care delivery although its cost-effectiveness needs to be proven. The incumbent clinician needs to have an open mind to critically assess the safe application of such innovations in real-time practice.

8.8 Benefits of IoT in the healthcare system

IoT is quickly gaining popularity in all spheres of healthcare systems. Following are the advantages of IoT implementation in healthcare:

1. **Improved patient care services:** IoT enables healthcare professionals to be more watchful and connect with the patients proactively. Data collected from

IoT devices can aid physicians to identify the best treatment process for patients and obtain the expected outcomes. Real-time remote monitoring via connected IoT devices and smart alerts can diagnose illnesses earlier, treat diseases, and save lives in case of a medical emergency.

2. **Reduction of healthcare costs**: IoT is cost-effective as it decreases the need for a prolonged hospital stay of the patients and monitoring as remote monitoring is possible with technology. It reduces the economic burden due to the illness per se and the treatment cost.
3. **Medical data accessibility**: Accessibility of EMRs by physicians can be of immense help for them to take timely decisions and provide appropriate management. EMRs in hospitals have revolutionized the healthcare system by providing quality care.
4. **Quality improvement and patient safety**: The convergence of Internet-based technologies can avoid unnecessary waiting time for sick patients in the waiting rooms, improving the prognosis. IoT can be an integral part of the hospital infection control protocol as IoT-enabled infection control devices can prevent vulnerable patients from acquiring life-threatening infections. IoT devices help in ensuring patient safety by notifying intake of correct drugs at the appropriate time, thereby reducing medical error.
5. **Improved healthcare management**: In a hospital ecosystem, the real-time location of emergency medical equipments, like wheelchairs, defibrillators, nebulizers, oxygen pumps, and ventilators, can be tracked using IoT devices tagged with sensors. This can improve healthcare delivery. IoT devices also help in asset management in hospitals with a stock management registry of all the equipment and their maintenance.
6. **Disease prevention**: The new Internet tools help to measure and improve the healthy behavior of humans. The application of these devices can prevent most chronic diseases through lifestyle modifications in the long run.
7. **Research**: The data collected through IoT devices can be cumulated and can be utilized for medical research in the long run. The unique feature of IoT is that it can store massive data rather than paper-based case records. The data analyzed in research can help in formulating protocols for evidence-based medicine.

8.9 Challenges with IoT in healthcare

Even though the IoT can greatly benefit healthcare, there are still many challenges to address before its full-scale implementation in all healthcare management systems. Hence, it is crucial to identify the risks and challenges so that preventive and corrective action can be taken. Security remains a potential threat in using IoT technology for medical purposes, as there is a chance for data being hacked or breached. as healthcare monitoring solutions have the potential to be breached or hacked. The data about the patient needs to be confidential as per medical ethics, hence the worry of leakage of sensitive information about the patient's health. This can result in grave consequences. Hence physicians with little knowledge of technology will be hesitant in using it for their patients. The question arises among the stakeholders regarding who owns the data, which deems to be confidential. Another major concern is the failure of the system due to any bugs or power failure. This can affect the performance of sensors and connected equipment placing healthcare operations at risk.

Although the application of wearable sensor devices in health has significant potential to positively transform health outcomes, there are certain disadvantages. One major challenge of wearable devices in the long term is the motivation of the patients and the users to use the device continuously. Several studies have proven that counseling, personalized monitoring, the feasibility of the device to be used in day-to-day life, and the design of the device being user-friendly can keep the participants motivated. The users must be given proper advice on the benefits of the device and the need for its use.

A qualitative research analysis revealed that healthcare staff is already overloaded with data, and incorporation of the data from additional devices could unnecessarily burden the already-busy professionals. This could affect patient management. One more concern is that healthcare providers feel becoming increasingly distanced from the patient and that the doctor–patient relationship is affected. However, there are still grey areas in its implementation by the physician, which can be addressed by knowledge dissemination and further research.

Most of the IoT-based system in pregnancy care was limited to the monitoring of maternal medical parameters rather than the fetal parameters and prediction of complications. Hence IoT devices customized to pregnancy care, involving obstetricians, fulfilling their needs, have to be developed. Sometimes the abnormalities in the parameters may be missed because of participants not wearing the smart wristbands and other devices, which can be detrimental. The hesitancy to wear a sensor with a fear of the baby being affected is yet another possibility. While IoT promises to reduce the cost of healthcare in the long term, the initial cost of its implementation in hospitals and staff training is quite high.

The implementation of IoT in healthcare is possible with a clear protocol for the management of data, privacy, confidentiality, and security. If the above-said issues are addressed, IoT shall be put in routine medical practice with confidence. Future innovations should be on IoT technology for a specific health condition, user-friendly and high level of security, addressing the health needs of that particular population.

References

[1] Lapao LV. The future impact of healthcare services digitalization on health workforce: the increasing role of medical informatics. *Studies in Health Technology and Informatics*, 2016;228:675–679.

[2] Dimitrov DV. Medical Internet of Things and big data in healthcare. *Healthcare Informatics Research*, 2016;22(3):156.

[3] Azodo I, Williams R, Sheikh A, and Cresswell K. Opportunities and challenges surrounding the use of data from wearable sensor devices in health care: qualitative interview study. *Journal of Medical Internet Research*, 2020;22(10):e19542.

[4] Minerva R, Biru A, and Rotondi D. *Towards a Definition of the Internet of Things (IoT)*. vol. 1 Torino, Italy: IEEE Internet Initiative, 2015, pp. 1–86.

[5] Haghi M, Thurow K, and Stoll R. Wearable devices in medical Internet of Things: scientific research and commercially available devices. *Healthcare Informatics Research*, 2017;23(1):4–15.

[6] Anliker U, Ward JA, Lukowicz P, *et al.* AMON: a wearable multiparameter medical monitoring and alert system. *IEEE Transactions on Information Technology in Biomedicine*, 2004;8(4):415–427.

[7] Bertolotti GM, Cristiani AM, Colagiorgio P, *et al.* A wearable and modular inertial unit for measuring limb movements and balance control abilities. *IEEE Sensors Journal*, 2016;16(3):790–797.

[8] Chen M, Ma Y, Song J, Lai CF, and Hu B. Smart clothing: connecting human with clouds and big data for sustainable health monitoring. *Mobile Networks and Applications*, 2016;21(5):825–845.

[9] Sanfilippo F and Pettersen KY. A sensor fusion wearable health-monitoring system with haptic feedback. In: *Proceedings of 11th International Conference on Innovations in Information Technology (IIT)*, Dubai, UAE, November 1–3, 2015, pp. 262–266.

[10] Dimitrov DV. Medical Internet of Things and big data in healthcare. *Healthcare Informatics Research*, 2016;22(3):156–163.

[11] van Uem JMT, Maier KS, Hucker S, *et al.* Twelve-week sensor assessment in Parkinson's disease: impact on quality of life. *Movement Disorders*, 2016;31(9):1337–1338; [Epub]. http://dx.doi.org/10.1002/ mds.26676.

[12] Tung CE, Su D, Turakhia MP, and Lansberg MG. Diagnostic yield of extended cardiac patch monitoring in patients with stroke or TIA. *Frontiers of Neurology*, 2015;5:266.

[13] Nugent R. Chronic diseases in developing countries: health and economic burdens. *Annals of New York Academy of Sciences* 2008;1136:70–79.

[14] Sinnapolu G and Alawneh S. Integrating wearables with cloud-based communication for health monitoring and emergency assistance. *Internet of Things*, 2018;1:40–54.

[15] Saravanan M, Shubha R, Marks AM, and Iyer V. *SMEAD: A Secured Mobile Enabled Assisting Device for Diabetics Monitoring*. USA: IEEE, 2017, pp. 1–6.

[16] Dadkhah M, Mehraeen M, Rahimnia F, and Kimiafar K. Use of Internet of Things for chronic disease management: an overview. *Journal of Medical Signals and Sensors*, 2021;11(2):138–157. doi:10.4103/jmss.JMSS_13_20.

[17] Chinnasamy A, Malavizhi N, and Manzoor SR. Evaluating the use of IoT for healthcare providers in treating elderly NCD patients in Malaysia. *European Journal of Molecular & Clinical Medicine*, 2020;7(8):696–704. ISSN 2515-8260.

[18] Hammad Memon M, Li JP, Ul Haq A, Hunain Memon M, and Zhou W. Breast cancer detection in the IoT health environment using modified

recursive feature selection. *Wireless Communications and Mobile Computing*, 2019;2019:1–19.

[19] Van Calster B and International Ovarian Tumour Analysis Group. Evaluating the risk of ovarian cancer using the ADNEX model. *BMJ*, 2014;349:g5920.

[20] Edoh T. *Internet of Things in Emergency Medical Care and Services*. http://dx.doi.org/10.5772/intechopen.76974.

[21] WHO. *Global Status Report on Road Safety: Time for Action*. Geneva: World Health Organization, 2009. Available at: www.who.int/violence_injury_prevention/road_ safety status/2009.

[22] Allahem H and Sampalli S. Framework to monitor pregnant women with a high risk of premature labour using sensor networks. In: *Proceedings of the 2017 IFIP/IEEE Symposium on Integrated Network and Service Management (IM); Lisbon, Portugal*, 2017, pp. 8–12.

[23] Marko KI, Krapf JM, Meltzer AC, *et al.* Testing the feasibility of remote patient monitoring in prenatal care using a mobile app and connected devices: a prospective observational trial. *JMIR Research Protocols*, 2016;5: e200. doi: 10.2196/resprot.6167.ay 2017; pp. 1178–1181.

[24] Grym K, Niela-Vilén H, Ekholm E, *et al.* Feasibility of smart wristbands for continuous monitoring during pregnancy and one month after birth. *BMC Pregnancy Childbirth*, 2019;19:34. doi: 10.1186/s12884-019-2187-9.

[25] Sasubilli SM, Kumar A, and Dutt V. Improving health care by help of Internet of Things and bigdata analytics and cloud computing. In: *2020 International Conference on Advances in Computing and Communication Engineering (ICACCE)*, 2020, pp. 1–4, doi: 10.1109/ICACCE49060.2020.9155042.

[26] Kotturu PK and Kumar A. Comparative study on machine learning models for early diagnose of Alzheimer's disease: multi correlation method. In: *2020 5th International Conference on Communication and Electronics Systems (ICCES)*, 2020, pp. 778–783, doi: 10.1109/ICCES48766.2020.9137872.

[27] Fagerberg J and Kurkinen L. mHealth and home monitoring. *Berg Insight Report*, December 2015. www.berginsight.com, Accessed October 2015.

[28] Yu S-Y, Yang JH, Kim Y, Kwak H-W, and Blumenkranz M. Reliability of the smartphone-based electronic visual acuity testing. *ARVO*, 2014;55(13): 5598–5598.

[29] Myung D, Jais A, He L, Blumenkranz MS, and Chang RT. 3D printed smartphone lens adaptor for rapid, high-quality retinal imaging. *Journal of Mobile Technology in Medicine*, 2014;3(1):9–15.

[30] Myung D, Jais A, He L, Blumenkranz MS, and Chang RT. Simple, low-cost smartphone adapter for rapid, high-quality ocular anterior segment imaging: a photo diary. *Journal of Mobile Technology in Medicine*, 2014;3(1):2–8.

[31] Toy BC, Myung DJ, He L, *et al.* Smartphone-based dilated fundus photography and near visual acuity testing as inexpensive screening tools to detect referral warranted diabetic eye disease. *Retina*, 2016;36(5):1000–1008.

Chapter 9

Smart healthcare monitoring framework using IoT with big data analytics

S. Usharani[1], P. Manju Bala[1], T. Ananth Kumar[1], R. Rajmohan[1], A. Balachandar[1] and A.S. Adeola[2]

Wireless networking systems, including the Internet of Things (IoT), have effectively expanded through different aspects of our lives as costs have decreased and functionality has improved. Healthcare is one of the most rapidly evolving and challenging technology fields. From digital healthcare tracking facilities for elderly people and assisted treatment, IoT would have an effect on the medical environment, from treating and handling chronic conditions to delivering customized medicine. The IoT in medical uses, the different ways in which it has penetrated the healthcare industry, and future developments in its growth such as nano-IoT and bio-IoT are all discussed. When processing and executing data analytics on the large volumes of data generated by wearable body sensor networks, a fusion of cloud and IoT architectures is used to make sensible medical systems that can support real-time applications. Wireless body sensor networks (WBSNs) are wireless communicating instruments that are mounted in or on a patient's body. This chapter discusses the design and deployment of a particular medical application based on a WBSN. The suggested method relies on the patient's heartbeat, plethysmogram, and absolute oxygen ratio data. Using IoT technologies, the recorded data is transmitted from the WBSN to the centralized database. Data recorded are analyzed by exploratory data analysis and data mining algorithms. So, if the patient has any abnormal symptoms then the system network will send the data report to the physician. This has been tested and simulated using TensorFlow and the system's performance is determined on the basis of network resilience, as well as the accuracy of the recorded data in different communication networks, as well as system consistency and efficiency in the management.

[1]Department of Computer Science and Engineering, IFET College of Engineering, India
[2]Computer Engineering Department, Ladoke Akintola University of Technology, Nigeria

9.1 Introduction

The recent growth of advancements in big data analytics and the Internet of Things (IoT) for biotechnology and health systems is based on actual cutting-edge research work conducted globally. The IoT and big data analytics are intended to change several areas, together with biotechnology and healthcare systems, as researchers distinguish them in this day and age. The demand for the healthcare system has grown significantly faster than that of other ventures. Because an aging population has an impact on the perception of skills and administrations, social welfare analysts should be prepared to express the appropriate social service management to meet the needs of the patient, wherever but whenever they arise. Many advanced frameworks have indeed been constructed by highly skilled people who rely on the finest and computerized processes with the aim of improving present clinical care methodologies and providing innovative capabilities in light of the massive data upset which ensures a smart framework has been established. As a result, the underpinning process of necessary requirement investigation for the smart workflow of the treatment facility is an action research and an innovative smart healthcare monitoring system (SHMS) for smart treatment facility based on a few interpretative methods utilized for social event necessities. Furthermore, this work provides a better technique to understand the mind-boggling healing center framework and simplifies the medical clinic process flow. The simulation results showed that the suggested SHMS is faster than current process management.

The growth of the IoT and big data, as well as a widespread concept of small wearable biosensors, has provided new possibilities for positive environment checking. Big data applications are used in various fields, the healthcare sector being one of the most important. Many challenges are currently unable to be redirected to make an accurate decision and to generate a modeling approach for well-being and consciousness checking. Big data research on IoT-based smart healthcare monitoring framework is suggested, and it uses an IoT-based healthcare monitoring model that provides a "wireless body sensor network." These devices generate massive amounts of data that the doctor is unable to handle. The doctor's problem is that they should make basic decisions regarding their patients' health based on these massive amounts of health information. Researchers separate the data pertaining to one particular patient from the flood of care mind data originating from a number of hospitals. The Intel Galileo Arduino IDE functions as an IoT operator, transferring patients' medical data to the cloud. This might deal with the large demand for health data, cleverly access the information all over healthcare frameworks, and make big data examinations feasible. Internet-based notifying of patient data is a crucial big data process that is essential in several suggested works. The cloud system is used to evaluate large amounts of health sensor data. Because the proposed model has a shorter response time, it is also appropriate for continual expressing concern. The primary aim of this chapter is to investigate the crucial role of IoT and big data in the healthcare industry. It conducts a study and showcases the big problems regarding healthcare services in a contemporary technologically advanced environment, with a focus on IoT and big data analytical methods.

This chapter focuses mainly on the following services: telemedicine monitoring, health information technology, healthcare data analysis, and appropriate treatment using suitable decision-making technology. This topic specifically explains information exchange in a wireless body sensor network. Smart applications for healthcare services are outlined, and difficult issues are also addressed in this chapter.

9.2 Related work

Big data analytics and IoT, in general, are important twofold technologies that will revolutionize the biomedical industries and improve society. Big data is used in various fields, including healthcare, agriculture, and social networking. Data analytics is determined by the amount of data collected, the quantity of information received (i.e., various categories), data generation speed, variability (i.e., date inconsistency), and incomplete inadequacy in certain data [1–3]. Case-by-case and large data analytics could be rather complicated when these dependent dimensions and many other aspects are taken into account. Data analytics can handle complicated and vast datasets created incredibly quickly in a useful way for healthcare or diagnostic service providers to understand the existing technologies in biomedical care [4–7]. Healthcare and biomedical-related information are produced at an unparalleled rate from diverse resources in today's clinical system, where everything is computerized. Medical data, photos, and records uploaded from various sensors in integrated systems, operations, pharmaceutical data, and other sources make up the majority of healthcare and biomedical data. There are various good evaluations on big data in healthcare that can be found here [8–12]. The data science for health conference is focused on developing developments in big data in healthcare and biomedical informatics, such as translational bioinformatics, sensor informatics, and imaging informatics, as well as an inclusive system [13–16]. This chapter focuses on information storage and implementation of a wide range of data from multiple sources and resources, such as genomics, proteomics, metabolomics, imaging, clinical diagnostics, physiological sensing, and so on.

9.3 Overview of IoT and big data

The IoT is a game-changing technology for digital security architectures. In many ways, business and private security products were at the forefront of IoT and continue to share numerous critical characteristics of the subject. Overall, the rapid decline in the price for IoT devices, as well as the millions and millions of innovative IoT devices expected to be incorporated in the preceding 4–8 years, make it a force to be reckoned with in terms of how we perceive digital security architectures and the industry as a whole. IoT greatly enhances business sector tasks since the most frequently mentioned customer benefit of IoT in the home is for security systems. In general, the rise of IoT increases the benefit for safety purchasers because the economic side of the IoT ecosystem will adapt to decreasing expense weight on surveillance sectors, sometimes as they become more smart and provide high features. Because of disenfranchised care information systems, emerging economies were related to healthcare data and were not entirely used to help and improve healthcare outcomes. The Indian healthcare event, ground-up capabilities, data innovation, and the challenges that they face were all addressed in this chapter. Our nation also has a smart healthcare framework, with an accessible healthcare component that works directly with such a covert treatment sessions arrangement. The information and records tightly pertinent to the medical information systems (MIS) have been collected, designed, and classified into three primary categories: collected, stocking up, and monitored, and for now carried out in various MIS that are split and imploded. As a result, there is a need for an organized MIS. Researchers explore two potential methods and techniques and corresponding test technology to their challenges. Another approach to dealing with enriched iterator design constraints has already been suggested, which makes use of managed design and configuration management capability and ideologies. The power of artificial intelligence has indeed been tremendously felt in changing various endeavors and care services technologies. This chapter discusses the use of evolving technologies like IoT and big data analytics in the management of a smart healthcare monitoring framework. Upon being inspired by the above-mentioned innovations in this chapter, The SHMS based on wearable body sensor networks was suggested and designed to simulate utilizing the TensorFlow simulation model. The simulation results demonstrate that performance is measured by moving forward with the project. The correctness of captured data in various communications systems, and also framework consistency and monitoring efficacy are demonstrated.

9.4 Data sources for healthcare

Datasets collected in medical care areas include both primary data and secondary data. Primary data is quantifiable information used for comparative analysis. For instance, temperature, age, weight, and any other distinct parameters. Secondary data is non-numerical information used to depict fitness issues. Gender and habits are some instances. Scientific information and medical evidence are two kinds of data sources in the field of medicine. Clinical evidence includes information derived from medical surveys or observational data. Information collected to bench fields of science is referred to as scientific data. The data documented and gathered in the medical area are both qualitative and quantitative in nature. The term "qualitative information" relates to the data collected and analyzed by an entire demographic. This information gathered can be used to address the research questions. Quantitative information is dependent on the underlying data that is used for certain purposes. Such data are being used to solve study queries. Figure 9.1 depicts the diverse data resources of healthcare systems.

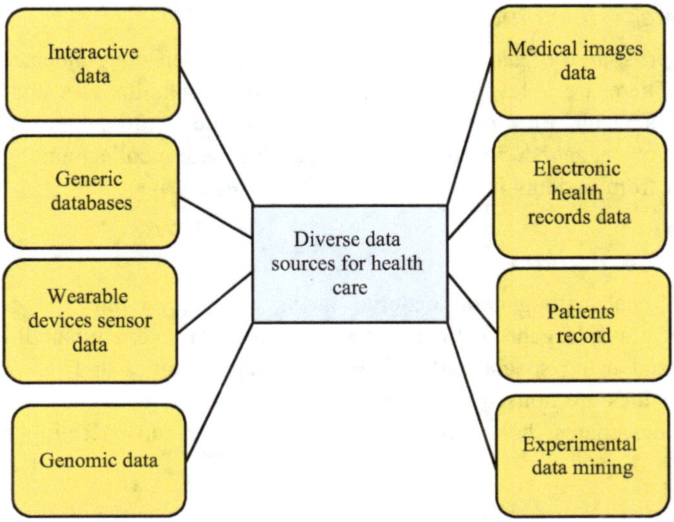

Figure 9.1 Diverse data sources for healthcare

9.4.1 Electronic health records (EHR) data

It is a vital source of information in the healthcare field. The term "electronic health records" refers to patients' electronic records. The information can be processed from almost anywhere at any time. It can be either saturated or unsaturated. All documents in datasets are supervised and categorized in a database. Unstructured records, on the other hand, are hazy and incoherent, and thus are presented in the form of web documents of medical information. For example, pdf, email messages, digital photos, sound recording, or streaming video. Such data soon evolved into structured records.

9.4.2 Medical images data

X-rays, magnetic resonance images (MRI), and computed tomography (CT) scan images are examples of large amounts of data. Photo archival and communication system is a system for storing and retrieving diagnostic data records. Images are dumped into a medical images data source during the diagnostic search process. These image-based information require a lot of memory and are difficult to handle.

9.4.3 Experimental data mining

Healthcare data can be organized, unorganized, or contain text-based data. Data mining could be used to obtain benefits and make it feel information from large amounts of actual data. Some data mining methods include classification and sentiment analysis. It can be used for optimized drug leveling, accurate illness diagnosis, and effective clinical therapy. Natural language processing has the potential for use in healthcare data mining.

9.4.4 Interactive data

The acquired source of data is from maneuverability sensors connected to social networks. There are a few social media sites that track illnesses and their side effects. As a result, they provide a more appropriate treatment procedure for a medical history engaged. Sensor nodes can also be used to collect and agglomerate illness data from patients in the healthcare amenities.

9.4.5 Genomic data

These data deal with genetic material facets in the systemic and consecutive structure of multiple gene features and functionality. These data should be stored and processed using specialized software. Human genomes and genome-related association rules are housed in a repository known as the genomic database. This repository designates the genetically identical side effects affecting health and illness.

9.5 Big data's evolution in IoT

The healthcare industry is a mash-up of various industries. Medicines, preventative with the analysis working of the possessions, and socialization of care are all included in the fields. Various nursing homes, medical trials, subcontracting, health coverage, telehealth, and voluntary organizations are also included in the national healthcare businesses. Nowadays, the health sector is seen from a business standpoint. People benefit from the company's earnings and other benefits. In addition, information and communication technology has indeed been incorporated. The information and communication technology (ICT) organization helps to afford the healthcare industry with various jobs that have helped to strengthen the business. The system desired is able to assist the current system in avoiding medical errors. Data comes from various places in the system. Clinical trials, medicine, fitness, fluctuating symptoms, prescription, laboratory report, insurance data, and numerous additional patient and doctor-related information are among the data sources. The volume of data in the healthcare industry is increasing exponentially, with existing data sizes in the petabyte range. This massive increase of data has resulted in a slew of issues with data storage, transport, and analysis. Conventional database systems can be used to examine and process this type of data. Furthermore, typical database systems can only process data that is organized. The data saved in the form of big data, on the other hand, is unstructured. ICT will assist society in a better storage with the introduction of novel and effective mechanisms for the stowing and retrieving of information. eHealth refers to the process of integrating ICT into the healthcare industry. As a result, its adoption in the medical industry would aid in the handing out of data, subsequent analysis, and the enhancement of the decision-making process for the gathering of improved treatment solutions for disease symptoms. The abundance of data is one of the most notable elements of the healthcare industry's use. The medical industry has been utilized rapidly to advance the division in recent decades, thanks to recent developments in diagnostics and treatment processes.

The term "big data" refers to a technique for storing and analyzing huge datasets like petabytes, terabytes, exabytes, and zettabytes. One such measure of data is typically beyond human skill to retain data in a physical effort; yet, gathering and evaluating the data can provide substantial value to the commercial and decision-making process. As a result, nowadays businesses are utilizing big data technology to store, process, and analyze data. Volume, velocity, and variety are three fundamental properties of big data, called the three Vs. Structured, unstructured, or semi-structured data are all possibilities. Structured data is the data that has been organized in a specific way; audio files, video files, text records, web sheets, computer programs, and online activities are examples of unstructured data that do not have a predetermined organization. Semi-structured data is neither totally structured nor fully unstructured, with unstructured data accounting for 80% of all data. Big data is linked to analytics, and analytics is a well-known term. The systematic techniques that have been in practice for numerous years are regression methods, simulation, and machine learning algorithms. It is understandable that the above-mentioned techniques were used to analyze unstructured data such as email, computer programs, and papers. However, as previously said, advanced sources in computer technology and software, such as social networking sites, like Facebook, blogs, Twitter, and various other commercial prospects, are making massive amounts of data every minute. As a result, a trend toward sophisticated tech known as big data analytics has emerged, which supports a wide range of data analysis technologies. Apart from processing large amounts of data, big data applications offer a slew of other advantages.

9.6 Recent trends in big data analytics and IoT

9.6.1 Specialized medical envisioning

Presently, image processing simply uses two-dimensional X-rays, ultrasonic, MRI, and CT scans, and several other techniques, but investigators are coming up with innovative and enhanced image processing options such as multiple kinds of electromagnetic imaging, electron beam or ultrasonic image analysis, ultrasonography, and clinical artificial intelligence, which would provide more efficiency and computational results. With the assistance of big data analytics and IoT, various diagnosis and analysis methods exist, which are explained in-depth.

9.6.2 Telehealth

Telehealth contributes significantly to technical advancement in fields such as computing, healthcare, molecular biology, telecommunication, machine intelligence, simulated appearance, information technology, and automation. Using machine learning, the telehealth system improves healthcare effectiveness while lowering costs.

9.6.3 Portable gadgets and the IoT

A new wave of portable technology has arisen that is aimed at biological and healthcare systems over simply recording your daily activities. Such devices can monitor accurate data about human healthcare while also functioning as a wristwatch or an easy adapter for our smartphones. The data was gathered and processed using normal methods by artificial intelligence and the ability to hunt for probable health-related concerns to anticipate. Beyond such prosthetics techniques, architecture, computers, and automation have improved to the point that microprocessors or supplementary actuators or robots equipment-based assistance methods are now available.

9.6.4 Biological IoT

Medicinal manufacturing processes, integrated healthcare gadgets, intelligent prostheses, operational helpers, and clinical decision support are all examples of biological IoT. Various biomedical breakthroughs are highly published and have the potential to change the way we think about healthcare services. Newer methods of patient details, nanotechnology employed in targeted drug discovery, or distribution of other materials to specific types of cells, such as tumor cells, could be designed to quantify cancer treatment. Further, brain imaging is the latest thing in bioengineering, with terms like "cardiology," "transcutaneous," and "ultrasound scan" are showing up in scientific report names more frequently. Some researchers are working on ways to restore brain activity after mental disorders using various forms of activation, whereas others are planning to operate on neural emerging technologies to power artificial limbs.

The internet of nanoobjects (IONO) is essentially more than the connecting of nanostructured materials with existing infrastructure. As a result, it ushers in state-of-the-art developments in the field of electromagnetic information sharing among nanoscale objects. A nanomachine combines tiny parts to execute various activities. It works in the same way that the IoT does, but with one important difference: it can interconnect nanoelements which the IoT cannot. Because the IONO can identify nanomaterials, a modest biosensor can identify a gas explosion and provide us with an instantaneous alarm so that we can take any action. However, the IONO is still in its initial phases, and this idea will be soon used in the same manner that the IoT devices are being used in various applications. The idea of the IONO may be used in various sectors to improve the functioning of IoT devices. Nanomaterials, developed in conjunction with the notion of the IONO, are now assisting in various beneficial applications by detecting nanomaterials that are impossible to detect with conventional sensors. Here are several examples:

- Domestic animals are being tracked using nanomaterials. This alleviates a major source of concern for farmers, as agricultural livestock such as cows, pigs, and sheep frequently run and hide when grazing. The domestic animals are furnished with nanomaterials, which are controlled by a centralized personal computer.
- Nanosensor systems are used to assess the status of plants growing. The researcher will gather reliable information about the condition of plants and crops and recommends corrective action.

- Nanomaterials can identify viruses that are difficult to detect with conventional sensors. There are wristbands that may be worn like a regular bracelet which would give us data when we come around items that are infected with viruses or germs.
- We have meters and detectors that monitor temperature, but nanomaterials are a point ahead of them. Nanomaterials can detect and record the heat of live cells. As a result, it enables doctors and researchers to make more precise judgments while prescribing drugs or performing experiments.

9.7 Big data challenges in healthcare

The trials can be classified in two ways.

9.7.1 Challenges relating to budgetary and economic considerations

Medical professionals can only be compensated if they have a face-to-face connection with their clients in the era of payment for performance. It operates as a roadblock to the development of new innovations that allow customers to return without having to be physically present. Furthermore, as we move away from direct communication models, where economic costs are higher, there is more opportunity to use modern technology to minimize needless face-to-face encounters. Face-to-face encounters with patients are costly in these situations, whereas advanced technology has a favorable effect on health benefits.

9.7.2 Challenges relating to expertise

The condition of medical information is one of the most significant technological obstacles to achieving this objective. Medical data records are heavily segregated into institution-based divisions, thanks to the EHR system. Through the use of standardized medical coders and communication format, every effort is made to cope with the interchange of specific data files between divisions. However, it does not address the problem of data segmentation. People in the medical sector have recently realized that the promise of medicine is information gathering, not merely exchanging copies of patient information. Information can only be addressed directly and used if it can be acquired from various sources, normalized, and resolved using specific medical IDs. Descriptive statistics has two major advantages:

1. It overcomes the compatibility problem. Data links and data conversions across enterprise applications are no longer expected by organizations. All they have to do now is link sources of data to a single application programming interface package. Successful machine intelligence technology is based on this data collection.
2. It offers maximum time, enabling machine learning and deep learning to function implemented in real.

9.8 IoT challenges in healthcare

The primary goal of reputable IoT companies is to deliver easy and powerful applications of IoT and information processing solutions. It aids in the creation of data analytics apps, visualization platforms, and healthcare IoT applications by developers. The following are some of the essential skills that IoT businesses must have:

Networking is simple: An optimal IoT company should be able to connect computers quickly and easily, allowing device administration to be simplified.

- **Simple device leadership:** It allows for increased usage of numerous assets and resources, resulting in increased productivity and lower operating costs.
- **Data consumption:** In the IoT, advanced analytics processing and preservation are critical. Data is imported from many sources of data, and then pertinent information is retrieved using predictive analytics.
- **Contextual intelligence:** For efficient decision-making and seamless processes, accurate raw data analysis is critical. It is utilized in real-time analytics and tracking current situations so that appropriate responses may be made. Furthermore, a straightforward interface makes it easier to comprehend and use.
- **Lower risk:** Use a unit dashboard to act on alerts and isolate actions gathered elsewhere in the company.

9.8.1 *IoT and portable gadgets*

Figure 9.2 shows the different Portable gadgets in IoT. A new wave of electronic sensors has developed that are aimed toward biological and healthcare uses beyond simply counting your daily activities. These gadgets can gather extensive information about our healthcare while also acting as a wristwatch or a basic connection for our cellular telephones. The data was gathered and processed using conventional

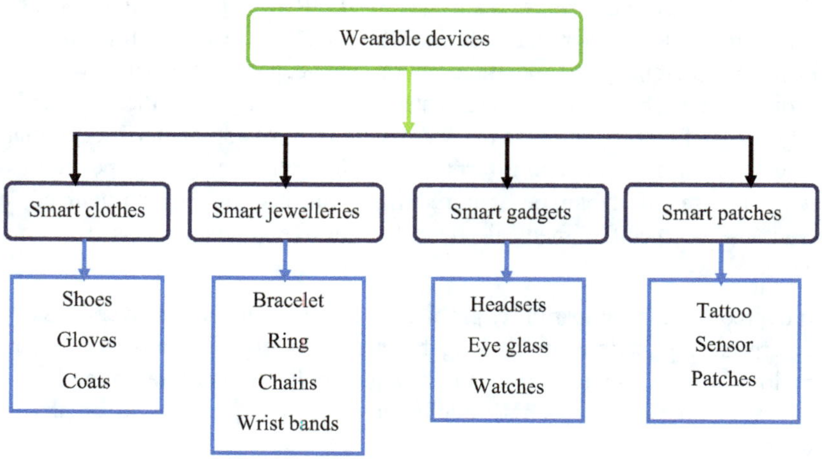

Figure 9.2 Classification of wearable devices

methods by artificial intelligence in order to seek for any healthcare concerns to forecast. Beyond these advancements in components and research, prosthetic technology has made tremendous progress in recent years. Chip-enabled prostheses with increased movement and adaptability, as well as supplementary actuators which can help give more speed and courage, as well as more robotic systems, are the future, blurring the borders between medicinal and supporting equipment. Wearable gadgets are mostly used for object detection and detecting what we are doing and where we are doing it.

9.8.2 Modes of communication in wearable devices

Figure 9.3 shows the different modes of communication in wearable devices. However the gadgets are tiny and run on a rechargeable generator to keep power usage low. They have their own energy restrictions, such as restricted processing, caching, and computational resources, which results in limited power consumption. We have categorized the items based on geographical availability:

- **Short range:** Radiofrequency communication is a mechanism that has been developed for extremely short-distance transmission. It requires very little power to send a small quantity of data over a limited duration.
- **Medium range:** Bluetooth is ideal for communicating over short distances. The gadgets are supplied by a rechargeable battery, which has a low budget and low energy consumption. This portable gadget can possibly send limited data over a distance of up to 100 m. In Bluetooth, a total of seven devices can connect; one acts as a virtual machine, while the other six function as virtual machines. It is not suitable for multimodal transmission due to the restricted data transmission.
- **Long range:** When data must be transmitted over a greater distance with a tolerable latency, Wi-Fi and mobile networks are the best alternatives. In comparison to smaller distributed systems, the gadget has higher power consumption, processing capabilities, and storage authority. Multimodal data such as speech and video is sent using these communication methods.

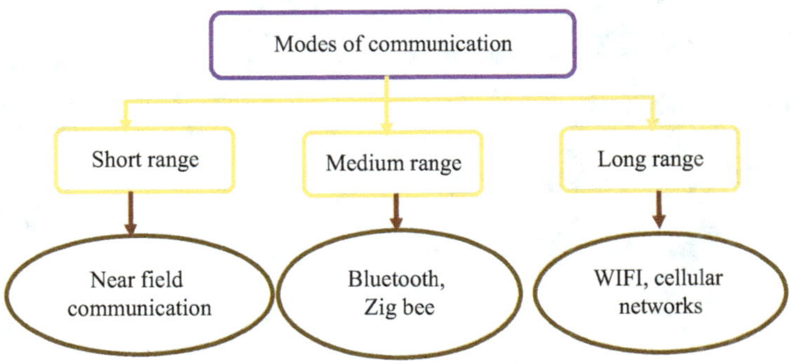

Figure 9.3 Modes of communication in wearable devices

9.8.3 Smart healthcare monitoring frameworks

Figure 9.4 illustrates the SHMS network framework. As universal healthcare has been made safe utilizing sophisticated security techniques in contemporary digital technology, a sophisticated fog services platform is now being used for simple access to health services and improved device using novel encryption standards. An observable information holding plan will be used to protect the encoded data on the cloud. Security approach is used as a component of protecting social assistance information to increase reliability and privacy. Enhanced security is provided by the suggested structure. The correctness of management tools information is assured using this proposed technique, as well as visible evidence of archive component proximity in the database. This technique also emphasizes information integration, modification, and termination while attempting to reduce the server's computation time. The surprise attack uses encrypted information, and the findings demonstrate that the suggested plan outperforms current techniques in terms of information accumulation security. Healthcare monitoring systems are progressively progressing, with the potential to transform how common insurance is now delivered. The SHMS, such as the one used in this study, automates the process of interpreting and monitoring activities. We created and developed a GPRS-based spectacular wearable architecture with a three-sensor, three-lead electrocardiogram (ECG) monitoring framework and continuous scrutiny framework in this study. The device is capable of detecting unexpected falls, cardiac irregularities, and hyper/hypertension, making it appropriate for continuous monitoring, self-conclusion, and distant location. The framework may also be used to transmit alerts to social insurance specialists as soon as any of the previously stated healthcare irregularities are discovered so that necessary steps can be taken. It is a clever, perceptive, and ubiquitous

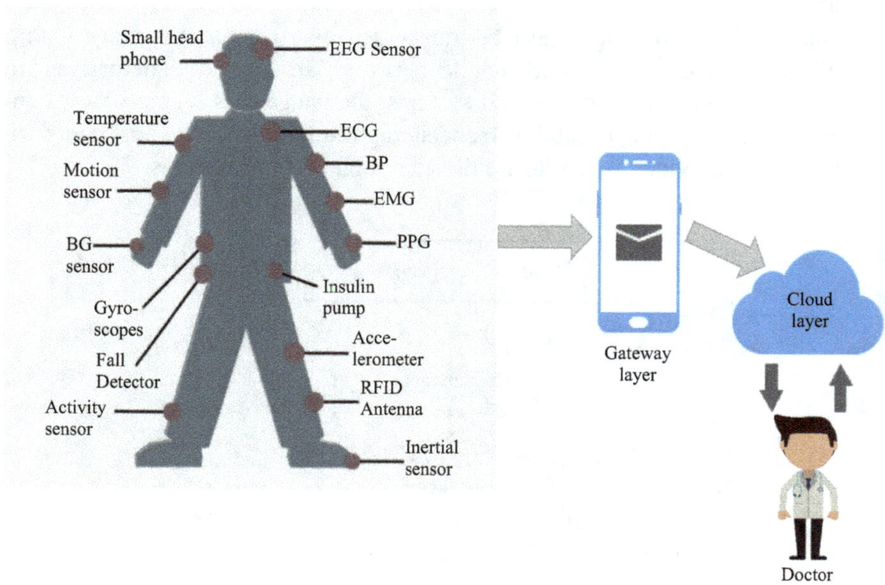

Figure 9.4 SHMS network framework

health-monitoring system that is of enormous importance not only to the aging and post-agent victims but also to people who live in rural regions and are unable to afford the costs of social security in metropolitan clinics or institutions. Current manifestations are indicative of how the usual age of the human population has stretched through, and the most numerous causes of mortality are heart disease and cardiac difficulties. A few researchers are pushing the need for acceptable preventive cardiology today and in the future. The most fundamental aspect of intensive predictive cardiac is the collection, screening, and separation of specific health data. The attempts correspondingly provided to organized therapy care professionals have constantly been successful surveillance and interpretation of information presently protected. Despite the fact that the associated facts must be scrutinized more carefully, the job objectives are invariably unquestionable. Fatigue and overburdening workloads are two probable causes of delayed emergency response, which might have harmed patients' chances of life (Figure 9.4).

9.8.4 SHMS principles in the IoT

Figure 9.5 illustrates the three-layer architecture of SHMS. The operating principles of SHMS in the IoT are divided into three layers of computing:

1. **WBSN layer:** It houses the data-sensing wearable sensors used in the framework. Here we examined the healthcare use in our diagram, where the temperatures and pulse are measured and transferred to an upper layer.

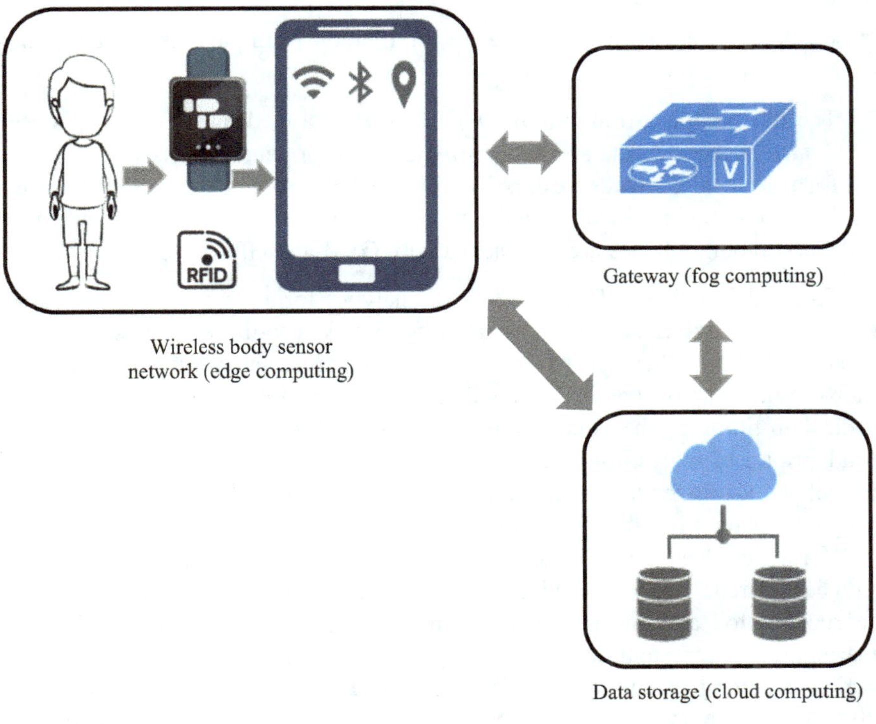

Figure 9.5 SHMS three-layer architecture

2. **Gateway layer:** The raw sensor data are gathered, analyzed, and communicated to cloud services using different wireless technologies such as Bluetooth, ZigBee, RFID, and Wi-Fi for remote access.
3. **Cloud layer:** Through communication technology, data from worn sensors is stored and accessed from a cloud service. In the remote system, it was also used to evaluate data and monitor the program.

9.8.5 Implementation of SHMS with big data analytics

Complex IoT-enabled technologies are beneficial to medical professionals in monitoring large amounts of healthcare information records in order to maintain monitoring of their patient's health status. The health state of patients is continually followed using the signals sent by the IoT system, and if any anomalies are detected, alarm signals are sent to doctors. For instance, if a patient's glucose level falls, the system will send an intimate signal to the doctor, who will be able to communicate with the patient and take the necessary action. This information is gathered from patients and stored in the cloud. Data can be gathered and aggregated more quickly and precisely in the cloud environment. To obtain the required value from the acquired medical data records, the proposed system model includes various elements such as medical utilities, doctors, important health services, patients, and other technology.

9.8.6 Proposed model

Sensors should be used to gather health information from patients, which is then processed and communicated.

- The Intel Galileo Arduino IDE, which serves as a medical IoT broker, is used to analyze sensor data and save it in the cloud for data analysis.
- Data analytics of records are done by using the Apache Spark approach in the cloud.
- Medical monitors manage graphical user interfaces for seamless advancement.
- Smart mobile phones are entrenched with GSM and GPRS capabilities.

Figure 9.6 depicts IoT-based big data analytics for a smart healthcare framework. The IoT broker health service provider is linked to the various sensors. The system model's medical metrics are used and incorporated within the cloud. To evaluate such massive medical data collected from multiple patients across various medical institutions, the cloud paradigm is used. With the collected data in the cloud, prior data analysis of health information for patients is possible. The doctor's smartphone is utilized to connect remotely to the IoT broker through Arduino IDE and gather patient health information.

If a patient has a dangerous medical condition, immediate medical assistance may be offered. The proposed smart healthcare monitoring framework, which incorporates IoT and big data analytics, may also be directly incorporated with a smartphone. The system architecture depicted in the figure provides cloud connection, resulting in end-to-end customer satisfaction. The Apache Hadoop distributed file system and the Apache Spark technique are used to analyze sensor data in the cloud. It is used to store and process sensor data linked to healthcare.

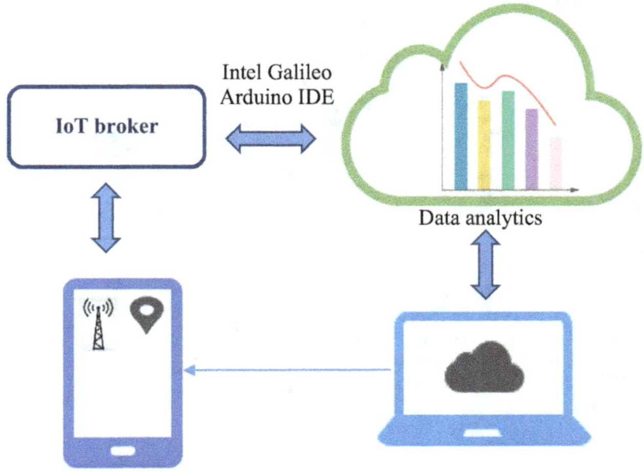

Figure 9.6 Smart healthcare framework using IoT with big data analytics

9.8.7 Case study

Using fog computing and data mining techniques, we can remotely monitor individual's health system and discover the cause of arrhythmia. Rather than analyzing data in the cloud, as in previous models, the goal is to deliver solutions near to consumers to enhance quality and minimize the delay. In particular, the alarm system immediately informs the patient or doctor in an urgent situation, such as when a patient suffers a heart attack. Eventually, the information collected from the fog layer is sent to the cloud layer through communication networks for stable storage in the patient's computer files as shown in Figure 9.7. Due to numerous issues with the normal functioning of the heart, heart disease is India's second-largest cause of mortality. One of these issues is cardiac arrhythmia, which, if left untreated, can result in irreversible consequences such as a heart attack. An ECG of the patient is one approach to diagnose this illness. To suggest a method for monitoring a patients health, in this chapter we consider cardiac arrhythmia patient monitoring system using IoT. The system to monitor heartbeat and perform ECG was tested and ran using an Arduino electrical board and an AD8232 sensor module. As a result, one of the most widely used data mining method, the deep neural network (DNN) algorithm, is utilized to identify and validate the kind of cardiac arrhythmia.

9.8.8 Performance evaluation of data analysis

The DNN algorithm was used to test the accuracy of arrhythmia diagnosis by analyzing and extracting patient data from the Google repository. The different kinds of arrhythmias for each class are defined by the total number of patients, as shown in Table 9.1 of the Google repository database. The pattern and shape of the pulses, which may be referenced as health values, are included in the aforementioned values in this table, and the system analyzes and classifies the input data with this information using the DNN algorithm. As a result, the primary patterns extracted from this database serve

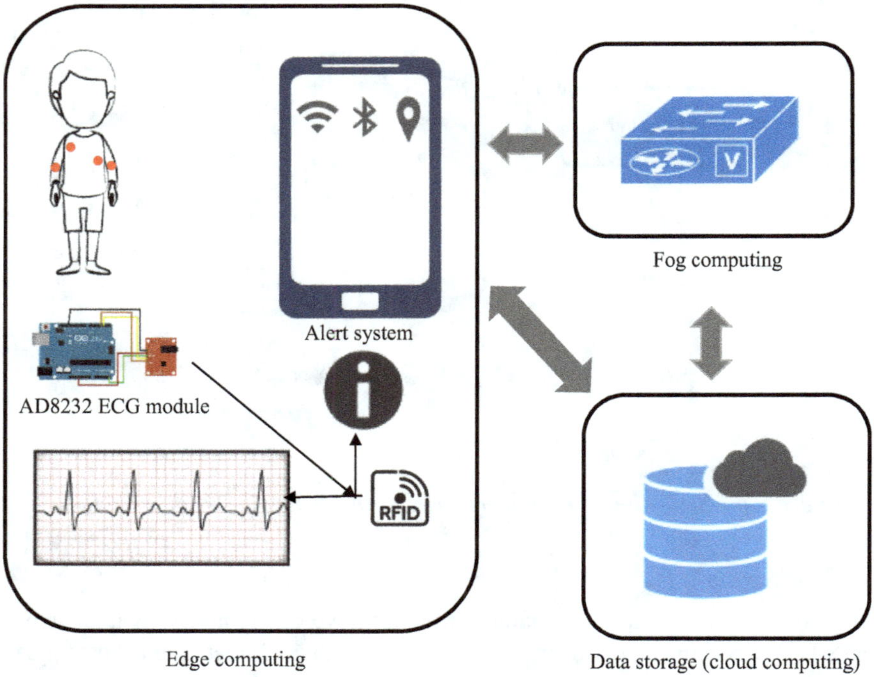

Figure 9.7 SHMS for arrhythmia patients

Table 9.1 Google repository database arrhythmias

S. no.	Classes	Types	Cases
1.	Normal	Normal	240
2.	Tachycardia	Tachycardia	13
3.	Bradycardia	Bradycardia	15
4.	Supraventricular arrhythmias	Premature atrial contractions	2
5.		Paroxysmal supraventricular tachycardia	5
6.		Accessory pathway tachycardias	0
7.		AV nodal re-entrant tachycardia	0
8.		Atrial tachycardia	0
9.		Atrial fibrillation	2
10.		Atrial flutter	3
11.	Ventricular arrhythmias	Premature ventricular contractions	3
12.		Ventricular tachycardia	4
13.		Ventricular fibrillation	2
14.		Long QT	0
15.	Bradyarrhythmias	Sinus node dysfunction	0
16.		Left bundle block	9
		Right bundle block	50

as the foundation for our accuracy. In TensorFlow, two subroutines have been built in the data mining area. The data in a CSV file will be first sorted randomly to check the findings in the first subroutine. The remaining data is divided into training and test categories, with 80% and 20% of the total being used for each category. Owing to non-uniform data sequence, the determined values for the trained and test data are different each time the program is run, and hence the outcomes will be varied. We use the DNN algorithm technique to categorize the training and test data with the DNN classify command. The results will be compared with the intended values and the overall accurateness of the results were calculated in the next subroutine, charted by the computation of root mean square error in Table 9.2.

The DNN regression figures resultant from the program output in TensorFlow are presented in Figures 9.8 and 9.9.

Table 9.2 Rate of accuracy

Accuracy			
Type1	100%	**Type9**	86.2%
Type2	87%	**Type10**	100%
Type3	80%	**Type11**	100%
Type4	76%	**Type12**	73%
Type5	75%	**Type13**	NAN
Type6	NAN	**Type14**	NAN
Type7	NAN	**Type15**	23%
Type8	NAN	**Type16**	80%

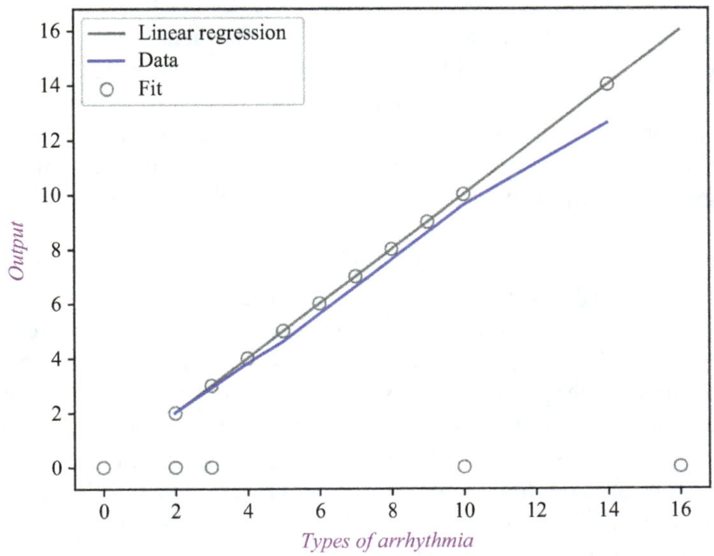

Figure 9.8 DNN regression on training data

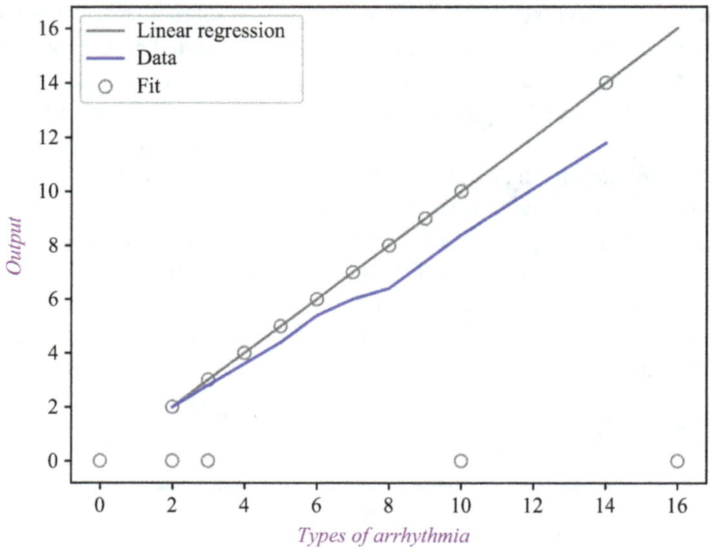

Figure 9.9 DNN regression on test data

9.9 Conclusion

Several new technologies are constantly being used in the medical sector, such as equipment and systems that check vital signs on a regular basis, as well as other gadgets that track actual clinical data. Patients and physicians are utilizing mobile applications to manage the health needs as data rates have increased and smartphones have become more widely available. In the healthcare industry, combining big data analytics with IoT technologies is critical. We have done a detailed discussion and analysis of the definition and impact of IoT and big data analytics technology in the medical industry in this chapter. Following that, a case study on the SHMS framework with data analytics is demonstrated, as well as its implementation. A large number of patient cardiac ECG signal data records are acquired, and a medical specialist is necessary to sort the data according to the patient. The IoT broker utilized in the overhead solution to connect scientific data of troubled patients with the cloud is Intel Galileo Arduino IDE. The proposed framework is used to process the large amount of patient data stored in the cloud. It was discovered that the responses time was significantly reduced when there was no major delay. It can also be used in a real-time context to keep track of patients' health. Clinical metrics are coordinated using the Intel Galileo Arduino IDE GPRS/GSM connection capabilities and alarm signals. Doctors can benefit from this well-developed healthcare system by giving the right information to the right patient at the right time. The DNN algorithm was also used to analyze the presence or absence of cardiac disease, which was then split into six categories of 16 different types of cardiac arrhythmias. For each category of cardiac arrhythmias, great accuracy was established after numerous studies. As a result, patients can receive effective and quick diagnoses in a short period of time. Different IoT monitoring system for various health problems using IoT are prototyped in the wireless framework [17–24].

References

[1] Mishra, S., B. K. Mishra, H. K. Tripathy, and A. Dutta. "Analysis of the role and scope of big data analytics with IoT in health care domain." In: *Handbook of Data Science Approaches for Biomedical Engineering*, Academic Press, 2020, pp. 1–23.

[2] Rath, M., and V. K. Solanki. "Performance improvement in contemporary health care using IoT allied with big data." In: *Handbook of Data Science Approaches for Biomedical Engineering*, Academic Press, 2020, pp. 103–119.

[3] Banerjee, A., C. Chakraborty, A. Kumar, and D. Biswas. "Emerging trends in IoT and big data analytics for biomedical and health care technologies." In: *Handbook of Data Science Approaches for Biomedical Engineering*, Academic Press, 2020, pp. 121–152.

[4] Moghadas, E., J. Rezazadeh, and R. Farahbakhsh. "An IoT patient monitoring based on fog computing and data mining: cardiac arrhythmia usecase." *Internet of Things*, 2020;11: 100251.

[5] Pavithra, M., R. Rajmohan, T. Ananth Kumar, and R. Ramya. "Prediction and classification of breast cancer using discriminative learning models and techniques." *Machine Vision Inspection Systems, Volume 2: Machine Learning-Based Approaches*, 2021; 241–262.

[6] Gokulan, S., S. Narmadha, M. Pavithra, R. Rajmohan, and T. Ananthkumar. "Determination of various deep learning parameter for sleep disorder." In: *2020 International Conference on System, Computation, Automation and Networking (ICSCAN)*, IEEE, 2020, pp. 1–6.

[7] Narmadha, S., S. Gokulan, M. Pavithra, R. Rajmohan, and T. Ananthkumar. "Determination of various deep learning parameters to predict heart disease for diabetes patients." In: *2020 International Conference on System, Computation, Automation and Networking (ICSCAN)*, IEEE, 2020, pp. 1–6.

[8] Jeyaranjani, J., T. Dhiliphan Rajkumar, and T. Ananth Kumar. "Coronary heart disease diagnosis using the efficient ANN model." *Materials Today: Proceedings*, 2021; 1–10.

[9] Rajmohan, R., M. Pavithra, T. Ananth Kumar, and P. Manjubala. "Exploration of deep RNN architectures: LSTM and Gru in medical diagnostics of cardiovascular and neuro diseases." *Handbook of Deep Learning in Biomedical Engineering and Health Informatics*, 2021; 167–202.

[10] Manju Bala, P., S. Usharani, R. Rajmohan, and G. Leemaroselin. "Applications of machine learning and deep learning for maintaining

electronic health records." *Global Journal of Internet Interventions and IT Fusion*, 2020;3(1):1–10.

[11] Usharani, S., P. Manju Bala, R. Rajmohan, and M. Pavithra. "Application of ML/DL for improved radiotherapy." *Global Journal of Computing & Software Testing*, 2020;2(1):1–10.

[12] Manju Bala, P., S. Usharani, G. Leema Roselin, A. Balachandar, and M. Pavithra. "Lung cancer diagnosis using deep learning methods in health care systems." *International Journal of Recent Development in Computer Technology & Software Applications*, 2021;5(1):11–18.

[13] Usharani, S., P. Manju Bala, and G. Leema Roselin. "Applications of deep learning in bio informatics and biomedical images." *Global Journal on Innovation, Opportunities and Challenges in Applied Artificial Intelligence and Machine Learning*, 2021;5(1):21–29.

[14] Balachandar, A., E. Santhosh, A. Suriyakrishnan, N. Vigensh, S. Usharani, and P. Manju Bala. "Deep learning technique based visually impaired people using yolo v3 framework mechanism." In: *2021 3rd International Conference on Signal Processing and Communication (ICPSC)*, IEEE, 2021, pp. 134–138.

[15] Swarna, S. R., A. Kumar, P. Dixit, and T. V. M. Sairam. "Parkinson's disease prediction using adaptive quantum computing," In: *2021 Third International Conference on Intelligent Communication Technologies and Virtual Mobile Networks (ICICV)*, 2021; pp. 1396–1401. doi:10.1109/ICICV50876.2021.9388628.

[16] Sasubilli, G. and A. Kumar. "Machine learning and big data implementation on health care data," In: *4th International Conference on Intelligent Computing and Control Systems (ICICCS)*, 2021; pp. 859–864. doi: 10.1109/ICICCS48265.2020.9120906.

[17] Mamdiwar, S. D., S. Zainab, C. Utkarsh, S. Kathiravan, and C.Chuan-Yu. "Recent advances on IoT-assisted wearable sensor systems for healthcare monitoring." *Biosensors*, 2021;11(10):372.

[18] Chakraborty, C., and M. Rathi. "Smart healthcare systems using big data." In: *Demystifying Big Data, Machine Learning, and Deep Learning for Healthcare Analytics*, Academic Press, 2021, pp. 17–32.

[19] Parvathy, V.S., S. Pothiraj, and J. Sampson. "Automated Internet of Medical Things (IoMT) based healthcare monitoring system." In: *Cognitive Internet of Medical Things for Smart Healthcare*, Cham: Springer, 2021, pp. 117–128.

[20] Chakraborty, C., S. Roy, S. Sharma, T. Tran, P. Dwivedi, and M. Singha. "IoT based wearable healthcare system: post COVID-19." *The Impact of the COVID-19 Pandemic on Green SocietiesEnvironmental Sustainability*, 2021; 305–321.

[21] Ilmudeen, A. "Big data-based frameworks for healthcare systems." In: *Demystifying Big Data, Machine Learning, and Deep Learning for Healthcare Analytics*, Academic Press, 2021, pp. 33–56.

[22] Poongodi, M., S. Ashutosh, H. Mounir, M. Ma, and C. Naveen. "Smart healthcare in smart cities: wireless patient monitoring system using IoT." *The Journal of Supercomputing*, 2021; 1–26.

[23] Biswas, R. "BIG data analytics: a boon for SMART healthcare." *International Journal on Engineering Technologies and Informatics*, 2021; 2(1):15–16.

[24] Usharani, S., S. Prithivi, S. Sharmila, P. M. Bala, T. Ananth Kumar, and R. Rajmohan. "Mobile application for doctor appointment scheduling," In: *2021 International Conference on System, Computation, Automation and Networking (ICSCAN)*, 2021, pp. 1–6. doi: 10.1109/ICSCAN53069.2021.9526398.

Chapter 10

Experimental analysis and investigation of dementia detection framework using EHR-based variant LSTM model

P. Manju Bala[1], *S. Usharani*[1], *R. Rajmohan*[1], *T. Ananth Kumar*[1], *A. Balachandar*[1] *and S. Arunmozhi Selvi*[2]

In developing nations, dementia is one of the major factors of death. While promising findings have been documented using computer-aided models in the study, there is no functional diagnostic tool used in clinical practice. Deep learning models have grown rapidly in recent years, particularly when working with photographs. Since 2013, deep learning has got a lot of interest in dementia detection science, with the number of articles published in this field rapidly increasing since 2020. When compared with traditional machine learning (ML) methods, deep learning models are increasingly effective in detecting dementia. Dementia diagnosis remains difficult, and identification necessitates a highly discriminative characteristic classification to distinguish related brain patterns. This chapter evaluates the success of a deep learning model to that of a conventional ML model in enhancing the precision of a dementia predictive model. The model can be used as a first-round diagnostic method for medical follow-up, involving neurological assessment, neuropsychological evaluation, screening and clinical study recruitment if it is accurate enough. Seven datasets with 2 years of data were established, three to eight times previously to the time of diagnosis, as well as an incident dataset. For each dataset, multiple trained models were constructed and their output was measured using verification and testing dataset: enhanced nodes, feed-forwarding model, convolutional neural network and fuzzy system with pre-trained weights. The current study developed a model for estimating the number of dementia cases centred on the organisation of the long short-term memory (LSTM) network which takes into account the impression of oldness and gender on dementia susceptibility. From 1997 to 2020, data collection was performed to gather the estimated metrics of patients identified with dementia. For predicting the

[1]Department of Computer Science and Engineering, IFET College of Engineering, India
[2]School of Computer Science & IT, DMI St. John the Baptist University, Africa

severity of dementia, the analysis indicates that the LSTM network has a better prediction performance than the other three forms of the model. We build an electronic health record (EHR)-based LSTM architecture that develops to detect dementia and implicitly learned functionality from communicative extracts. When categorising participants into dementia and control classes, our method achieves an accuracy rate of 93%, establishing a cutting edge in dementia identification using the EHR dataset.

10.1 Introduction

Dementia is a disorder that mainly affects senior citizens and is a massive strain on families and communities due to their high level of cognitive incapacity. There is presently no effective dementia drug that can slow or stop the progression of the disease. It is impossible to exaggerate the emphasis placed on the disease's early stages, fast medication and postponement. Clinical diagnoses of dementia are made based on a client's and people's thorough medical history, clinical findings and cognitive tests. Further examinations, such as haematological, computed tomography and magnetic resonance imaging, are required to rule out other reasons for dementia. Neurocognitive tests are essential for diagnosing problems in the human 'brain regions'. Despite the existence of diverse diagnostic indicators for the early identification of dementia, there is still a great deal of uncertainty. The development of enhanced analysis techniques is critical. Outcomes result from proper cognitive decline diagnosis, but it is also vital for healthcare. The manual identification of cognitive decline, which may involve numerous items of evidence such as a cognitive test score, scientific research findings, informed witness accounts and so on, is time consuming in medical diagnosis.

The physician's insightful details determine the effectiveness and correctness of the diagnosis. It will be far more complicated to classify and diagnose dementia in some isolated places where expert technicians are scarce. Machine learning (ML) is a cutting-edge computational technique that can enhance the interpretation of healthcare data and make decisions for patients instantly. The goals of this paper are to (1) apply optimisation methods to improve or even minimise the amount of cognitive impairment needed to diagnose dementia sufferers and (2) construct and evaluate an appropriate prediction system based on the clinical data of registered participants.

Dementia is characterised by a decline in intellectual capacity. People who suffer from this illness have difficulty in socialising and going about their everyday life. Cognition, communication skills, problem-solving, spatial cognition, consciousness, and the capacity to pay concentration and awareness are all harmed or lost as a result of the illness. Many dementia patients lose control of their feelings and their identities shift. Dementia is the most common disease, accounting for an estimated 65–85% of case [1]. Dementia affects more than 48.5 million people globally, with 7.9 million new cases diagnosed annually. The prevalence of dementia in the normal community aged more than 65 years is around 5–9%. In regards to medicare and social assistance, dementia is typically more expensive than cancers, hepatitis and coronary cardiovascular disease combinedly. By 2025, it was anticipated that there would be 76 billion people with dementia, with an estimated cost of $3 trillion for medical services [2]. Although there is no proper treatment for the disease earlier, the lives of those who undergo can be assisted and enriched. The goals of dementia care are to provide caretakers with awareness and long-term assistance, improve brain, exercise, personal functioning or wellness, recognise and manage behaviour and emotional problems [19], and diagnose dementia earlier. When it comes to diagnosis of dementia, the quicker a patient is precisely diagnosed, the quicker they can seek health care. Unfortunately, since there is no standardised diagnosis for identifying dementia, treating it can be difficult [3]. On the other side, the latest theory predicts that 38% of dementia occurrences may be avoided if customised danger could be predicted in advance, allowing physicians and individuals to make better-informed decisions earlier. As a result, in addition to strategies to achieve dementia identification, scientists have lately focused on establishing methods for forecasting dementia risk. One of the most critical places for ML techniques is medical care. It is, nonetheless, among the most complicated [4] and difficult topics, above all in the fields of classification and prediction [5].

10.2 Related work

In 2020, more than 65 million individuals globally have dementia, with that figure anticipated to rise to 155 million by 2040. The most frequent form of dementia affects an anticipated 5.9 million individuals of different ages in the United States, with that figure anticipated to climb to 14.8 million by 2040 [6]. In 2019, the overall cost of dementia-related medical services and long-term care in the United States was $280 billion [7]. Physicians, public health professionals and biologists will require tools to help them detect people in danger of dementia and treat the problem. Prediction methods have been introduced in recent decades to aid in the identification of patients with severe dementia before the initiation of the illness. These devices can be classified into two categories. The first set of scientists choose potential risks based on medical feedback and use quantitative measures to estimate their impact [8]. Neural network (NN) statistics strategies have been introduced [9] to detect and monitor risk variables from massive healthcare datasets in the former method, which is significantly smaller than in the first. Convolutional neural network (CNN) techniques, a subclass of artificial intelligence [10], have shown particular potential in harnessing the energy of enormous procedural records and electronic medical record (EMR) collections, as well as improved technology to make more effective healthcare predictive analytics. With a c-statistic of 0.8 and 0.66, in [11] the authors create deep neural networks (DNNs) using EMR data for identifying activities such as customer's actual release diagnosis and 1-month unscheduled repatriation. Their systems used a series of occurrences building up to an indicator item as parameters to recurring algorithms with long short-term memory (LSTM) [12], which can use the order and duration among events. Create a contextual prediction model using gated recurrent units (GRUs) [13] that leverage medical records (treatment, procedures and medicine) and time among appointments to forecast diagnostic and prescription orders, as well as the time to another appointment, in [14]. When comparing estimates using logistic regression analysis, their results revealed a 9% enhancement in the region under the curve (RUC). Choi *et al.* [15] used GRUs to construct a neural interest framework [16] that not only delivers a value for the estimation method but also interprets the result (using time order) by highlighting the relevance of the elements that contribute to it. Focus methods have been applied efficiently in a number of disciplines, including natural speech processing and image processing [17]. These methods have the advantage of regulating response time as the duration of the image input grows longer, which is a disadvantage of CNN, but at the expense of acquiring to learn more components. For diagnosing illnesses like dementia, it takes decades after the onset. Sequencing sizes are less useful because the amount of information generated declines significantly. LSTM-based methods were used by Kim *et al.* [18] to predict patients who might be at a chance for dementia diagnosis.

10.3 Materials and methods

10.3.1 EHR datasets

The OASIS-2 data was used in this investigation, which included concealed procedural records and EMR information. Continuous brain imaging, medical and cognitive sets of data of healthy aged and dementia people are all included in the repository. OASIS-2 is a retroactive collection of information of about 1500 individuals acquired over a period of 35 years by the WUSTL Knight ADRC via numerous ongoing initiatives. Patients' age ranges between 45 and 96 years and the sample comprises 610 typically developing people and 490 people with varying stages of cognitive impairment. All patients were given a randomly generated identity, and all periods were deleted and normalised to match the number of days since they first entered the research. The majority of the magnetic resonance imaging (MRI) scans contain spatially differentiating documents created by sequence adaptive multimodal segmentation processing. Over 1,600 unprocessed MRI images from three different devices, PIB as well as the corresponding post-processed data from the OASIS-2, were considered.

Simulations can be fed data in two contexts: as a grid or as credit value. It is generated as the range of discrete days where a person had a score in systems that employ a grid. The median of 61 values derived using the training samples was used to limit (winsorize) the absolute value in the grid. The highest value of each column was used to scale the sparse values, while the median and normal deviation were used to normalise the impenetrable columns. For systems that use code series of people, 350 codes from the appropriate 24 months data-collecting period nearest to the indexing date are used, with no more than 10 instances of a character. This enables the learning to go quickly without compromising precision. A mobile device was used to capture the azimuth statistics. The preceding quasi-static data was logged using an approach proposed and assembled on the mobile device: time and date (ms), velocity effort over the a, b and c axes, such as gravitational force (m/s^2), a positive linear correlation of the gadget (m/s^2), gadget direction or the slope among true poles and the gadgets' *b* coordinate axes (degree). The information was sampled at a rate of 25 Hz.

The study enlisted the help of 15 non-dementia volunteers who have been psychologically stable and capable of autonomous gait. Every participant in the primary group was informed of the research aims and has given their consent to take part in the study. Topics were described and samples of each sort of travel behaviour were presented before the conservation area. Even before a real gathering, each participant was asked to repeat each sequence once to ensure that they understood it. All of the participants held the smartphone in their hands so that they could readily engage with the user experience on the mobile device to begin and finish capturing data. The individuals were asked not to vary the position in which they held the phone throughout data recording in order to maintain data integrity. The data collectors first notified the subjects of the traffic patterns that were assessed to be acquired during this authorised signatory. After that, the participants were asked to walk in the same fashion. During the walk, the participants were required to carry a smartphone with an information storage application installed. Whenever the individuals began walking, they press arrow buttons on the smartphone camera to initiate data acquisition, which began trying to log the accelerometer data for that trip cycle. The subjects rotated the switch to cease data analysis for that travelling event when they started walking. The smartphone would then immediately save the accelerometer data. The participants' actions during their trip were constantly watched and monitored to maintain system quality and accuracy. If the individuals' traffic volumes did not match our data gathering plan, we rejected the information.

10.3.2 ML models

Figure 10.1 shows the proposed framework for the development of ML models using characteristics taken from original time series as input. As ground truth destinations, sources, and their respective labels, and the collected characteristics from the sample set series data are inputted into the trained ML classifiers for assessment, which is subsequently describes the outcome of the traveling trends.

The tsfresh library package is used to extract the 15 characteristics listed in Table 10.1 for each time series. The linear trend characteristic, in particular, is used to describe the linear trend of straightforward trends and has four characteristics extracted: interception, *p*-value, slope and standard error. The final three elements (from the bottom up) in Table 10.1 are added to illustrate the repeated characteristics of pacing and lapping patterns. The remaining characteristics are statistical characteristics that are frequently used in classification tasks. Five ML categories from the Spark MLlib and Numpy libraries are used to train and evaluate

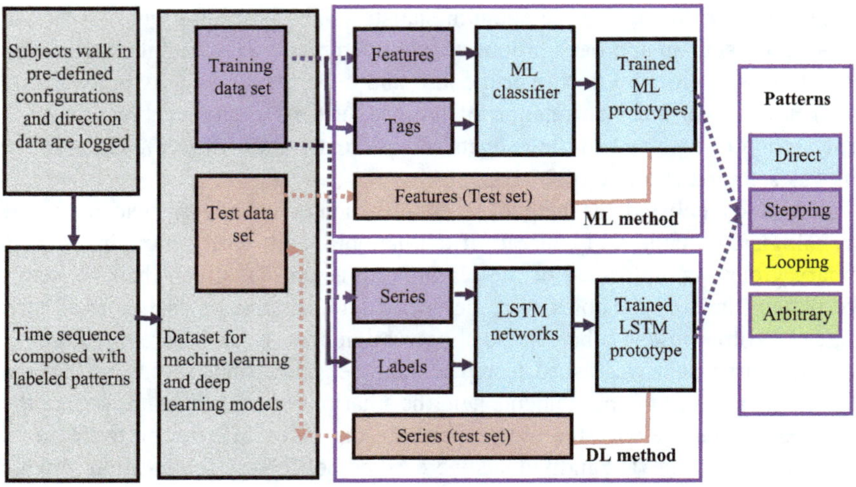

Figure 10.1 Proposed framework

Table 10.1 Characteristics of time series

Extracted characteristics of time series			
Average	Median	Kurtosis	Length
Autocorrelation	Maximum	Minimum	Standard deviation
quantile_q_0.62	peaks	linear_trend	quantile_q_0.26
Value of correctness number to the length of time series			
% of recurring values to all values			
% of recurring data elements to all data elements			

classification models, notably artificial NNs, decision tree, random forest (RF) and support vector machine (SVM).

10.3.3 Approach to deep learning

LSTM-based deep networks are used to identify trip trends. To store temporal characteristics based on the memory cells with a limited number of gates and to retain valuable data that have lengthy relationships use memory cells with a small proportion of valves in LSTM. Given time series data, LSTM has been used to classify various human activities and has performed well. LSTM is a suitable option for this issue since the alignment indicators of dementia-related mobile arrangements are distinctive time series with temporal and spatial dependence. Figure 10.2a depicts the general LSTM network construction.

In the meantime, we utilise bi-directional LSTM, an enhanced form of LSTM, that includes a backward and forward layer (Figure 10.2b) to account for both former and future data. The training LSTM models are then used to forecast mobile

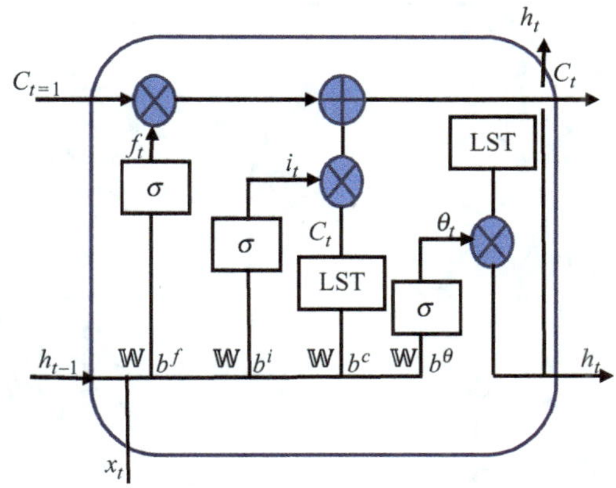

(a) LSTM network Construction

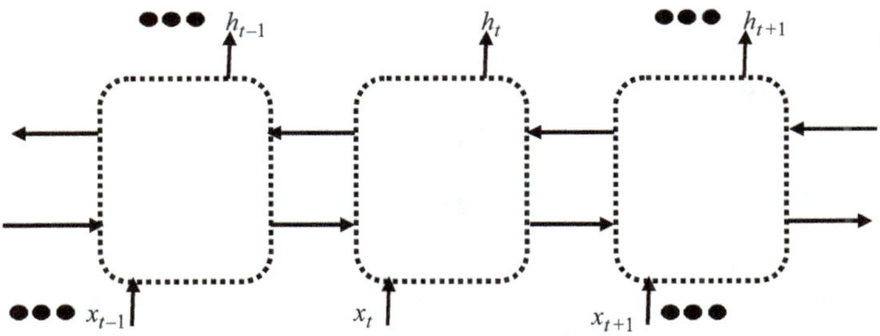

(b) Bi-directional LSTM network

Figure 10.2 LSTM network. (a) LSTM network construction. (b) Bi-directional LSTM network

arrangements based on code, similar to ML methods. Time series segments from the test set is worth noting that both the instruction sets and trail sets are pre-processed to guarantee that they have a similar arrangement length before being fed into the LSTM models.

10.3.4 Analysis of models

The researchers utilised two distinct DNN architectures: a feed-forwarding network and a recurrent neural network (RNN). The design of a deep feed-forward NN using matrix input is shown in Figure 10.3. The network includes a number of secret layers of varying sizes, each of which has its output passed through a rectified linear unit activation function as shown in Figure 10.4. A sigmoid or logistic activation function has been used in the output layer. Between each of the secret

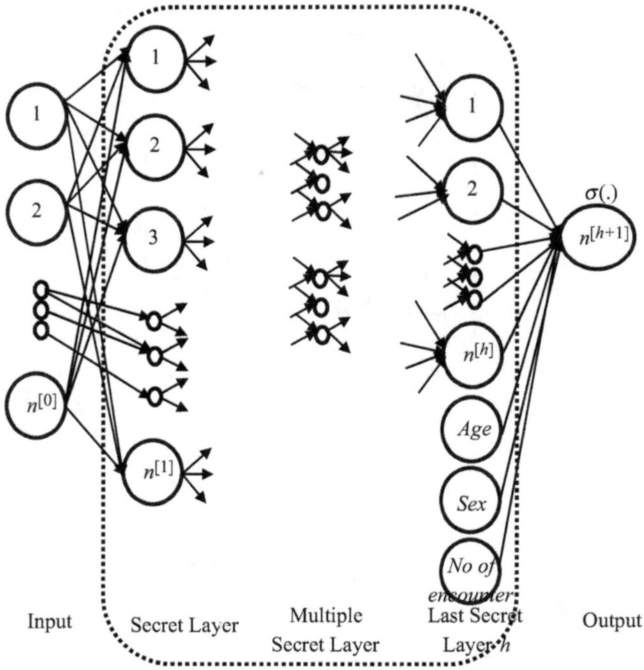

Figure 10.3 Design of a deep feed-forward NN

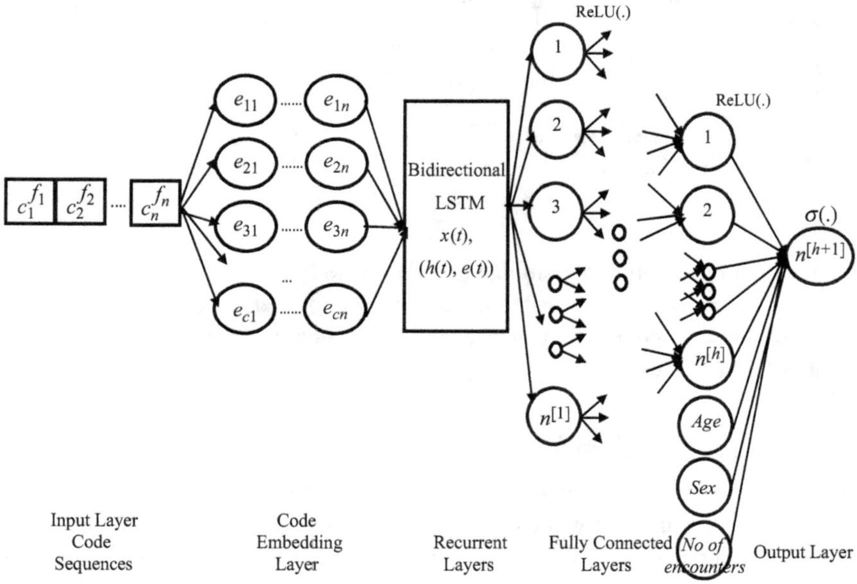

Figure 10.4 NN with ReLU activation function

layers, as well as between the source and the first secret layer, a dropout layer is used for generalisation to avoid the fitting problem.

The inclusion of the characteristics such as age, gender and number of encounters in the final secret layer, which is directly linked to the output layer, is a significant innovation in the design. Because there is no dropout layer between this and the output layer, each of these important characteristics has a greater impact. In the presence of certain characteristics, the other weights are modified appropriately. The structure was compared against a network that includes age, gender and the number of interactions in addition to all other input layer characteristics. In all cases, the update with the three characteristics in the final secret layer performed significantly better; therefore it is the only one shown here.

The RNN-based architecture has a visual programming language as a source for each individual in the cohort. This is being sent via an embedding layer, which produces N-dimensional vectors by each coding. These vectors are processed through a bi-directional LSTM in a standard format, which summarises the codes from both the past and future to provide a more enriched secret state. After that, the data in the secret state is transferred via a series of fully connected layers with ReLU activation and dropouts between the levels before being sent to the output layer with sigmoid activation. Several LSTM variants, including two LSTMs with the output of one feeding into the other, a single LSTM that only utilises the previous states, and recurrent neural units, were tested before agreeing on the above design, which steadily improved the others.

10.3.5 Proposed methodology

The proposed methodology for classifying patients is classified into two categories: with dementia and without dementia. Transcripts of the clients' voiced discussions are fed into the structure. The translations are around the same length as the originals (the average participant spoke 73 words across 16.8 words). Rather than dividing the discussion into individual statements, we evaluate all participant speech in a single block, enabling the model to evaluate the whole discussion circumstances in a way comparable to a reality diagnostic situation.

10.3.6 Model architecture

The suggested model structure is a deep CNN-LSTM with a few tweaks. At the conclusion of the LSTM layer, we added a thick NN to take into account additional factors. Linguistic characteristics are important in prior studies. Humans used a bi-directional LSTM instead of a traditional unidirectional LSTM and added an attention mechanism to the secret states of the LSTM. We anticipate our success in this manner. A model for detecting certain linguistic forms is linked to dementia. Furthermore, when considering lengthy sequencing, the attention mechanism has been shown to enhance performance. To adjust for the dataset mismatch, we added class weights to the loss function during training.

As the design shown in Figure 10.5, we remove interviewer comments and truncate the remainder texts to 73 words before pre-processing each complete

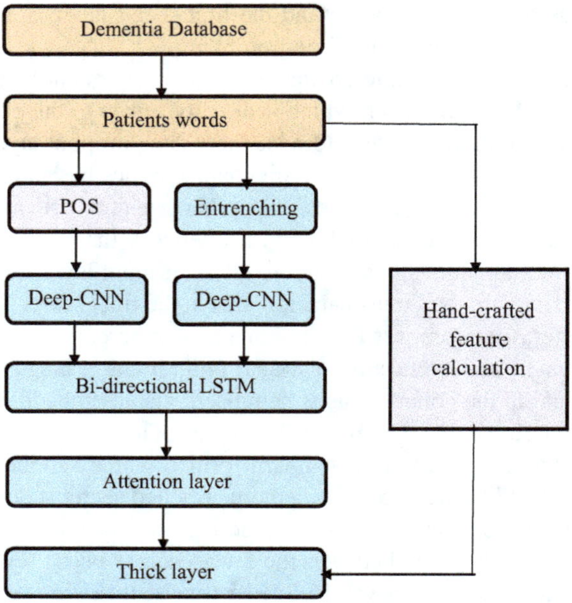

Figure 10.5 Model architecture

interview transcript from DementiaBank. This is done such that (a) each instance has consistent font sizes and (b) the situations are of reasonable length, allowing for sufficient information to evaluate the patient's condition. The median (participant-only) interview length is 73 words; therefore, 50% of occurrences contain the whole interview (padded if required), while 50% are shortened to the first 73 words. The interviews are inputted into separate tokens, and each token's POS tags1 are calculated. The tokenized interview and the associated POS tag list are sent to the system. The pre-trained 300-dimensional GloVe embedding for the conversation text tokens is generated using the Wikipedia 2014 and Giga word 5 datasets. Each word's POS tag is provided as a single-hot encoded vector. The terms embedding and POS are used interchangeably.

Matrices are inputted into two separate CNNs with identical architecture, and the output of the two CNNs is compressed and input into a bi-directional LSTM with a focus mechanism. The output of the bi-directional LSTM is then fed into a layered NN, which additionally takes into account language characteristics that have previously been shown to be beneficial, as well as certain demographic characteristics. The model's ultimate result is achieved using a single sigmoid activation function neuron at the finish of the impenetrable layer. Keras is used to implement the model. The benefit of our hybrid design, which takes into account both implicitly learned and manufactured characteristics, is that it may combine information that may be valuable but is secret from the human observer with information that explicitly conveys clinical and psycholinguistic results.

10.4 Dataset for the suggested method

The suggested method is based on a dataset from the Organisation for the Advancement of Science and Technology (OASIS); oasis-brain.org is a website dedicated for the study of the brain. The Brain dataset is a collection of open-source neuroimages. It is home to a large number of people. It is useful in studies since it includes important interpretations. OASIS brain image is made up of three types of neuroimaging datasets: OASIS 1, 2 and 3. The OASIS-1 and OASIS-2 databases are used in this study. In OASIS-1, 416 cross-sectional MRI data are available. During the course of 434 sessions, these individuals are of certain age. Both males and women between the age of 18 and 96 years are included in this dataset. Women have to keep track of vital information regarding their brain sickness. OASIS-2 is a dataset of longitudinal MRI data from 150 people with 375 gatherings in total. These individuals are of age between 60 and 80 years. Grey matter is given special attention in the OASIS-1 dataset. There are map pictures and medical dementia ratings.

In dementia patients, there is a substantial loss of grey matter, having a clinical dementia grade of more than 0. The OASIS-2 dataset contains the gender, age and years of schooling which are highlighted. In order to get experimental findings using these methods—mini-mental state, socio-economic status, minimum mean square error, total intracranial volume estimation, atlas, normalised whole brain volume, clinical dementia rating and atlas scaling factor—an end-to-end gratification is, as suggested in this article, needed. Some of the necessary libraries and tools, such as NumPy, Pandas, Keras, Matplotlib, Sklearn and TensorFlow, are available for implementation. Computer hardware such as a graphical processing unit (GPU) or terminal is required for access and graphics processing.

10.4.1 Dataset pre-processing

Grey Matter Axonal Connectivity Maps comprise the input data in the cross-sectional dataset. Every data point in the set is made up of 90 slices of 110×90 pixels MRI scan pictures of the individuals, each of which represents a different representation of the brain. To remove any haze or blurriness, every data point in the image is smoothened. The smooth image function from the Nilearn library is used for this. Each part in the piece of image is resized to 50×50 pixels to ensure homogeneity. To create model input and output data, the continuous dataset needed to train the RNN and LSTM models is standardised. The missing hypotheses are recognised and switched with the average mode of the complete column. The sexual category property is changed from ‘M/F’ to ‘1/0’ for simplicity. The output data is created using the clusters in the database file from the second dataset. In demand to construct a bi-classifier, the participants are labelled as ‘converted’ and are regarded as insane. The words ‘demented’ and ‘non-demented’ are used to relate the findings to the DNN model. Both input data and resultant data are labelled and converted using the LabelEncoder from the TensorFlow toolkit.

The MRI dataset comes with a lot of missing and null values. To understand the variable relations, co-relation and dispersion matrix were generated. The aim is to understand the variable dependency in non-standardised data. In order to reduce the noise in the data, we followed a set of pre-processing steps such as data normalisation as follows. The missing values found in the dataset are assigned as 0. The demented value in the dataset is labelled as 1, non-demented is labelled as 0 and the converted is labelled as 0.5. The MRI Id from the dataset is also converted to float to overcome the basic problem of string to float data conversion in TensorFlow.

10.4.2 Parameters of the CNN model

Values and bias vectors in the CNN model are specified in four layers, with each and every row includes the value and bias of each secret unit. The first layer values are a 4×4×4 block, 1 input and 32 characteristics to be calculated in the value vector. V1 = [2,4,34] is the value vector for this level. The value vector for the second level is V2 = [4,4,4,64,128], and the values are a 4×4×4 block, 64 inputs and 128 characteristics. The values for the third secret unit are the custom product value for the reshaping algorithm, which is used on the output tensor of the second secret layer, and the number of secret elements in this layer, because the third level has 248,770 neurons. As a result, V3 = [248768, 1024] is the value vector of the third step. The number of secret elements in the preceding layer and in the input layers is used to calculate the values for the secret layers, which are VW4 = [1024, 3]. The selection of attributes to be calculated or the amount of secret elements in that layer is the biases in the bias vector. B1 = 64, B2 = 128, B3 = 1024 and B4 = 8 are the biases for the four layers, respectively.

The input vector is initially convolved with values V1 and bias B1 with a wastage of 0.6, before the softmax procedure is performed on it by the first layer of the CNN model. With values V2 and bias B2, the second level does something similar to the first layer’s results. The result of the two layers is rearranged to 248,770 in the third layer, and then the softmax activation function is performed to the regression results of the vector with values V3 and bias B3 in the fourth layer. The vector linear regression model with values V4 and bias B4 is used in the last layer to estimate output. The rate of the error between the definite and anticipated outputs is determined after the model has been trained, and the model is optimised with a computed value of 1e−3. A total of 1,717 iterations have been developed. The data for training and validation is divided into a 0.9 proportion. Each iteration loss and correctness is kept in an ‘out.txt’ file when the training is completed.

10.4.3 Parameters of the RNN model

The train and test fragmented the dataset in the proportion of 0.3 with a stochastic state of 32. Verification data refers to the first 50 items in the test data, whereas test data refers to the remaining entries. Following the division, there are 262 statistics for training, 48 for verification and 73 for testing processes. When creating the simulation, a sample size of four is chosen. A sequential model class is created using a predefined location, a bi-directional LSTM layer and two impenetrable layers. The input size in the secret layers, which is 262, determines the number of inputs for model training. The output size is 20 and the source distance is 10 because the input data comprises nine characteristics. In the bi-directional LSTM layer, the number of secret layers is supposed to be 62. The third secret layer is the impenetrable layer, which contains 62 secret elements and uses the ReLU stimulation function. The last layer is an impenetrable layer with 2 units and a sigmoid stimulation function, in addition to consolidating the output between 0 and 1. During model development, the loss is handled as 'binary-cross variance', and the training rate is set to 1e−4. The prototype is refined after 48 iterations.

10.4.4 Parameters of the LSTM model

The train and test fragmented the dataset in the proportion of 0.2 with a randomised state of 32. Validation data refers to the first 50 items in the training data, whereas test data refers to the remaining entries. Following the division, there are 262 statistics for training, 48 for verification and 73 for testing processes. When creating the simulation, a sample size of four is chosen. A sequential model class is created using a predefined location, a bi-directional LSTM layer and two impenetrable layers. The input size in the secret layers, which is 262, determines the number of inputs for model training. The result size is 410 and the input length is 10 since the input data comprises nine characteristics. In the LSTM layer, the amount of secret units is estimated to be 110. The last layer is an impenetrable one-unit layer with a sigmoid activation function to bring the output between 0 and 1. Through design generation, the loss is handled as a 'binary-cross probability'. The prototype has been trained after 12 iterations.

10.5 Dementia detection and prediction model

According to investigations, brain alterations related to dementia disease can develop more than 20 years earlier than symptoms. The method of MRI is used to scan the structure of the brain. Dysplasia is assessed by the decrease of density in a specific brain area between two images. It logically follows that within the sequence-dependent temporal context, various regions of the brain have varied illness diagnosis skills. The LSTM is a type of RNN that may be capable to link prior data to specific activity. To store the chronological relationship between inputs and the next phase of dementia, an LSTM system with max-pooling and activator layers is constructed.

The LSTM is meant to circumvent the long-term dependence concern by using a sequence of repeated NN components. In an LSTM deep network, there are four communication infrastructure layers: a forget gate, an input gate, an update gate and an output gate. The 'forget gate', indicated in (10.1), agrees whether the data is discarded away from the initial state.

$$f_a = \sigma\left(L_f.[g_{t-1}, y_t] + C_f\right) \tag{10.1}$$

where L_f is the load matrix, C_f is the cross vector and f_a is a value between 0 and 1, where 0 indicates forget and 1 denotes keep.

Change of data is determined by the input gate, which has a sigmoid function and a tanh function. Equations (10.2) and (10.3) are demonstrated as follows:

$$i_t = \sigma(L_i.[g_{t-1}, y_t] + b_i) \tag{10.2}$$

$$A_t = \tanh\left(L_C.[g_{t-1}, y_t] + b_C\right) \tag{10.3}$$

where L_i and L_C are weight matrices, b_i and b_C are bias variables and i_t, A_t are outcomes of algorithms.

In (10.4), the update gate replaces the past cell state value obtained from the input gate. Furthermore, the output gates in (10.5) and (10.6) determine which value should be produced from the layers.

$$A_t = g_t * A_{t-1} + i_{t*}A_t \tag{10.4}$$

In (10.4), g_t determines which information should be ignored and i_tA_t determines how many entries for the cell should be changed.

$$i_o = \sigma(L_o.[g_{t-1}, y_t] + b_o) \tag{10.5}$$

$$g_t = o_t * \tanh\left(A_t\right) \tag{10.6}$$

Wherever the quantity of i_o in (10.5) determines which amount of the cell condition will be the output, o_t*tanh(A_t) in (10.6) yields g_t, which is the outcome of the components i_o, by multiplying the new device state A_t by o_t and using the tanh.

Because LSTM may learn long-term interconnections, a framework based on LSTM was created to explain the development of dementia. In this chapter, a framework that uses time series data to forecast dementia 4 months in advance has been shown. The dementia prediction model has three layers: the partial integrated layer, cellular layer and fully connected layer. One activation function and a softmax function make up the pre-completely coupled layer; one LSTM layer and a dropout wrapper make up the cells layer and one fully connected layer and a softmax layer make up the post-fully attached layer. During the training stage, characteristic extraction sequential information with sample points is given to the prototype, and the model predicts the state of the next 7 months. The first two steps of R1 time consecutive data from 6 to 12 months are supplied to the model as shown in Figure 10.3. The analysis shows that the 18th month will be dementia, 6 months after the 12th month. When the model is tested with the subject for characteristics in the 18th and 24th month, the result is a forecast of the subject's condition in the 30th month.

The RUC is distinct as the region under the receiver operating characteristic (ROC) curve. It is one of the most important assessment metrics for evaluating the efficiency of classification models. It calculates the trade-off between true positives and false positives rate. Regions with several classes are also available. Under the ROC (mRUC), which is agnostic of group sizes, each class is assigned an overall level of classification ability. The RUC=mRUC method is used to assess our system, with RUC accounting for the binary-class and mRUC accounting for the multi-class. The following is a list of the metrics: RUC is for the region underneath the curve, and RUC(c_i; c_j) stands for the RUC of class c_i compared to another class c_j, as demarcated in (10.7):

$$RUC(c_i, c_j) = \frac{R_i - \frac{m_i(m_i+1)}{2}}{m_i m_j} \tag{10.7}$$

where m_i and m_j denote the total number of points in classes i and j, correspondingly. The total of the rankings of the probability belonging to class i is referred to as R_i. The average RUC(c_i; c_j) for classes i and j is RUC(c_i; c_j). As shown in (10.8), the mRUC is a multi-class area that denotes the ROC curve as follows:

$$mRUC = \frac{2}{M(M-1)} \sum_{i=2}^{M} \sum_{j=1}^{i} RUC(c_i, c_j) \tag{10.8}$$

RUC(c_i; c_j) signifies the average RUC(c_i; c_j) for classes i and j, and M denotes the number of classes. We can see how effective our strategy is by utilising MRI data to test it. When brain region characteristics are submitted to the prototype as time series, pre-full-service retrieves the most associated combination of attributes. After that, the cell layer selects time-sensitive properties. When characteristics are submitted to a fully connected layer, a mixture of time-sensitive properties is engaged to forecast dementia. The following parameters are fixed in trials based on prediction to change the network topology of our model: group size is set to 262, learning proportion is set to 4, cell type is LSTM cell, number of partially connected cells is set to 524, number of fully connected cells is set to 3 and the probability is set to 0.78. In order to discover the best values, we change our model using the following parameters: the number of entirely connected cells, the number of LSTM cells and the number of LSTM layers. Experimental results on dementia, mild cognitive impairment (MCI) and normal cognition (NC) reveal that the best mRUC is achieved when the number of totally linked cells is 524 and the number of LSTM layers is 2. Our prototype attains the finest mRUC when the number of LSTM cells is raised to 131 cells, as shown in Table 10.1.

10.6 Experimental results

According to research, dementia-related brain changes can begin more than two decades before symptoms appear. Multiple symptoms are a reflection of the degree of cell loss in diverse parts of the brain, which varies from person to person. The

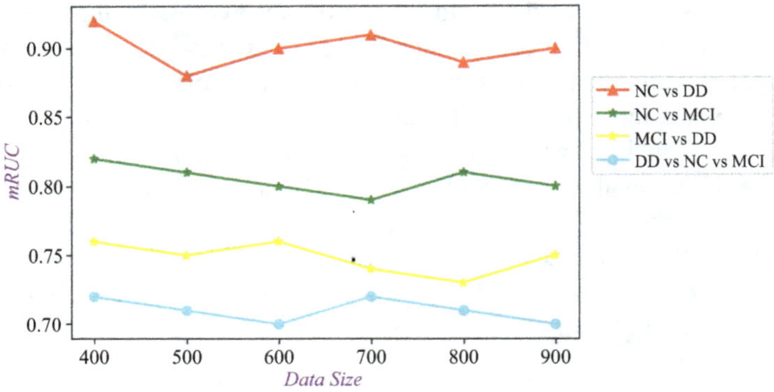

Figure 10.6 Data size variation in various prediction models

severity of the complaints varies. In this study, we focus on the 6-month predictive states of dementia detection. As a result, the predicted status refers to the patient's 6-month state. To evaluate the models' performance (Figure 10.6), binary prediction tasks and multi-prediction tasks are used. After preparing the data using intervals, we used cross-validation methodology for entire procedures. We separated the data into five subgroups at irregular periods in the past. Then we choose one subset to test and the other four to learn from. Method 1 uses classifier self-representation sparse regression to identify characteristics and an SVM classifier to classify them. Method 2 selects characteristics using the RF approach and categorises the ailment using the DNN method. The DNN is utilised in combination with principal component analysis–least absolute shrinkage and selection operator in method 3 to select attributes, and the SVM is used to categorise the ailment. The TensorFlow application and a single GPU 1060i Nvidia card in the target PC were used for testing and cross-validation for all of the strategies. The outcomes display that our prototype surpasses all other prediction strategies by a wide margin.

An experimental result demonstrates that our algorithm produces the best RUC/mRUC in the greatest of the estimates grounded on all the parameters of MRI data, including 900 delay discounting (DD), MCI and NC, when compared to other techniques. DD vs NC has an RUC of 0.935, DD vs MCI has an RUC of 0.798 and DD vs NC vs MCI has an RUC of 0.777. Because the NC vs MCI dataset cannot distinguish between MCI progressing to DD and MCI remaining stable, our approach scores lowest in the estimate for MCI against NC, with an RUC of 0.697. The suggested method goal is to identify time-relative biomarkers connected to the sickness state. Five biomarkers that have been recognised by medical technology are as follows: surface region (SR), cortical thickness average (TA), WM parcellation volume (PV), neuronal parcellation volume (NV) and thickness standard (TS). One of the five characteristics of TA is the detection of significant atrophy in the prefrontal cortex and some subcortical areas. In neuroscience, dementia is defined as the loss of neuronal cells in the cerebrum and some subcortical locations, which leads to a massive shrinking of the affected regions. The RUC/mRUC comparison with various characteristics categories is considered in Figures 10.7–10.12, based on various data sizes, from 400 to 900. The outcomes show that the algorithm is reliable in many of the characteristics, particularly the TA characteristic, which ensures the highest achievement in RUC/mRUC among such

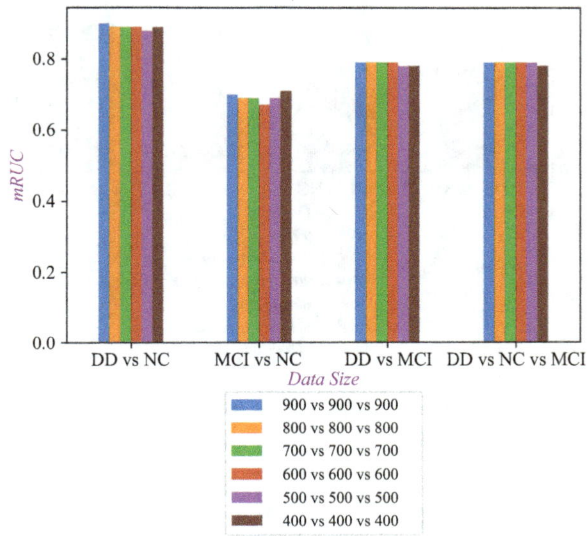

Figure 10.7 Data size variation on NV characteristics

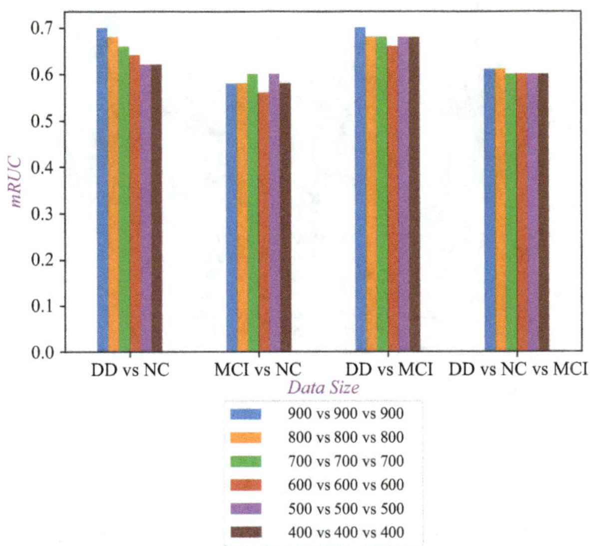

Figure 10.8 Data size variation on SR characteristics

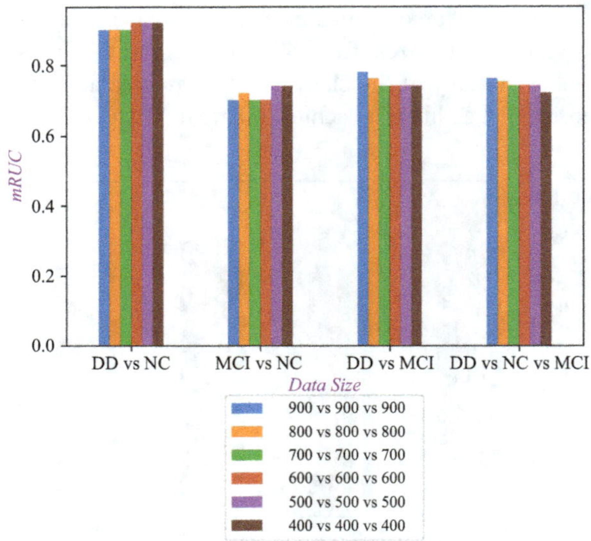

Figure 10.9 Data size variation on PV characteristics

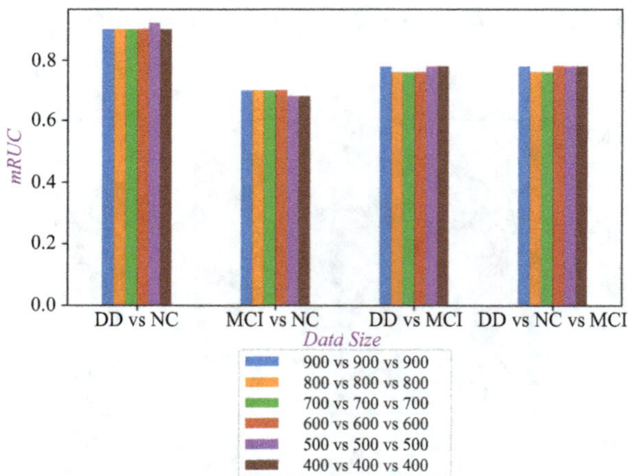

Figure 10.10 Data size variation on TA characteristics

characteristics. The DD is associated with widespread shrinkage of the cerebrum and some specific brain regions, which is characterized by the TA characteristic, according to neurosciences.

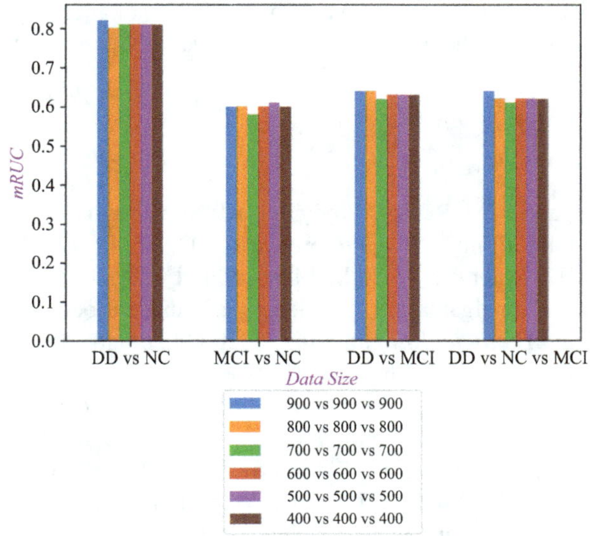

Figure 10.11 Data size variation on TS characteristics

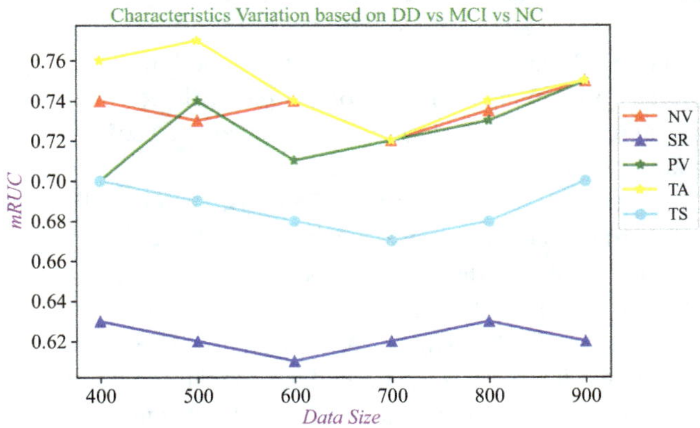

Figure 10.12 Characteristics variations based on prediction models

10.7 Conclusion

This chapter introduces a deep learning model for forecasting the progression of dementia disease. Because the condition is fundamentally progressive, the model takes into account the temporal data gathered from the cases. In contrast to existing methodologies, our model can predict the disease's future condition rather than just categorise the state of a present diagnosis. Experimental results have displayed that our prototype outclasses the vast majority of existing techniques. Furthermore, this method contains a table for various data sizes. Simultaneously, the findings demonstrate that the TA characteristic is a substantial predictor of DD progression.

References

[1] Zedan, M. and F. M. Ramo. "Predicting Alzheimer's disease using grey wolf intelligent algorithm." *Computer Science*, 2020;15(4): 993–1003.

[2] Dua, M., D. Makhija, P. Y. L. Manasa, and P. Mishra. "A CNN–RNN–LSTM based amalgamation for Alzheimer's disease detection." *Journal of Medical and Biological Engineering*, 2020;40(5): 688–706.

[3] Vuong, N., Y. Liu, S. Chan, *et al.* "Deep learning with long short-term memory networks for classification of dementia related travel patterns." In: *2020 42nd Annual International Conference of the IEEE Engineering in Medicine & Biology Society (EMBC)*, IEEE, 2020, pp. 5563–5566.

[4] Miah, Y., C. N. E. Prima, S. J. Seema, M. Mahmud, and M. Shamim Kaiser. "Performance comparison of machine learning techniques in identifying dementia from open access clinical datasets." In: *Advances on Smart and Soft Computing*, Singapore: Springer, 2021, pp. 79–89.

[5] Bansal, D., R. Chhikara, K. Khanna, and P. Gupta. "Comparative analysis of various machine learning algorithms for detecting dementia." *Procedia Computer Science*, 2018;132: 1497–1502.

[6] Enshaeifar, S., A. Zoha, A. Markides, *et al.* "Health management and pattern analysis of daily living activities of people with dementia using in-home sensors and machine learning techniques." *PLoS One*, 2018;13(5): e0195605.

[7] Rishad Ahmed, M. D., Y. Zhang, Z. Feng, B. Lo, O. T. Inan, and H. Liao. "Neuroimaging and machine learning for dementia diagnosis: recent advancements and future prospects." *IEEE Reviews in Biomedical Engineering*, 2018;12: 19–33.

[8] Rodman Shankle, W., S. Mani, M. J. Pazzani, and P. Smyth. "Detecting very early stages of dementia from normal aging with machine learning methods." In: *Conference on Artificial Intelligence in Medicine in Europe*, Berlin, Heidelberg: Springer, 1997, pp. 71–85.

[9] Spooner, A., E. Chen, A. Sowmya, *et al.* "A comparison of machine learning methods for survival analysis of high-dimensional clinical data for dementia prediction." *Scientific Reports*, 2020;10(1): 1–10.

[10] Shah, S. A., J. Ahmad, F. Masood, *et al.* "Privacy-preserving wandering behavior sensing in dementia patients using modified logistic and dynamic Newton Leipnik maps." *IEEE Sensors Journal*, 2020;21(3): 3669–3679.

[11] Usharani, S., P. M. Bala, R. Rajmohan, T. A. Kumar, and S. A. Selvi. "Pregnancy women—smart care intelligent systems: patient condition screening, visualization and monitoring with multimedia technology." In: Tyagi, A. K., A. Abraham, and A. Kaklauskas (eds). *Intelligent Interactive*

Multimedia Systems for e-Healthcare Applications, Singapore: Springer, 2022, https://doi.org/10.1007/978-981-16-6542-4_9.

[12] Swarna, S. R., A. Kumar, P. Dixit, and T. V. M. Sairam. "Parkinson's disease prediction using adaptive quantum computing." In: *2021 Third International Conference on Intelligent Communication Technologies and Virtual Mobile Networks (ICICV)*, 2021, pp. 1396–1401, doi:10.1109/ICICV50876.2021.9388628.

[13] Sasubilli, G. and A. Kumar. "Machine learning and big data implementation on health care data." In: *4th International Conference on Intelligent Computing and Control Systems (ICICCS)*, 2020, pp. 859–864, doi: 10.1109/ICICCS48265.2020.9120906.

[14] Aschwanden, D., S. Aichele, P. Ghisletta, *et al.* "Predicting cognitive impairment and dementia: a machine learning approach." *Journal of Alzheimer's Disease*, 2020;75(3): 717–728.

[15] Castellazzi, G., M. G. Cuzzoni, M. C. Ramusino, *et al.* "A machine learning approach for the differential diagnosis of Alzheimer and vascular dementia fed by MRI selected features." *Frontiers in Neuroinformatics*, 2020;14: 25.

[16] Pavithra, M., R. Rajmohan, T. Ananth Kumar, and R. Ramya. "Prediction and classification of breast cancer using discriminative learning models and techniques. Machine vision inspection systems." *Machine Learning-Based Approaches*, 2021;2: 241–262.

[17] Gokulan, S., S. Narmadha, M. Pavithra, R. Rajmohan, and T. Ananthkumar. "Determination of various deep learning parameter for sleep disorder." In: *2020 International Conference on System, Computation, Automation and Networking (ICSCAN)*, IEEE, 2020, pp. 1–6.

[18] Narmadha, S., S. Gokulan, M. Pavithra, R. Rajmohan, and T. Ananthkumar. "Determination of various deep learning parameters to predict heart disease for diabetes patients." In: *2020 International Conference on System, Computation, Automation and Networking (ICSCAN)*, IEEE, 2020, pp. 1–6.

[19] Usharani, S., S. Prithivi, S. Sharmila, P. M. Bala, T. Ananth Kumar, and R. Rajmohan. "Mobile application for doctor appointment scheduling." In: *2021 International Conference on System, Computation, Automation and Networking (ICSCAN)*, 2021, pp. 1–6, doi: 10.1109/ICSCAN53069.2021.9526398.

Chapter 11

An intelligent agent-based distributed patient scheduling using token-based coordination approach: a case study

E. Grace Mary Kanaga[1], M.L. Valarmathi[2], J. Dhiviya Rose[3] and Lincy Grace Kanagaselvi[4]

Patient scheduling is the process of sequencing and scheduling the patients for different resources in the healthcare domain. This problem is highly complex because multiple constraints and multiple objectives need to be achieved. The schedules may be more or less congested based on the number of available resources. In this Covid-19 pandemic situation, we could witness this. In countries like India in particular, demand often exceeds the availability of resources and greatly increases the waiting time of many patients with medical procedures. Based on the World Health Organization statistics, it is observed that only one doctor is available for every 2,500 people in India, which is far below the international standards. Also, the availability of diagnostic equipment like magnetic resonance imaging scan, computerized tomography scan, ultrasound, electrocardiogram, lab, and other facilities like beds, oxygen, and ventilators are not on par with its population size when compared with other developed countries. However, even in this case, effective scheduling could reduce the patient waiting time and improve hospital resource utilization. Hence, in this chapter, inspired by the real needs in hospital environments an attempt has been made to develop algorithms for effective scheduling and to reduce the waiting time of patients and hence saving their life.

This chapter discusses a new intelligent agent-based distributed patient scheduling using a token-based coordination technique to reduce the waiting time of the patients. This novel approach generates efficient scheduling with reduced message passing and computational complexity in a distributed pattern. JADE is used as an

[1]Department of Computer Science and Engineering, Karunya Institute of Technology and Sciences, Coimbatore, India
[2]Department of Computer Science and Engineering, Dr. Mahalingam College of Engineering and Technology, Pollachi, India
[3]Cybernetics Cluster, School of Computer Science, University of Petroleum and Energy Studies (UPES), Bidholi, India
[4]Department of Information Technology, University of Technology and Applied Sciences-Salalah, Sultanate of Oman

implementation tool and the performance is evaluated on the data set with 10 departments and a varying number of patients. It shows that the total weighted waiting time and the total weighted completion time of the agent-based patient scheduling (APS)-token are reduced by 57.91% and 12.57%, respectively, in comparison with APS-auction. The number of late patients is reduced by 35.71–39.02%. The total idle time of all the resources is less in the token-based approach. There is a drastic reduction in the execution time in the proposed method. and the average reduction factor of execution time is 34.89%. It clearly shows that the APS-token approach gives better performance in less computation time.

11.1 Introduction

Sequencing and scheduling the patients for different departments like labs, computerized tomography (CT), magnetic resonance imaging (MRI), electrocardiography (ECG), electroencephalography, etc. in the healthcare domain is referred to as patient scheduling. By implementing a good patient scheduling support system, the hospital administration can afford its patients a comfort zone by reducing the waiting time and improving the utilization of critical resources. This patient scheduling system is implemented by tracking the availability of the various resources and services provided by the hospital toward the need of utilization by every patient in a dynamic manner. The patient scheduling problem is highly complex because multiple constraints and multiple goals need to be achieved. The constraints include the availability of the resources like doctors and nurses, usage of various medical equipment, usage of laboratory facilities, time gap required between each sample given for testing in labs, etc. The schedules may be more or less congested based on the number of available resources. In countries like India in particular, demand often exceeds the availability of resources and greatly increases the waiting time of many patients with medical procedures. A statistical study states that in a real-time scenario with patient scheduling the time between the appointments and the actual time of arrival difference is larger and unpredictable [1,2]. Predictive scheduling algorithms are often used for static environments, but resequencing is often done to help in recovering the system from outages. Reactive scheduling in dynamic environments often uses various priority rules [3].

11.1.1 Brief introduction to agent paradigm

Agent paradigm provides a new means to implement flexible and intelligent communication systems with an autonomous ability to sense, understand, negotiate, decide, and act. Agent technologies are a natural extension of current component-based methodologies, and this area of computer science is now one of the most vibrant and fascinating fields of research. Software agents offer a powerful new way to build next-generation information systems. Agent-based computing is a potential way of creating complicated domain applications.

Agents have a wide range of applications in this fast-paced technological era. In most circumstances, a single agent will not be able to accomplish the complete goal on their own. As a result, a multi-agent system (MAS) is more appropriate for complicated applications. In most cases, MASs are employed to solve problems that are too big for a single centralized agent to handle. MASs are environments in which interacting intelligent agents or actors exist. MASs have the following characteristics:

- Each agent has inadequate information or capabilities for solving the problem, resulting in a limited viewpoint.
- There is no system-wide control.
- Data is decentralized.
- Computation is asynchronous.

11.1.2 Patient scheduling

Patient scheduling is an inevitable aspect of hospital management. Figure 11.1 depicts the typical scheduling pattern in the hospital. The patient flow is an iterative model and hospitals need to have coordination among the staff and resources in an effective manner to meet the need of a patient. Scheduling is a primary area in the healthcare domain that operates toward improving facilities for the patients by reducing the waiting time and proper usage of healthcare resources like doctors, labs, stores, testing centers, etc. In a normal scenario, a patient after getting suggestions from the doctor in the primary checkup inside the hospital is taken to multi-counters wherein every counter he/she has to wait and go with individual order if the patient scheduling is not handled properly.

Patient scheduling is stochastic by itself as scheduling the current patients on day to day basis along with appointments scheduled for the future period which in turn is timebound [4]. Yan *et al.* [5] proposed a system for managing the scheduling of patient appointments. Extending the consideration of the future visit and doctors'

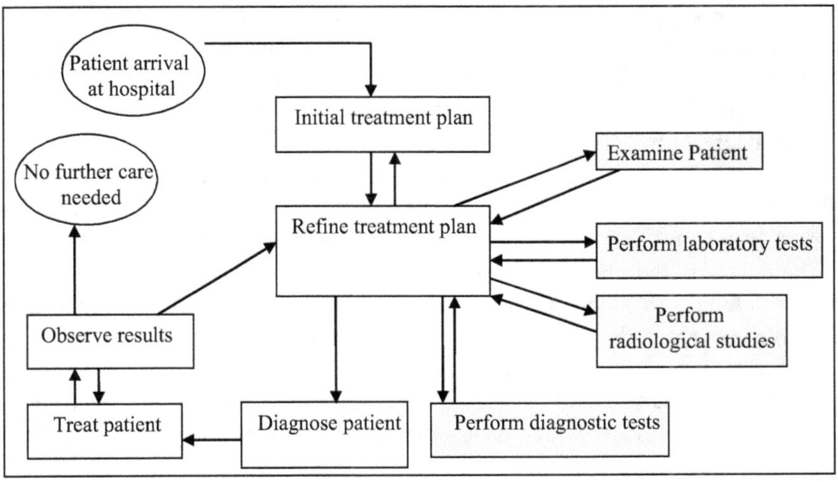

Figure 11.1 A model for patient care

preferences has been a promising insight in the development that eases humans by reducing the waiting time. According to statistics, World Health Organization (WHO) said that every 2,500 people in India have a doctor, which is far below the international standards. Also, the availability of other medical equipment like MRI, CT scans, ultrasound, ECG, lab, and other diagnostic equipment is not on par with its population size when compared with other developed countries. Not only the various hardware resources are to be considered for scheduling, but the human resources like nurses, attenders, doctors, etc. are also one of the high priorities in this COVID pandemic season, which made mankind towards more improvements in the healthcare domain [6]. As scheduling is a highly time complex problem in the hospital domain, it is impossible to make a quantitative and qualitative schedule by hand or even with exact methods. Hence a patient scheduling system to find an adequate solution in a limited time is a chief requirement in this domain. However, even in this case, effective scheduling of patients and resources is used to reduce the patient waiting time and improve hospital resource utilization.

11.1.3 Agent-based patient scheduling

Scheduling in general refers to the notion of assigning time slots and resources to a given set of tasks. Scheduling theory is concerned with allocating limited resources to activities in the most efficient way possible over time. Jobs must be scheduled on machines while adhering to particular constraints in order to optimize some objective functions. A hospital is an institution where the sick or injured are given medical or surgical care. Because of the fundamentally dynamic and distributed characteristics of hospitals, as well as the ineffectiveness of numerous electronic scheduling techniques, current research has expanded on the idea of using software agents to tackle the scheduling problem and it has been adopted by Vermeulen *et al.* [51]. Agents are an obvious choice since they are well-known for accurately representing a dispersed environment and continuously adapting to varied schedules in response to a changing environment. Agents can also express unique limits and preferences, which is important in a lot of hospital scheduling issues. Decker and Li [52] provide an agent-based strategy to deal with preferences and constraints in hospital scheduling.

Real-time hospitals adopt manual scheduling that includes open booking, double booking, and time-specified scheduling for booking the slots for patients based on first come first serve algorithms. In appointment scheduling, the patients visit the doctor in the given timeslot as per the appointment, fixed by the personal assistant to the doctor. However, this becomes ineffective if some high-emergency critical patients reach the hospital and need immediate attention. In such a scenario, the problem is to be revamped and circulated again to all the stakeholders. Moreover, the integration of the hospital scenario into the computerized system started two decades back, which paved the way for the use of various recent technologies in achieving efficiency [7].

Recent research has given new life to the idea of using software agents to solve the scheduling challenge in the healthcare industry. In the scheduling problem, a schedule must be created in a specified period of time, and all of the agents involved must agree with the schedule that is created. Agents are a natural choice because they are recognized to accurately reflect a distributed environment and can also coordinate with one another, which is useful in a dynamic environment [8]. The social abilities allow agents to engage with one another in order to achieve their own goals, which aids in the realization of adaptability and dynamism. Specific limitations and preferences can be represented by agents as goals and plans, which play an important part in many hospital scheduling issues. Agents with this attribute produce highly fault-tolerant and flexible systems. Agents can agree through auctions, negotiations, or reasons, or a combination of these methods [9]. The environment in our problem domain is dynamic, distributed, and decentralized, so agents need to work together and be coordinated to arrive at a solution [3]. To achieve coordinated distributed problem-solving, agents collaborate by exchanging tasks or results. The nonlinear programming-hard scheduling has also been proved to be solved with the help of software agents [10]. The agents' environment is classified into fully observable vs. partially observable depending on the input data, deterministic vs. stochastic depending on the outcome of an action, episodic vs. sequential depending on the dependency between actions, static vs. dynamic depending on the predictability, and single agent vs. multi agent depending on the number of agents in action.

The in-demand computer science technology MASs have the capability of social coordination with the interacting nature [7]. This software program is capable of acting on behalf of the user. The MASs work in a heterogeneous environment where various types of software agents collaborate and communicate. This chapter describes the various application of agents in the healthcare sector with the involvement of the same in scheduling. An algorithm for an effective scheduling case study in hospitals with the metric of evaluations is to reduce the waiting time of patients and gain patient satisfaction. It discusses the originality of the auction based on tokens used to manage the divorce resource and optimize the processing time for the patient agents (PAs) and the resource agents (RAs). The simulation runs conducted to prove the effectiveness of the proposed system with the various metrics used in our approach are also discussed.

11.2 Context of study and problem description

There is growing interest in the problem of agent-based technologies applied to the medical field. Across the globe, many researchers have discussed the applications of agents in healthcare systems [8,11–13], and various agent-based approaches to solving hospital scheduling with preferences and certain constraints have also been developed [14–18] over a decade. The scheduling problem in hospitals is decentralized and heterogeneous. Among the various techniques adopted, the auction-based approach has given a high positivity rate in such distributed environments but with the message and computational complexities [19–22].

11.2.1 Application of agents in healthcare

Agents work by allowing people to assign their tasks to them. Software agents are programs that help and act on behalf of people. They are capable of working with various computer operating systems, networks, and databases, among other things. MASs have been mostly utilized in healthcare to handle real-time scheduling, modeling, and planning difficulties. Its nature-like distributed, artificial reasoning and negotiation with the potential to govern the system toward the goal make the system unique and powerful [23]. Several authors have proposed effective systems using multi-agents in healthcare that have decreased the life complexity of patients in real-world survival. An intelligent ambient system is addressed by creating a good ambiance in the hospitals based on the input from various sensors and controlling through the MASs [24]. The sensor combined with multi-agents has been used in intelligent crop irrigation systems [25] and other systems [26]. MASs is a system created by Jemal *et al.,* where the agents in the teamwork exchange data to find a suitable schedule to fix the appointments [27]. Aldea *et al.* found a system to search for a suitable donor for the transplantation of organs and streamline the process of donation and transplant donor using MASs in Spain [28]. An organization has developed a system AIDA that helps in monitoring the health informatics that manages through the agents [29] which has equipped itself with intelligent agents to solve the interoperability problems. In Kenya, an agent-based surveillance system has been proposed that helps in keeping track of the patients' records and continuous surveillance that had led to the improvement in the healthcare scenario [30]. In France, an agent-based orchestration system HOST has been proposed and found to be a success in controlling the patient flow in hospitals [31].

Two decades before when the research on the MASs started, a system for managing the patients and the consultant was developed and proved to be efficient for managing the diabetic patients [32]. The planning of the schedule is performed using the MASs across the various nodes using the coordination mechanism, which is extended by providing a coordination mechanism for resource sharing across nodes. A roster-based schedule is prepared for the nurses in the hospital which gave high feedback on the system [26]. In this paper, a bidding strategy is used as the coordination mechanism for making the global schedule after matching with the local schedule [33]. In healthcare, agents are not only used for record-keeping but also used for creating intelligent systems with the help of sensors as input. The VEPR system proposed has integrated the data with the help of agents technologies where the database is able to be mined for a year [33]. This type of system can be enhanced for future research by using the current data analytics algorithm to find out various new patterns of the patients and schedule them accordingly.

In the system framed by Spyropoulus *et al.*, the hospital infrastructure is classified into four groups. Employees, which include physicians of various specialties, nurses, specific technicians, and other support personnel such as nurse scheduling and rostering, are the first group. When it comes to nurse-rostering issues, the nurses' preferences, legal rules, personnel policies, and other hospital-specific criteria should all be taken into account, since these constraints are much more difficult to fix. Critical care units are the second type, which includes the extensive infrastructure needed to support the monitoring and treatment of intensive care patients. Several factors should be examined in this group, including the number of critical care units, emergency patient arrivals at random times, and the length of stay of patients. Surgical operation units are the third group with specialized equipment, tools, and procedures for various surgeries. One of the primary goals is to reduce surgical procedure waiting lists. Each department also has its own regulations and preferences for patient scheduling and resource allocation [34].

11.2.2 Application of agents in scheduling

MAS comprises many independent intelligent agents that cooperate, coordinate, and plan to achieve a single goal. While coordination, negotiation is a process that needs some optimization in achieving a common goal where every agent will go through a situation in which it has to coordinate and come to conclude with what to do and make plans on which it can act social responsibility. According to Stefano *et al.,* coordination can be best explained as the process by which they manage interdependencies between the agent's activities with autonomy [35]. A value-driven framework for the coordination of various intelligent agents was proposed where based on the values introduced by the local agent the overall system generates its plan focusing on the goal [36]. Recent advances in computing technology have made it possible to implement more sophisticated algorithms for production optimization and performance. Few researchers suggest various algorithms like genetic algorithms [37,38], ant colony optimization, swarm intelligence, and many more are used in solving search problems based on biological evolution at the background framework of MASs to achieve a good schedule [39].

The scheduling agent develops an optimal schedule for the various proposals by obtaining the treatment plan for each patient from the common agent (CA) in a short period of time. Arthur and Sankaranarayanan [40] have also used fuzzy logic in developing MAS that functions on preference rules, to make a proper decision in scheduling the appointments of the patient with the doctors. Many heuristic optimization techniques have benefited time management in constructing an efficient schedule in such very large combinatorial search problems as patient scheduling [41]. For resolving the conflicts among the paths, the usage of heuristics algorithms has shown good results in MASs [42,43]. It has also been proved that using this algorithm in the dynamic domain for generating optimal schedules with minimal time and hardware/software requirements has improved the quality of livelihood.

11.2.3 MAS toward coordination

Coordination is the process of mediation that the heterogeneous agents take to achieve a common task. If the agents work in a distributed environment many factors influence the coordination mechanism leading to technical challenges. The process of coordination is considered a two-step process where during the first step the dependencies are identified followed by actions and plans that can be shared or common [44]. The coordination of a MAS includes signal broadcasting, simple auction, and negotiation. In signal broadcasting, the schedule is communicated to all members and each agent can plan accordingly to the broadcast schedule. However, this technique is less expensive and lacks dynamic coordination among the agents. Auction is another coordination mechanism where bidding takes place with a utility score and the agent that owns the bid is responsible for generating the plans. Further, in research, this auction is added with various negotiation algorithms that gain improvement in the performance of the system. Multi-agent negotiation is one of the coordination mechanisms in which the heterogeneous agents follow to arrive at and complete a particular task with a win–win strategy among the agents. The important metrics used in finding the best negotiation methods are with less execution time and resource consumption in achieving a task [45]. Paulussen *et al.* developed MAS which has been tested in a German hospital and it has shown a significant time for negotiations at 1.500 ms with a standard deviation of 5 ms even at the time of high load and medium load [46]. The token-based approach has been identified as one of the effective distributed coordination mechanisms for a scalable environment that reduces the routing time [16].

11.3 Related work

11.3.1 Token as a coordination mechanism

The study on the recent successful token-based algorithms for task allocation has been developed for the effective coordination of various resources in the hospital and to allocate patients to these resources in an optimal way [17]. Token-based systems are used in the supply chain to manage the demands to reduce the bullwhip effect using multi-agents [47]. Traffic scheduling systems are also developed with the help of tokens to avoid congestion in MASs [48]. In power supply energy management, MASs are used and tested for coordinating among the production unit, distribution, and demand units. The demand units reciprocate their value as the token has yielded a good response in the bulk supply chain management [42,49].

With this objective, the system has proved to eliminate myopia of decision-making while maintaining the response of the system to a real and dynamic hospital environment. The token-based approach increases the visibility of the RA by incorporating useful information in the token based on the global and future conditions. Based on the study made on the recent successful token-based algorithms for task allocation, this chapter discusses a new distributed token-based coordination technique using multi-agents. The proposed token-based coordination mechanism adopts the conventional way of scheduling used in hospitals based on a market-based strategy [50]. This approach generates efficient scheduling with reduced message passing and computational complexity in a distributed pattern. This system operates on top of the conventional process of hospital scheduling that increases the RA's visibility by incorporating useful information in the token based on global and future conditions. The chief objective is to develop an efficient distributed dynamic patient scheduling system using a token-based coordination approach. This framework aims to minimize the stay time/waiting time for patients in the hospital. This approach also strives to optimize resource utilization in hospitals without violating the resource constraints. The coordination mechanism aims to be effective in minimizing the communication overhead.

11.3.2 Agent-based patient scheduling using token-based coordination

The framework of the token-based approach for patient scheduling is depicted in Figure 11.2. Traditional interfaces, which rely on users to transfer manual data and execute any problem-solving step by hand, would be unable to keep up with the increasing complexity of scenarios faced by both patients and health care practitioners. The suggested method explains how the elements will result in a new generation of patient programming that rationalizes user interfaces and enables more complicated communication and problem-solving. Intelligent agent software is a new type of computer software that intends to bring aspects like lesson, energy, heuristic problem-solving, goal action, and communication to the table [3]. Diagnostic units, X-rays, and other testing labs are shown as agents which collaborate round the clock to ensure efficient use of hospital resources for enhancing patient care and minimizing anxiety and conflicts among staff and family members.

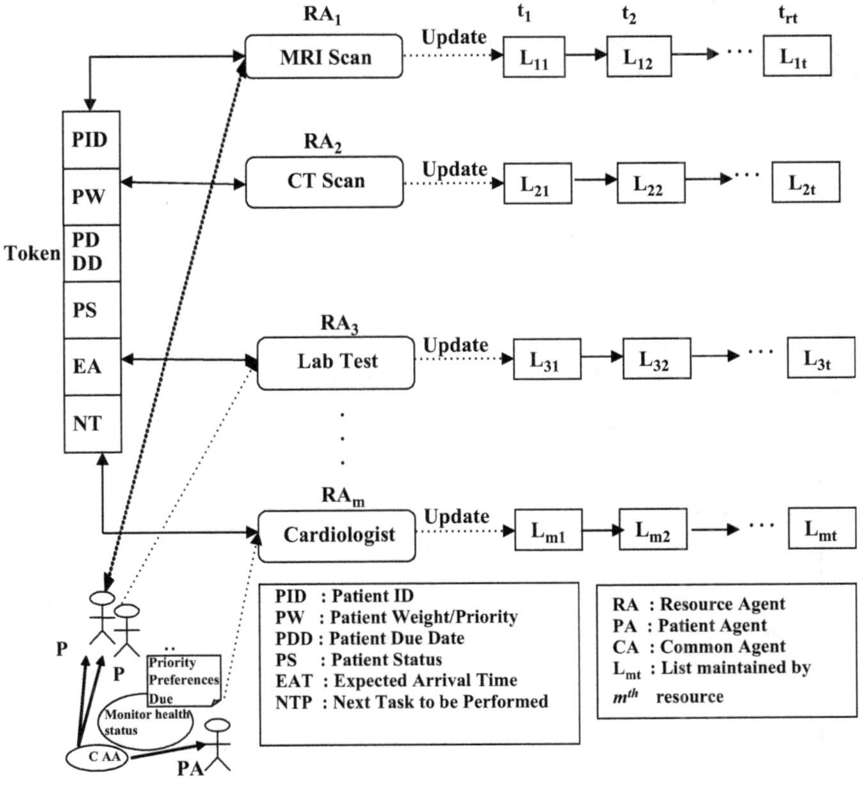

Figure 11.2 Framework of token-based approach for patient scheduling

The new multi-agent paradigm has a lot of potential for assisting hospital administration and patients in resolving conflicts that develop during patient scheduling.

This framework contains three types of agents, namely, PA, RA, and CA. But this approach gives a new way of coordination based on tokens. The token contains the following information like patient ID, patient priority, due date, the status of the patient, expected arrival time, and next task to be performed. The above-said information is encapsulated in the token. These tokens are passed on to the corresponding resource where the patient needs to be processed next. Each RA m maintains a priority list, L_{m1}, $L_{m2}, \ldots, L_{mt}$, that is sorted with a timeline, and m and t are the indices of the resource and time slot, respectively. Each priority list is composed of a list of patients ordered based on their priority/weight. The weight of the patients varies from 1 to 5. Higher weight values are assigned to the patient with higher priority (i.e., in a poor health state). The weight values are assigned by the CA. Once a RA receives a token, it updates its list based on the information present in the token. The information present in the token gives a look-a-head view of each RA, leading to the better allocation of patients to the resources. The description of a single priority list is shown in Figure 11.3.

11.3.3 Algorithm for updating the nonlocal viewpoints of the resource

The RA_m updates its nonlocal viewpoint by receiving tokens from other resources and modifies the priority list L_{mt}. The algorithm for updating the nonlocal viewpoint of the agent is shown in Figure 11.4.

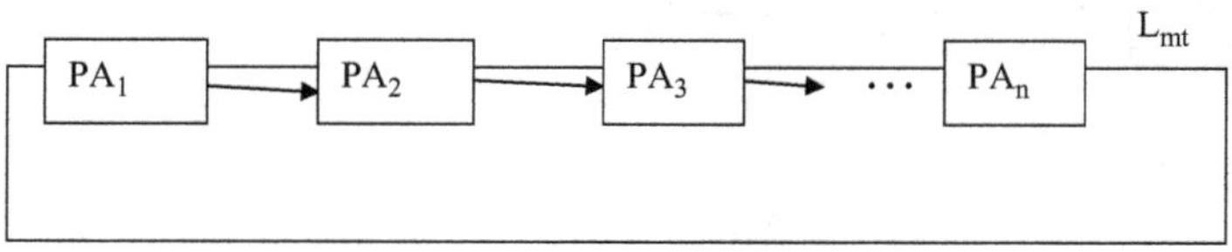

Figure 11.3 Priority list maintained by RA m at time slot t

Step 1: Receive the tokens

Step 2: Extract the information from the tokens

Step 3: Update the priority list based on the expected arrival time and priority

Step 4: Process the task of the patient based on the updated priority list

Step 5: Once the process started, update the expected due date, status and the next task to be performed fields in the token.

Step 6: Send the token to the resource where the next task to be performed.

Figure 11.4 Algorithm for token-based approach

11.4 Model implementation and validation

The proposed solution is implemented in the JADE platform, which is a Java-based software framework. It facilitates the development of MASs by providing a middleware that adheres to the foundation for intelligent physical agents guidelines, as well as a set of tools for debugging and deployment. All simulations are run on a system that has the following configurations: 2.2 GHz Intel Core 2 Duo processor, 2 GB DDR2, and 320 GB hard drive.

11.4.1 Performance metrics

When it comes to scheduling issues, there are various performance metrics to consider. The objective of the patient scheduling problem is to reduce patient waiting times and increase hospital resource usage. The performance measures listed below can be used to assess the patient scheduling issue. All of these metrics values will be effectively minimized by a suitable scheduling approach.

- **Total weighted waiting time** ($\sum w_i T_i$) is the most important metric for assessing the patient scheduling issue. It displays whether the patient's waiting time is reduced by considering the patient's priority.
- **Total waiting time** ($\sum T_i$) is the overall total waiting time of all the patients.
- **Maximum waiting time** ($T_{\max}$) refers to the maximum lateness of all patients with a value equal to the completion time minus the due time for each patient. It illustrates the worst-case scenario for how long a patient might be delayed in the worst-case scenario.
- **Maximum completion time or makespan** ($C_{\max}$) is the maximum completion time for all the patients. In terms of patient scheduling, it refers to the time the last patient checks out and the time all the tasks in their diagnosis are completed.
- **Total number of patients scheduled late** ($\sum U$) is a good measure of how many patients have to wait beyond their scheduled time.
- **Total completion (flow) time** ($\sum C_I$) refers to the maximum lateness of all patients, with a value equal to the completion time minus the due time for each patient. It illustrates the worst-case scenario for how long a patient might be delayed in the worst-case scenario.
- **Total weighted completion (flow) time** ($\sum w_i C_i$) includes the priority of the patient when calculating the total completion time.

11.4.2 Comparison of results

The agent-based patient scheduling (APS) with token-based coordination is tested on the set of data used with 10 resources and a varying number of patients. The computational results of the APS-token are given in Tables 11.1 and 11.2 and compared with the APS-auction-based technique proposed in the literature (Grace Mary Kanaga *et al.*, 2009, 2010). The performance metrics based on waiting time are given in Table 11.1 and the performance metrics based on completion time are shown in Table 11.2.

A perusal of Figures 11.5 and 11.6 clearly shows that the total weighted waiting time and the total weighted completion time of APS-token are reduced by 57.91 and 12.57%, respectively, when compared to APS-auction. But when the maximum waiting time is considered, it gets increased in a token-based approach. Table 11.3 and Figure 11.7 give the comparison of the number of late patients for APS-auction and APS-token methods.

There is a considerable percentage of reduction in the number of late patients in the token method when compared to the auction mechanism. There is a reduction

Table 11.1 Comparison of performance metrics based on waiting time

Performance metrics (in min)	Scheduling techniques	
	APS-auction	APS-token
Max. waiting time	40	56
Total waiting time	645	474
Total weighted waiting time	1542	649

Table 11.2 Comparison of performance metrics based on completion time

Performance metrics (in min)	Scheduling techniques	
	APS-auction	Token
Max. completion time	87	85
Total completion time	2393	2171
Total weighted completion time	6564	5739

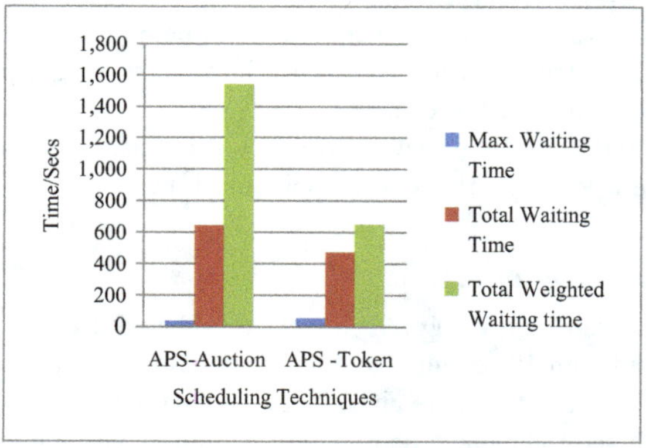

Figure 11.5 Comparison of performance metrics based on waiting time

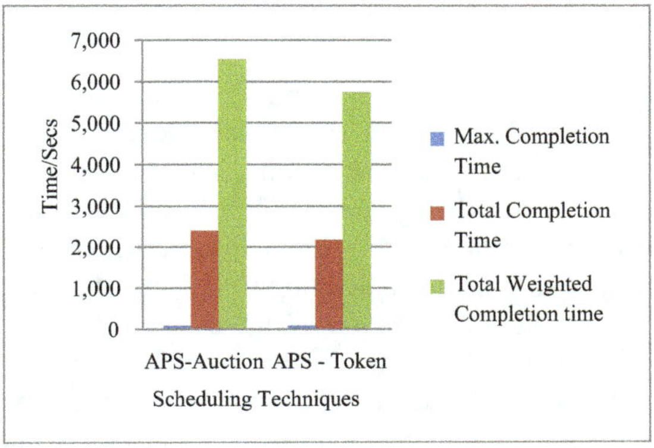

Figure 11.6 Comparison of performance metrics based on completion time

Table 11.3 Comparison of the number of late patients

Number of patients	**Number of late patients**		**Performance improvement in percent**
	APS-auction	**APS-token**	
10	2	2	0.00
20	14	9	35.71
30	22	14	36.36
40	34	19	44.12
50	41	25	39.02

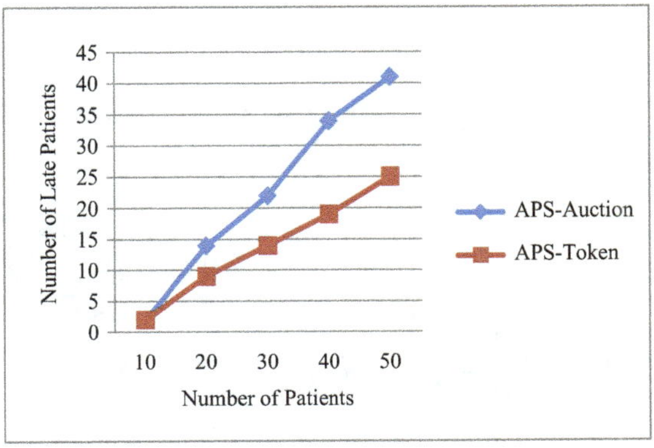

Figure 11.7 Comparison of the number of late patients

of 35.71% for 20 patients, and even for 50 patients, there is a reduction of up to 39.02%.

The idle time of each resource based on APS-auction and APS-token is tabulated in Table 11.4, and Figure 11.8 shows that the total idle time of all the resources is less in the token method when compared to the auction-based method. Even though the idle time of the resources R2, R7, R9, and R10 increases in the token-based method, the overall idle time of the resources gets reduced by 6.56% in the APS-token method, and hence in addition to reducing the waiting time of the patients, it also provides better utilization of the resources.

Table 11.5 and Figure 11.9 show the comparison of the execution time in auction-based and token-based patient scheduling. There is a drastic reduction in the execution time in the token method when compared to the auction-based technique. The reduction of execution time varies with the number of patients. The

Table 11.4 Comparison of resource idle time

Resources	**Resource idle time**	
	APS-auction	**APS-token**
R1	37	10
R2	51	54
R3	37	9
R4	10	1
R5	56	54
R6	9	10
R7	21	36
R8	24	13
R9	39	45
R10	36	67
Total idle time	320	299

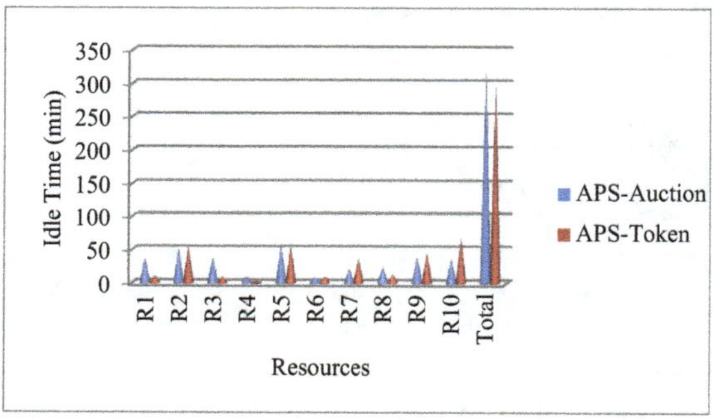

Figure 11.8 Comparison of resource idle time

Table 11.5 Comparison of execution time

Number of patients	Execution time in seconds		Performance improvement in percent
	APS-auction	APS-token	
10	31	12	61.29
20	47	23	51.06
30	70	56	20.00
40	80	63	21.25
50	120	95	41.67

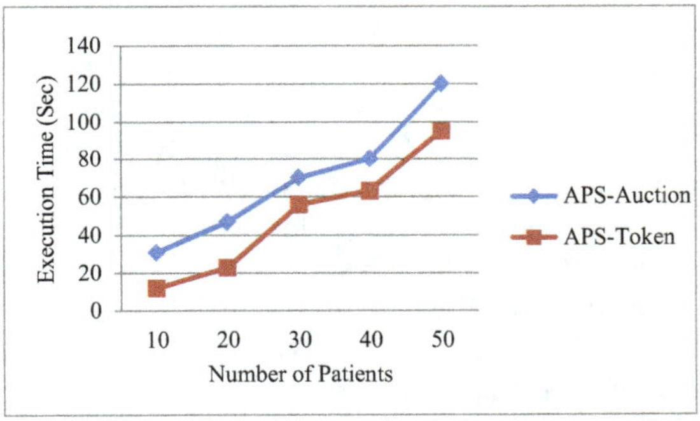

Figure 11.9 Comparison of execution time

average reduction factor of execution time is 34.89%. It clearly shows that the APS-token method gives better performance in less computation time.

11.5 Conclusion

A new token-based coordination technique for patient scheduling has been proposed and implemented in this paper. The results obtained show that this method gives better performance when compared to the traditional methods that are used currently. The chief performance metric is the total weighted waiting time and based on that metric, the token-based approach ranks the first. The prominent features of the token-based approach are:

- It handles the distributed system well.
- It handles emergencies and unforeseen disturbances in the hospital scheduling domain.

Patient scheduling is a dynamic problem, and a good solution will improve society's quality of life. A schedule that reduces the amount of time a patient has to wait could even save their life.

References

[1] X. Pan, N. Geng, and X. Xie, "Appointment scheduling and real-time sequencing strategies for patient unpunctuality," *Eur. J. Oper. Res.*, 295(1): 246–260, 2021, doi: 10.1016/j.ejor.2021.02.055.

[2] N. Benhajji, D. Roy, and D. Anciaux, "Patient-centered multi agent system for health care," *IFAC-PapersOnLine*, 28(3): 710–714, 2015, doi: 10.1016/j.ifacol.2015.06.166.

[3] D. Borrajo and S. Fernández, "Efficient approaches for multi-agent planning," *Knowl. Inf. Syst.*, 58(1): 425–479, 2019, doi: 10.1007/s10115-018-1202-1.

[4] A. Sauré, M. A. Begen, and J. Patrick, "Dynamic multi-priority, multi-class patient scheduling with stochastic service times," *Eur. J. Oper. Res.*, 280(1): 254–265, 2020, doi: 10.1016/j.ejor.2019.06.040.

[5] C. Yan, G. G. Q. Huang, Y. H. Kuo, and J. Tang, "Dynamic appointment scheduling for outpatient clinics with multiple physicians and patient choice," *J. Manag. Sci. Eng.*, 1–17, 2021, doi: 10.1016/j.jmse.2021.02.002.

[6] M. Qaisari, H. Abadi, S. Rahmati, A. Sharifi, and M. Ahmadi, "HSSAGA: designation and scheduling of nurses for taking care of COVID-19 patients using novel method of hybrid salp swarm algorithm and genetic algorithm," *Appl. Soft Comput.*, 108: 107449, 2021, doi: 10.1016/j.asoc.2021.107449.

[7] J. M. Corchado, D. I. Tapia, and J. Bajo, "A multi-agent architecture for distributed services and applications," *Int. J. Innov. Comput. Inf. Control*, 8 (4): 2453–2476, 2012.

[8] J. Xie and C. C. Liu, "Multi-agent systems and their applications," *J. Int. Counc. Electr. Eng.*, 7(1): 188–197, 2017, doi: 10.1080/22348972.2017.1348890.

[9] M. Mokni, J. E. Hajlaoui, and Z. Brahmi, "MAS-based approach for scheduling intensive workflows in cloud computing," *Proc. 2018 IEEE 27th Int. Conf. Enabling Technol. Infrastruct. Collab. Enterp. (WETICE 2018)*, pp. 1–2, 2018, doi: 10.1109/WETICE.2018.00010.

[10] S. R. Swarna, A. Kumar, P. Dixit, and T. V. M. Sairam, "Parkinson's disease prediction using adaptive quantum computing," *2021 Third International Conference on Intelligent Communication Technologies and Virtual Mobile Networks (ICICV)*, pp. 1396–1401, 2021, doi:10.1109/ICICV50876.2021.9388628.

[11] G. Sasubilli and A. Kumar, "Machine learning and big data implementation on health care data," *4th International Conference on Intelligent Computing and Control Systems (ICICCS)*, pp. 859–864, 2020, doi: 10.1109/ICICCS48265.2020.9120906.

[12] M. Reid, "Multi-agent system applications in healthcare: current technology and future roadmap," *Procedia Comput. Sci.*,52: 252–261, 2015, doi: 10.1016/J.PROCS.2015.05.071.

[13] E. Kiesling, M. Günther, C. Stummer, and L. M. Wakolbinger, "Agent-based simulation of innovation diffusion: a review," *Cent. Eur. J. Oper. Res.*, 20 (2): 183–230, 2012, doi: 10.1007/s10100-011-0210-y.

[14] I. Vermeulen, S. Bohte, K. Somefun, and H. La Poutré, "Multi-agent pareto appointment exchanging in hospital patient scheduling," *Serv. Oriented Comput. Appl.*, 1(3): 185–196, 2007, doi: 10.1007/s11761-007-0012-1.

[15] K. Mustapha, Q. Gilli, J. M. Frayret, and N. Lahrichi, *"Agent-Based Modelling and Simulation Framework for Health Care,"* Springer, vol. 676: pp. 171–197, 2018.

[16] A. Ghosh, E. A. Stepanov, J. M. Mayor Torres, M. Danieli, and G. Riccardi, "HEAL: a health analytics intelligent agent platform for the acquisition and analysis of physiological signals," *2018 IEEE 20th Int. Conf. e-Health Netw., Appl. Serv. Heal.*, 2018, doi: 10.1109/HealthCom.2018.8531120.

[17] D. G. P. Y. De Silva and P. Ravindra, "Intelligent agent to negotiate on goal oriented conversations," *Proc. Int. Conf. Image Process. Robot. (ICIPRoB 2020)*, 2020, doi: 10.1109/ICIP48927.2020.9367345.

[18] A. Degas, E. Kaddoum, M. P. Gleizes, F. Adreit, and A. Rantrua, "Cooperative multi-agent model for collision avoidance applied to air traffic management," *Eng. Appl. Artif. Intell.*, 102: 104286, 2021, doi: 10.1016/j.engappai.2021.104286.

[19] J. A. Arauzo, J. M. Galán, J. Pajares, and A. López-Paredes, "Multi-agent technology for scheduling and control projects in multi-project environments. An auction based approach," *Intel. Artif.*, 13(42): 12–20, 2009, doi: 10.4114/ia.v13i42.1042.

[20] S. Adhau, M. L. Mittal, and A. Mittal, "A multi-agent system for distributed multi-project scheduling: an auction-based negotiation approach," *Eng. Appl. Artif. Intell.*, 25(8): 1738–1751, 2012, doi: 10.1016/j.engappai.2011.12.003.

[21] C. Zeng, J. Tang, Z. Fan, and C. Yan, "Auction-based approach for a flexible job-shop scheduling problem with multiple process plans," *Eng. Optim.*, 51 (11): 1902–1919, 2019, doi: 10.1080/0305215X.2018.1561884.

[22] M. Asghari, D. Deng, C. Shahabi, U. Demiryurek, and Y. Li, "Price-aware real-time ride-sharing at scale: an auction-based approach," *GIS Proc. ACM Int. Symp. Adv. Geogr. Inf. Syst.*, 2016, doi: 10.1145/2996913.2996974.

[23] E. Shakshuki and M. Reid, "Multi-agent system applications in healthcare: current technology and future roadmap," *Procedia Comput. Sci.*, 52(1): 252–261, 2015, doi: 10.1016/j.procs.2015.05.071.

[24] R. Cruz-Correia, P. Vieira-Marques, P. Costa, *et al.*, "Integration of hospital data using agent technologies: a case study," *AI Commun.*, 18(3): 191–200, 2005.

[25] G. Villarrubia, J. F. De Paz, D. H. De La Iglesia, and J. Bajo, "Combining multi-agent systems and wireless sensor networks for monitoring crop irrigation," *Sensors (Switzerland)*, 17(8), 2017, doi: 10.3390/s17081775.

[26] S. J. Mason, R. R. Hill, L. Mönch, O. Rose, and M. Carlo *et al.*, "Proceedings of the 2008 Winter Simulation Conference," In: S. J. Mason, R. R. Hill, L. Mönch, O. Rose, T. Jefferson, J. W. Fowler (eds.), *Proc. 2008 Winter Simul. Conf.*, pp. 695–702, 2008.

[27] H. Jemal, Z. Kechaou, M. Ben Ayed, and A. M. Alimi, “A multi agent system for hospital organization,” *Int. J. Mach. Learn. Comput.*, 5(1): 51–56, 2015, doi: 10.7763/IJMLC.2015.V5.482.

[28] A. Aldea, B. López, A. Moreno, D. Riaño, and A. Valls, “A multi-agent system for organ transplant co-ordination,” *Lect. Notes Comput. Sci. (including Subser. Lect. Notes Artif. Intell. Lect. Notes Bioinformatics)*, 2101: 413–416, 2001, doi: 10.1007/3-540-48229-6_56.

[29] R. Sousa, D. Ferreira, A. Abelha, and J. Machado, “Step towards monitoring intelligent agents in healthcare information systems,” *Adv. Intell. Syst. Comput.*, 1161: 510–519, 2020, AISC, doi: 10.1007/978-3-030-45697-9_50.

[30] B. W. Peter and N. A. Wechuli, “Multi-agent based surveillance system for diseases,” *Int. J. Trend. Res. Dev.*, 3(6): 286–289, 2016.

[31] S. Ben Othman and S. Hammadi, “Driven workflow orchestration of patient pathway in hospital emergencies,” *Proc. of Inmeta*, pp. 1–11, 2016.

[32] G. Ganendran, Q. Tran, P. Ganguly, P. Ray, and G. Low, “An ontology-driven multi-agent approach for healthcare,” *Proc. of the Health Informatics Conference (HIC2002)*, pp. 1–7, 2002.

[33] U. Deshpande, A. Gupta, and A. Basu, “Coordinated problem solving through resource sharing in a distributed environment,” *IEEE Trans. Syst. Man. Cybern. B Cybern.*, 34(2): 1299–1304, 2004.

[34] C. C. Marinagi, C. D. Spyropoulos, C. Papatheodorou, and S. Kokkotos, “Continual planning and scheduling for managing patient tests in hospital laboratories,” *Artif. Intell. Med.*, 20(2): 139–154, 2000, doi: 10.1016/S0933-3657(00)00061-0.

[35] S. Mariani, G. Cabri, and F. Zambonelli, “Coordination of autonomous vehicles: taxonomy and survey,” *ACM Comput. Surv.*, 54(1): 1–34, 2021, doi: 10.1145/3431231.

[36] W. Jiao and Y. Sun, A values-driven self-organization mechanism for automating multiagent coordination, *Int. J. Comput. Intell.*, 34(2):635–678, 2018.

[37] M. Gen, J. Gao, and L. Lin, “Multistage-based genetic algorithm for flexible job-shop scheduling problem,” *Stud. Comput. Intell.*, 187: 183–196, 2009, doi: 10.1007/978-3-540-95978-6_13.

[38] V. Podgorelec and P. Kokol, “Genetic algorithm based system for patient scheduling in highly constrained situations,” *J. Med. Syst.*, 21(6): 417–427, 1997, doi: 10.1023/A:1022828414460.

[39] C. F. Chien, F. P. Tseng, and C. H. Chen, “An evolutionary approach to rehabilitation patient scheduling: a case study,” *Eur. J. Oper. Res.*, 189(3): 1234–1253, 2008, doi: 10.1016/j.ejor.2007.01.062.

[40] H. Arthur III and S. Sankaranarayanan, “Application of intelligent agents in hospital appointment scheduling system,” *Int. J. Comput. Theory Eng.*, 4(4): 625–630, 2012, doi: 10.7763/ijcte.2012.v4.545.

[41] E. Zitzler and L. Thiele, “An evolutionary algorithm for multiobjective optimization: the strength pareto approach,” Computer Engineering and Communication Networks Lab (TIK), Report no. 43, pp. 1–43, 1998.

[42] A. Felner, J. Li, E. Boyarski, *et al.*, "Adding heuristics to conflict-based search for multi-agent path finding," *Proc. Int. Conf. Autom. Plan. Sched. (ICAPS)*, vol. 2018(June), pp. 83–87, 2018.

[43] J. Li, E. Boyarski, A. Felner, H. Ma, and S. Koenig, "Improved heuristics for multi-agent path finding with conflict-based search: preliminary results," *Proc. 12th Int. Symp. Comb. Search, SoCS 2019*, pp. 182–183, 2019.

[44] H. Billhardt, R. Centeno, C. E. Cuesta, *et al.*, "Organisational structures in next-generation distributed systems: Towards a technology of agreement," *Multiagent Grid Syst.*, 7(2–3): 109–125, 2011, doi: 10.3233/MGS-2011-0168.

[45] B. Goertzel, C. Pennachin, and N. Geisweiller, "A formal model of intelligent agents." In: *Engineering General Intelligence, Part 1*. Atlantis Press, Paris, vol. 5, pp. 157–172, 2014.

[46] A. Zöller, L. Braubach, A. Pokahr, *et al.*, "Evaluation of a multi-agent system for hospital patient scheduling," *Int. Trans. Syst. Sci. Appl.*, 1(4): 375–380, 2006.

[47] T. Moyaux, B. Chaib-Draa, and S. D'Amours, "Multi-agent coordination based on tokens: reduction of the bullwhip effect in a forest supply chain," *Proc. Int. Conf. Auton. Agents*, vol. 2, pp. 670–677, 2003.

[48] A. K. Agogino and K. Tumer, "QUICR-learning for multi-agent coordination," *AAAI*, 1438–1443, 2006.

[49] H. V. V. Priyadarshana, M. A. K. Sandaru, K. T. M. U. Hemapala, and W. D. A. S. Wijayapala, "A review on multi-agent system based energy management systems for micro grids," *AIMS Energy*, 7(6): 924–943, 2019, doi: 10.3934/ENERGY.2019.6.924.

[50] T. H. Kyaw and N. L. Thein, "Token-based coordination in multiagent organization,"*Int. J. Adv. Comput. Sci.*, 5(5), 196–201, 2015.

[51] I. Vermeulen, S. Bohte, K. Somefun, and H. La Poutré, "Multi-agent pareto appointment exchanging in hospital patient scheduling," *Serv. Orient. Comput. Appl.*, 1(3): 185–196, 2007.

[52] K. Decker and J.Li, "Coordinated hospital patient scheduling," *Third Int. Conf. Multi Agent Syst. (ICMAS'98)*, pp. 104–111, 1998.

Chapter 12

Internet of Things (IoT) for the efficient healthcare system

Abhishek Choubey[1] *and Shruti Bhargava Choubey*[1]

Currently, we have major infections including Coronavirus (COVID-19), Chikungunya virus, Hantavirus, Malaria, blood glucose, chronic diseases like arthritis, hepatitis, cancer, AIDS/HIV, and Typhoid, which are currently widespread medical problems. Healthcare is a necessary module of human needs. It is energizing for personal growth as well as for the improvement of society. As a result, mental and physical well-being is critical because it has a significant impact on human society, public health, and the global economy, all of which are predictable to be pretentious in the near opportunity. Therefore, it is crucial to establish an additional intellectual and methodical application of non-pharmaceutical treatments to make the certain organization of illnesses such as COVID-19, Chikungunya virus, and Hantavirus with the least possible impact on our lives and society. The Internet of Things (IoT) concept necessitates the use of hardware strategies that screen or capture records and are associated with the open or private cloud, allowing them to naturally trigger specific events. Through smart cities, smart healthcare, and smart homes, the IoT technology is transforming our lives and resolving to construct a more appropriate and smarter community. In this chapter, significant emphasis has been given to IoT for an efficient healthcare system. The basic purpose of the chapter is to explore the numerous possibilities of IoT in the field of healthcare system.

12.1 Introduction

Many healthcare issues need the use of technologies and information solutions to improve and ease people's lives. Many diseases can only be treated or monitored if they are found or tracked at an early stage. As a result, recognizing diseases as soon as possible could save lives. People who are already ill, elderly, or visually impaired find it difficult to carry out normal things without the assistance of a caretaker. As a result, providing them with smart solutions based on the Internet of

[1]Department of Electronics Communication, Sreenidhi Institute of Science and Technology–Hyderabad, India

Things (IoT) would be an excellent way to identify their health issues or assist them in performing their daily routines without assistance. The advantage of IoT security is privacy. Notwithstanding action trackers, wearable biometric sensors, glucose and blood pressure, prescription allocators, and smart beds, the IoT applications in the medical services region use remote observing, brilliant sensors, and clinical gadget coordination [1]. If IoT is created and put resources into, the fate of medical care will be very encouraging [2]. This innovation will upgrade a component of medical care and adjust how it is directed. Since medical care requires IoT, numerous reactions present a reasonable and down-to-earth picture of the IoT's capacity in the field of medical services and change information into estimations. Since quantifiable healthcare can be better improved, quantifiable well-being will be the eventual outcome of medical services. As a result, it is reasonable to exploit quantitative healthcare innovation. Since information impacts execution, the ability to gauge people and track their health to create better outcomes is the reason behind IoT.

The healthcare business has witnessed tremendous expansion in recent years, resulting in a significant increase in revenue and employment [3]. Diseases and abnormalities in the human body could only be identified by a medical physical examination only a few years ago. The majority of the patients had to stay in the hospital for their treatment. This has resulted in higher healthcare costs and put a strain on rural and remote healthcare facilities. In recent years, technological advancements have permitted the diagnosis and monitoring of various illnesses using tiny devices such as smartwatches. Furthermore, technological advancements have had an impact on the healthcare industry [4,5].

Many clinical investigations, for instance, can be carried out at home even without the help of a medical professional (e.g., monitoring blood pressure, blood glucose level, pO2, level, and so on). Advanced communication capabilities may also be used to send diagnostic information from remote areas to healthcare institutions. The use of these telecom services in collaboration with rapidly expanding equipment (e.g., computer vision, predictive analysis, IoT, cordless sensing, smartphones, and cloud computing) has superior convenience to healthcare facilities. In order to optimize the usage of IoT in healthcare arrangements, many countries have taken up new technologies and laws. The purpose of this chapter is to provide a complete assessment of the technology, services, and applications that allow IoT-based healthcare systems, as well as to explain the development of high-tech research in this area.

IoT is an essential part of changing over medical care into keen medical care. The prerequisites are for better medical services as follows:

- The right actions should be given to the right individual at the right time.
- It enables professionals to accurately identify and treat patients' diseases.
- It enables all parties to interact and exchange information effectively.
- Data availability in an easily accessible location.
- Appropriate notification of patients to include them in their treatment process.

- As a result of its cost-effective approaches, healthcare may be brought to remote areas.
- It increases productivity by lowering waste and operating expenses [6].

12.2 Overview of IoT

The IoT technology is a network of networked devices that exchange data without the need for human–machine interaction. In other words, IoT is a network of interconnected technological devices. This is a fast-growing technology that brings together real-time connected gadgets and the Internet. It is popular in several areas due to its progression from a conventional item to a smart object. This will have long-term impacts on patient neurobiological data monitoring, administration, and healthcare services. Before being transferred to the health monitoring unit, patients are connected to sensors, and data is linked to controlling equipment. Data are sometimes stored in the cloud, which makes it easier to handle huge volumes of data while preserving security.

IoT demands the connecting of devices and the storage of sensitive data. Personal information security and confidentiality are extremely important to healthcare companies. Even if the data is sent across great distances, security and data encryption are required since healthcare data is sensitive. In order to maintain data security and personal safety, it is also important to be aware of the hazards posed by the addition of new devices to IoT. The summary of IoT applications is given in Table 12.1.

The idea of using computers, sensors, and networks to monitor and control objects has been around for decades. However, IoT is getting closer to mainstream use, thanks to a recent confluence of numerous technological sector breakthroughs. Ubiquitous connection, pervasive Internet protocol-based networking, computer economics, downsizing, improvements in data analytics, and the advent of cloud computing are

Table 12.1 Summary of IoT applications

S. no.	Application		Contribution	Reference
1	Commercial application	Medical and healthcare	IoT is impacting the healthcare industry by redefining the area of equipment and people interaction in offering healthcare solutions. In healthcare, IoT has applications that advantage patients, physicians, families, hospitals, and insurance companies	[7]
		Construction and home computerization	The use of IoT can be used to track and monitor hydraulic, industrial, and technological methods in various types of buildings' home automation and smart buildings (e.g., public and	[8]

(Continues)

Table 12.1 (*Continued*)

S. no.	Application		Contribution	Reference
			private, commercial, institutional, or residential) using sensors	
2	Industrial applications	Manufacturing	IoT will enable the seamless integration of diverse industrial equipment with sensing, detecting, processing, connection, automation, and communication infrastructure	[9]
		Agriculture	IoT solutions are intended to guide farmers in reducing the supply–demand gap by ensuring high yields, profitability, and environmental protection. Precision agriculture refers to the use of IoT technology to ensure the most efficient use of resources in order to maximize crop yields while lowering operational costs. IoT in agriculture technologies includes specialized equipment, wireless connectivity, software, and information technology services	[10]
		Energy management	Many energy-consuming appliances (such as switches, power outlets, lamps, televisions, and so on) now have an Internet connection, allowing them to interrelate with utilities to equilibrium power invention and energy use and get the most out of overall energy consumption	[11]
3	Military applications	Internet of Military Things	The application of IoT technology in the military sector for reconnaissance, surveillance, and other war-related purposes is known as the Internet of Military Things	[12]
		Internet of Battlefield Things	Battlefield Things Internet is a project started and implemented by the United States. Army Research Laboratory (ARL) focuses on basic science related to IoT that improves military capabilities	[13]
4	Consumer applications	Smart home	A smart home or computerized home possibly will be based on a mobile coordination and appliances proposal or hubs that control	[14]
		Eldercare	One main purpose of a smart home is to give a hand to disabled people and elderly people. Such home programs make use of assistive equipment to meet the unique handicaps of a human being	[15]

just a few of them. IoT should not be a constraint in terms of demand, resources, or infrastructure development. It is important to recognize barriers to building viable solutions. High-speed internet and high-performance computing facilities are driving the present research. It is important to prioritize that correspond with corporate objectives, as well as models that are compatible with healthcare concerns. Massive data storage devices and capabilities are required to minimize patient wait times, respond to emergency calls, and track hospital and patient movements through alerts.

IoT has given people more independence while also increasing their ability to interact with the rest of the world. IoT seems to have become a key contributor to global communication, thanks to future authentication schemes. It connects a wide range of devices to the Internet, including wireless sensors, domestic appliances, and electrical equipment [16]. IoT applications include agribusiness [17], vehicle [18,19], and home [20], in addition to healthcare [21]. IoT is gaining some traction due to its benefits of better efficacy, lower costs, and the skill to better envisage prospective events. Additionally, increased knowledge of software and applications, as well as improvements in mobile and cloud-based technologies, perspective emphasis on a network connection, and the expansion of the sharing transformation have all assisted the rapid IoT revolution [22].

Instruments, controllers, and other IoT devices have been integrated with other hardware objects to analyze and exchange data via Wireless, Zigbee, IEEE 802.11 (Wi-Fi), and many other communication networks. Devices are used in application scenarios to record sensor parameters from the body of the patient, such as heat, pulse rate, electrocardiograph (ECG), electroencephalograph, and so on, and are instead implanted or worn on the patient's psyche [23]. Warmth, precipitation, date, and time are all examples of environmental data that may be recorded. Information may be used to make accurate and relevant decisions regarding a patient's health. Information management and accessibility are extremely important in the IoT organization since a great quantity of data is obtained/verified from several resources (sensors, mobile phones, e-mail, software, and appliances). Doctors, professionals, and other authorized individuals receive data from either the aforementioned sensing devices. The capacity to communicate this information with health practitioners through cloud/server allows for the rapid scientific process and, if necessary, medical intervention. Users, patients, and the communication module must all work together to provide a successful and secure transfer. A large percentage of IoT applications have a graphical interface that acts as a screen for healthcare professionals, the ability to control, visualize, and evaluate data. There is a significant amount of research on the expansion of the IoT system in remote patient monitoring, management, privacy, and confidentiality in the literature [24]. The significance of IoT in the healthcare industry, as well as its bright future, can be seen in these accomplishments. The key challenge while developing an IoT mechanism is preserving the standard of examining prevailing conditions, and this includes the privacy of intelligence gathering, protection, price, reliability, and accessibility. Remote management in the healthcare business is now possible, thanks to IoT-enabled devices that can keep people healthy while also allowing doctors to deliver better care. Patients' engagement and happiness have improved

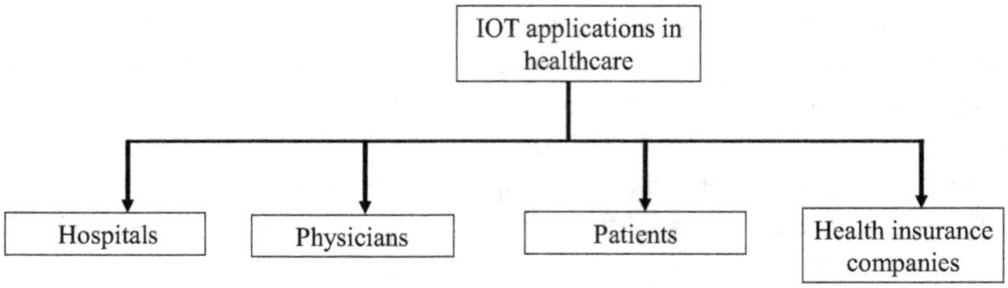

Figure 12.1 Application of IoT in healthcare

as communication with doctors became simpler and more efficient. Second, medical technology reduces hospital stays and prevents re-admissions. IoT has a significant impact on reducing healthcare costs and improving treatment outcomes. By reimagining the space of devices and human interaction in healthcare, IoT is altering the industry. The application of IoT in healthcare is shown in Figure 12.1.

12.3 Review of existing work

IoT, big data, and machine learning (ML)-dependent healthcare explanations are the trendiest research areas right now for improving healthcare services. Many research works based on data science, big data analysis, IoT devices, ML, and data mining approaches are available in the healthcare sector. Because of Chikungunya's rapid spread, an early diagnosis is a major problem for healthcare providers. The most essential step in managing the Chikungunya virus is to lower its death rate by recognizing the affected people early and offering preventative measures to those who are not afflicted. The infected individuals should be evaluated as soon as possible to ensure that they do not aid in the stretch of the Chikungunya virus. To prevent the extension of this illness, early identification and remote surveillance are now very desirable. Although diagnostic tests can aid in the proper diagnosis of viruses, most medical laboratories do not have these tests accessible right away. Because of a lack of staff and healthcare resources in hospitals, even patients are moved to another city. The Chikungunya virus has no particular antiviral medication or vaccine available to treat it. Chikungunya is one of the world's most prevalent vector-borne illnesses [25]. The summary of existing work using IoT to cure different diseases is shown in Table 12.2.

COVID-19: The relevance of IoT and additional digital technology in the encounter against COVID-19 or prospect pandemics has been underlined in numerous studies released in the previous year or two. Elansary *et al.* [26] suggest an overview of COVID-19-fighting technologies based on IoT. In [27,28], the authors propose the impact of the pandemic on various technology as well as the societal implications and present a thorough examination of digital health resolution carried out in countries in the middle of elevated COVID-19 occurrence. In [29], the authors presented COVID-19 diagnosis and treatment using IoT-assisted discussion of Chinese experts. In [30], the cost, time, and efficiency of possessions of IoT execution in healthcare are suggested.

Table 12.2 Summary of existing works

	Types of diseases	Article	Main research contribution
1	Chikungunya virus	[1]	In this article, the author proposed a novel framework for implementing a smart health system, for the Chikungunya virus. The proposed approach is based on the cloud that gathers data from mobile phones and sensors using edge servers. It makes use of edge computing and wireless communication technology.
		[2]	The use of a sensitivity factor with the time of various events for determining the severity of health is a key feature of this article. Emergency alerts are generated based on the severity of the health condition to deliver event information to users' mobile phones on time via the fog network.
		[3]	Wearable sensor technologies, a decision tree, and thin film transistor (TFT) are used to build a fog-assisted cloud-based healthcare system for forecasting and preventing Chikungunya infection. The users are divided into distinct groups using a J48 decision tree. The main goal of this article is to figure out how to calculate the likelihood of a health index based on the timing of numerous occurrences. Alerts are issued based on the possibility of a health index.
2	COVID-19	[13]	In this article, the authors proposed an integrated multi-criteria decision-making method under an uncertain environment. This method integrates the benefits of the best–worst method and technique in order of preference by similarity to the ideal solution (TOPSIS). To evaluate a set of alternatives according to decision criteria and proposed model effectiveness in detecting the COVID-19 in all five cases achieving detection accuracy of up to 98%.
		[14]	This article proposes a fog-cloud shared IoT platform for COVID-19 control and prevention by applying five non-pharmaceutical interventions, including COVID-19 quarantine monitoring, symptom diagnosis, social distancing, contact tracing and, COVID-19 outbreak forecasting, and SARS-CoV-2 mutation tracking.
		[15]	We offer an overview of IoT technology in the healthcare area in this article. First, enabling technologies for smart healthcare were briefly discussed, including sensors, cloud computing, fog computing, and wireless body area networks. Wearable sensors were given special attention since they are non-intrusive and popular with people. Pulse sensors, respiration rate sensors, body temperature sensors, blood oxygen sensors, and pulse oximeter sensors were among the wearable sensors examined. Second, real-time monitoring and alerts, telemedicine, chronic illness diagnosis and prevention, and home and senior healthcare were all mentioned as major IoT healthcare applications.
		[17]	In this article, the authors examine existing and evolving edge-computing architectures and techniques for smart healthcare,

(Continues)

Table 12.2 (*Continued*)

	Types of diseases	Article	Main research contribution
			as well as recognize the demands and challenges of various application scenarios. It also provides a brief overview of how IoT solutions are used in edge platforms for medical treatment and healthcare.
3	Diabetes diseases	[18]	In this article, a novel systematic strategy for diabetes illnesses and related medical data is developed using the UCI Repository dataset and medical sensors in order to forecast patients who are badly afflicted by diabetes. In addition, for identifying the disease and its severity, the authors proposed a novel classification method called fuzzy rule-based neural classifier. The experiments were carried out using the standard UCI Repository dataset as well as real health records obtained from various hospitals.
		[19]	This paper offers an automated identity system to assist individuals with chronic diabetes in effectively managing their blood glucose levels. The overall implementation results of the proposed self-care IoT platform indicate that it can seamlessly gather and send data when each device in the IoT pack is linked to a smartphone.

Applications of IoT, big data, and artificial sensors [31] proposed how artificial intelligence (AI) and blockchain can be used to lessen the impact of COVID-19. Rahman *et al.* [32] proposed a number of IoT technologies that might be useful in combating COVID-19. In various industries, according to Javaid *et al.* [33], 4.0 technologies (e.g., AI, IoT, virtual reality, etc.) can contribute to the prevention of disease transmission. Computer applications for COVID-19 were proposed in [34]. In [35], the authors proposed the COVID-19 pandemic in-depth, as well as the role of IoT, drones, AI, blockchain, and 5G in reducing its impacts. In [36], the authors proposed contact with tracking ought to be the accountability of services, and propose a fully automated contact tracing system that does not rely on user interaction. In [37], early detection, quarantine, and recovery as three different stages of IoT-based healthcare applications have been proposed. ML, AI, as well as other intelligent approaches for COVID-19 prognosis are discussed. In [38] the authors have compiled a list of possible COVID-19 IoT-based applications. They examine the competence of accessible IoT systems at several layers, like the perceptual layer, core network, fog layer, and cloud layer. They also talk about how IoT may be used to diagnose COVID-19 symptoms. Reference [39] proposes a four-layered structural design depending upon IoT and blockchain technology to aid in the fight against SARS COVID.

To protect the confidentiality and protection of physiological data transferred in the middle of IoT nodes, a blockchain-based approach is presented. It also includes a number of apps designed to detect and track prospective COVID patients. In [40], the authors examine the responsibility of IoT in the current digital healthcare infrastructure. It also looks at how data provided by IoT-enabled healthcare communications affect authorities' decisions. They have also mentioned existing enablers and roadblocks to IoT-based healthcare uptake. Reference [41] provides a comprehensive survey on the contributions of IoT in healthcare. This is a comprehensive investigation. It also compares different IoT system installation techniques both before and during the epidemic. Reference [42] provides a survey that enumerates early attempts to apply digital knowledge in healthcare to combat COVID, taking into account several classes such as finding, anticipation, and supervision. In [43], the authors emphasize the critical need of employing AI methods in the fight against upcoming pandemics. Finally, they argue that data science should be used in global health to create more accurate forecasts that can be used by policymakers. In [44], a comparison of digital technology adoption in various locations across the world is provided.

12.4 IoT architecture for Chikungunya and COVID-19

This research proposes a cutting-edge healthcare system based on IoT. The proposed system is a general healthcare system that includes a solution for Chikungunya virus, Hantavirus, Malaria, blood glucose, chronic diseases like arthritis, hepatitis, cancer, AIDS/HIV, and typhoid. Visually impaired, disabled patients and users can track their health and predict disease. The proposed work aims to create a real-time health prediction system based on IoT architecture. In the proposed system, the health parameters are sent to obtain the attributes from the data and to apply the proposed architecture to predict the healthcare dangers and send them as messages and suggestions to the patients and the healthcare contributor as well. Furthermore, the proposed systems included a speech recognition system and the best solution for managing smart home devices in the health sector. A block diagram of the IoT platform for Chikungunya prevention using a non-invasive wireless sensor is shown in Figure 12.2. The proposed prototype collects information from the wearable sensors and sends it over mobile networks to a data storage server. The proposed architecture provides the user with medical advice, risk alerts, and emergency on real-time detection.

COVID has worked as a catalyst for the adoption of e-Health, e-Education, and e-Commerce, according to the report. Several disease detection systems have been created and tested in various settings. Many different sensor types are used to monitor various health concerns, including blood pressure, heart rate, oxygen rate, accelerator sensors, and chemical sensors.

Smart gadgets (watches, phones, rings, and so on) have become commonplace in everyday life. The wearable wireless sensor measures a variety of environmental variables, such as temperature and humidity, as well as numerous human bodily metrics, such as heart rate, blood pressure, and other factors. Many suggestions have cited the body area network as the ideal method for transferring data from wearable medical and sensory equipment to cell phones [8], the second of which is a sensor gateway that stores data on the server, and the third of which is a warning system in the event of abnormal circumstances in a patient. Third, cloud computing will be used to store and monitor patient data, as well as provide online services for healthcare applications. Finally, apps for smartphones or tablets will be used to display data on the patient's health status. The system has shown to be effective. Symptoms of COVID-19 and SARS are shown in Figure 12.3.

Clinical decision-making is entirely based on the information available on various clinical devices, economic, and social variables. IoT medical devices supply healthcare service providers with a range of data that must be evaluated in order to anticipate unknown data. IoT wearable devices offer medical solutions based on data acquired from various sources, including patients. Smartphones and smart wearables assist consumers in carrying out their everyday tasks. They employ wireless connectivity to give healthcare systems data-driven advice for living a healthy life. Medical wearable devices with microprocessors, memory, sensors, and

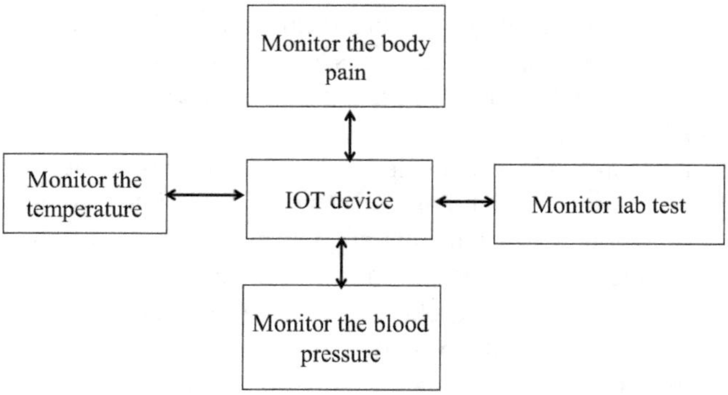

Figure 12.2 Block diagram of IoT platform for Chikungunya prevention

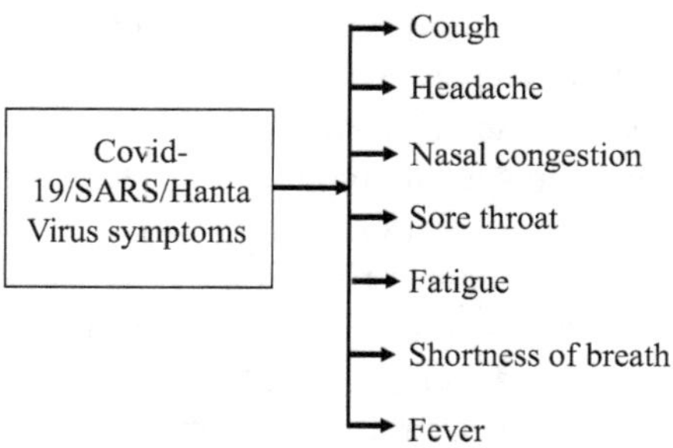

Figure 12.3 Symptoms of COVID-19 and SARS

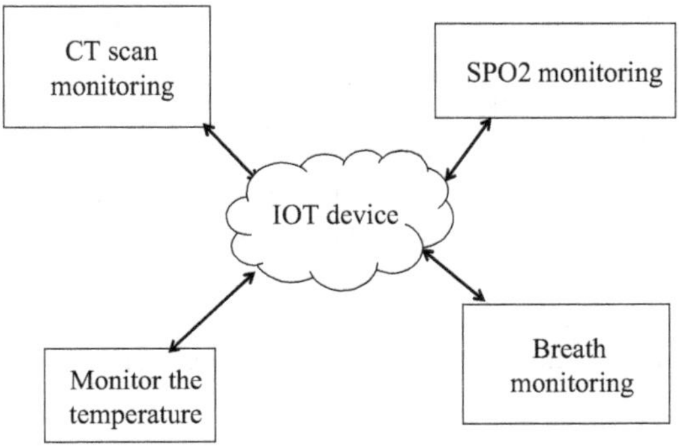

Figure 12.4 Block diagram of IoT platform for COVID-19 prevention

Table 12.3 Summary of sensors used in healthcare

S. no.	Sensor	Contribution	Reference
1	Heart rate sensor	It is a type of wearable sensor. Generally, ECG devices are recommended for monitoring heart rate including detailed analysis. However, because it is an expensive system and a prototype module, a simple heart rate sensor is used to explain the working principle, which generates a digital pulse for each heart function cycle.	[36]
2	Respiration sensor	To measure the amount of tumor necrosis factor, this sensor is linked to anti-tumor therapy.	[37]
3	Hall effect sensor	The hall sensor and a small magnet placed over an elastic belt can measure the rate of breathing.	[39]
4	Temperature sensing circuit	The purpose of this LM 35 circuit is to measure the temperature of the human body.	[40]
5	Pulse oximetry sensors	Portable devices can monitor the amount of oxygen in the blood, which is another important criterion for detecting diseases like hypoxia. These sensors use PPG impulses to determine blood oxygen levels.	[41]
6	Blood pressure sensors	It is used to measure one of the most routinely monitored health indicators, blood pressure. The monitoring of blood pressure can lead to the detection of hypertension, which can lead to heart attacks and other cardiovascular illnesses.	[42]

communication interfaces can be implanted or simply worn. They take measurements of the body's vital signals. They are great for measuring heart rates, as well as respiration, body temperature, and other factors. Wearable gadgets in healthcare can assist patients to have a better understanding of their health condition and provide recommendations for what they should do in their lifestyle. There are several sensors on the market now that can monitor health indicators and assist healthcare practitioners in keeping track of their patients. It's difficult to create a wearable gadget that can track health concerns. The wearable gadget should be light, have a long battery life, and have various medical sensors. The fitness tracker must have a communication method of analysis that integrates the device's readings to mobile phones or networking devices; the handheld technology must give the user astounding information about the health tracking, and the wearable technology must have a communication method of analysis that connects the Smartphone's accurate reading to smartphones or network servers.

12.5 Conclusions and future scope

In today's healthcare environment, health status monitoring is critical. It aids healthcare practitioners in the treatment of illnesses. Many systems provide health status monitoring strategies for obtaining information from wearable devices, fitness devices, magnetic resonance images (MRI), medical reports, texture analysis, and other sources. IoT technology is expected to be one of the fastest-growing sectors in the next years, with a significant impact on the healthcare industry. IoT enables individuals to connect via wearable devices to empower their health and wealth smartly. A new trend in IoT has emerged as a result of recent advancements in wireless sensor networks.

Smart health is a significant use in IoT. Patients with aberrant health problems may be monitored promptly and treated fast with the smart health care system. This sort of solution may be found in wearable devices that continually monitor the patient's activity and condition in a predictable manner. The major goal of this paper is to conduct a comprehensive study into sensor data collection, data analysis, and patient feedback based on numerous health indicators.

IoT is made up of modern fields such as embedded systems, wireless sensor networks, control mechanisms, automated (including remote monitoring and smart buildings), and others. In the consumer market, IoT technology is most closely associated with products that fall under the "smart home" definition, which encompasses devices and appliances (such as lighting fixtures, thermostats, home security systems, cameras, and other home appliances) that are connected to one or more widely accepted ecological systems and can be controlled using ecosystem-related devices like smartphones and smart speakers. There are quite various severe concerns raised regarding the hazards of IoT development, particularly in the areas of security and privacy, as well as industry and government initiatives to address these issues.

References

[1] S. Rani, S. H. Ahmed, and S. C. Shah, "Smart health: a novel paradigm to control the Chikungunya virus", *IEEE Internet of Things Journal*, 2019, 6(2), 1306–1311, doi:10.1109/JIOT.2018.2802898.

[2] Y. Dong and Y. -D. Yao, "IoT platform for COVID-19 prevention and control: a survey", *IEEE Access*, 2021, 9, 49929–49941, doi: 10.1109/ACCESS.2021.3068276.

[3] Z. Lou, L. Wang, K. Jiang, Z. Wei, and G. Shen, "Reviews of wearable healthcare systems: materials, devices and system integration", *Materials Science and Engineering: R: Reports*, 2020, 140, 100523.

[4] M. S. Hossain and G. Muhammad, "Cloud-assisted industrial internet of things (IIoT)-enabled framework for health monitoring", *Computer Networks*, 2016, 101, 192–202.

[5] J. Joonyoung, H. Kiryong, L. Jeonwoo, K. Youngsung, and K. Daeyoung, "Wireless body area network in a ubiquitous healthcare system for physiological signal monitoring and health consulting", *Journal of Image Process*, 2008, 47–54.

[6] J. Kim, A. Campbell, B. Ávila, and J. Wang, "Wearable biosensors for healthcare monitoring", *Nature Biotechnology*, 2019, 37(4), 389–406.

[7] G. Honan, A. Page, O. Kocabas, T. Soyata, and B. Kantarci, "Internet-of-everything oriented implementation of secure digital health (D-Health) systems", In: *IEEE Symposium on Computers and Communication (ISCC)*, IEEE, 2016, pp. 718–725.

[8] S. Sareen, S. K. Gupta, and S. K. Sood, "An intelligent and secure system for predicting and preventing Zika virus outbreak using fog computing", *Enterprise Information Systems*, 2017, 1–21.

[9] S. C. Weaver and M. Lecuit, "Chikungunya virus and the global spread of a mosquito-borne disease", *New England Journal of Medicine*, 2015, 372(13), 1231–1239.

[10] M. Page, T. Hassanalieragh, M. Soyata, B. Aktas, B. Kantarci, and S. Andreescu, "Conceptualizing a real-time remote cardiac health monitoring system", *IGI Global*, 2015, 1–34.

[11] D. Florez and J. Sepulveda, "Blooxy: On a non-invasive blood monitor for the IoT context", In: *2017 30th IEEE International System-on-Chip Conference (SOCC)*, IEEE, 2017, pp. 29–34.

[12] J. Zhu, P. Ji, J. Pang, Z. Zhong, H. Li, C. He, J. Zhang, and C. Zhao, "Clinical characteristics of 3,062 COVID-19 patients: a meta-analysis", *Journal of Medical Virology*, 2020, 92(10), 1902–1914.

[13] M. Mohammed, H. Syamsudin, S. Al-Zubaidi, *et al.*, "Novel COVID-19 detection and diagnosis system using IoT based smart helmet", *International Journal of Psychosocial Rehabilitation*, 2020, 24(7), 954–960.

[14] P. Tedeschi, S. Bakiras, and R. Di Pietro, "IoTrace: A flexible, efficient, and privacy-preserving IoT-enabled architecture for contact tracing", *IEEE Communications Magazine*, 2020, 59(6), 82–88.

[15] S. Dami and M. Yahaghizadeh, "Efficient event prediction in an IoT environment based on LDA model and support vector machine", In: *2018 6th Iranian Joint Congress on Fuzzy and Intelligent Systems (CFIS)*, IEEE, 2018, pp. 135–138.

[16] "Coronavirus disease (COVID-19)", World Health Organization, 2020. [Online]. Available at: https://www.who.int/emergencies/diseases/ novel-coronavirus-2019.

[17] V. Bianchi, M. Bassoli, G. Lombardo, P. Fornacciari, M. Mordonini, and I. De Munari, "IoT wearable sensor and deep learning: an integrated approach for personalized human activity recognition in a smart home environment", *IEEE Internet of Things Journal*, 2019, 6(5), 8553–8562.

[18] P. Kellam and W. Barclay, "The dynamics of humoral immune responses following SARS-CoV-2 infection and the potential for reinfection", *Journal of General Virology*, 2020, jgv001439.

[19] "Coronavirus disease (COVID-19) situation report−197," World Health Organization, August 2020. [Online]. Available: https://www.who.int/ docs/ default-source/coronaviruse/situation-reports

[20] S. Bagha and L. Shaw, "A real time analysis of PPG signal for measurement of SpO2 and pulse rate", *International Journal of Computer Applications*, 2011, 36(11), 45–50.

[21] S. Singh, A. Bansal, R. Sandhu, and J. Sidhu, "Fog computing and IoT based healthcare support service for dengue fever", *International Journal of Pervasive Computing and Communications*, 2018, 14(2), 197–207.

[22] S. K. Sood and I. Mahajan, "Wearable IoT sensor based healthcare system for identifying and controlling Chikungunya virus", *Computers in Industry*, 2017, 91, 33–44.

[23] P. d. S. Tavares and E. B. Rodrigues, "IoT-based architecture for data analytics of arboviruses in smart cities", In: *2018 IEEE Symposium on Computers and Communications (ISCC)*, IEEE, 2018, pp. 952–957.

[24] M. Pachetti, B. Marini, F. Benedetti, *et al.*, "Emerging SARS-CoV-2 mutation hot spots include a novel RNA-dependent-RNA polymerase variant", *Journal of Translational Medicine*, 2020, 18, 1–9.

[25] R. Yin, E. Luusua, J. Dabrowski, Y. Zhang, and C. K. Kwoh, "Tempel: Time-series mutation prediction of influenza a viruses via attention based recurrent neural networks", *Bioinformatics*, 2020, 36(9), 2697–2704.

[26] T. Avšič-Županc, A. Saksida, and M. Korva, "Hantavirus infections", *Clinical Microbiology and Infection*, 2019, 21, e6–16. DOI10.1111/1469-0691.12291.

[27] S. Liu, Y. Jiang, and A. Striegel, "Face-to-face proximity estimation using Bluetooth on smartphones", *IEEE Transactions on Mobile Computing*, 2013, 13(4), 811–823.

[28] F. A. Alaba, M. Othman, I. A. T. Hashem, and F. Alotaibi, "Internet of things security: a survey", *Journal of Network and Computer Applications*, 2017, 88(Suppl. C), 10–28. http://dx.doi.org/ 10.1016/j.jnca.2017.04.002.

[29] M. Ahmad Khan and K. Salah, "IoT security: review, blockchain solutions, and open challenges", *Future Generation Computer Systems*, 2018, 82, 395-411. ISSN 0167-739X.

[30] L. Atzori, A. Iera, and G. Morabito, "The internet of things: a survey", *Computer Networks*, 2010, 54(15), 2787–2805.

[31] J. Gubbi, R. Buyya, S. Marusic, and M. Palaniswami, "Internet of Things (IoT): a vision, architectural elements, and future directions", *Future Generation Computer Systems*, 2013, 29(7), 1645–1660.

[32] S. Milici, S. Amendola, A. Bianco, and G. Marrocco, "Epidermal RFID passive sensor for body temperature measurements", In: *2014 IEEE RFID Technology and Applications Conference (RFID-TA)*, IEEE, 2014, pp. 140–144.

[33] J. Xie, N. Covassin, Z. Fan, *et al.*, "Association between hypoxemia and mortality in patients with COVID-19", In: *Mayo Clinic Proceedings*, Elsevier, 2020.

[34] O. Elijah, T. A. Rahman, I. Orikumhi, C. Y. Leow, and M. N. Hindia, "An overview of internet of things (IoT) and data analytics in agriculture: benefits and challenges", *IEEE Internet of Things Journal*, 2018, 5(5), 3758–3773.

[35] S. Li, L. Da Xu, and S. Zhao, "The internet of things: a survey", *Information Systems Frontiers*, 2015, 17(2), 243–259.

[36] S. Selvaraj and S. Sundaravaradhan, "Challenges and opportunities in IoT healthcare systems: a systematic review", *SN Applied Sciences*, 2020, 2, 139.

[37] S. M. Sasubilli, A. Kumar and V. Dutt, "Machine learning implementation on medical domain to identify disease insights using TMS", *2020 International Conference on Advances in Computing and Communication Engineering (ICACCE)*, 2020, pp. 1–4, doi: 10.1109/ICACCE49060.2020.9154960.

[38] S. M. Sasubilli, A. Kumar, and V. Dutt, "Improving health care by help of internet of things and bigdata analytics and cloud computing", *2020 International Conference on Advances in Computing and Communication Engineering (ICACCE)*, 2020, pp. 1–4, doi: 10.1109/ICACCE49060.2020.9155042.

[39] P. K. Kotturu and A. Kumar, "Comparative study on machine learning models for early diagnose of Alzheimer's disease: multi correlation method", In: *2020 5th International Conference on Communication and Electronics Systems (ICCES)*, 2020, pp. 778–783, doi: 10.1109/ICCES48766.2020.9137872.

[40] S. K. Sood and I. Mahajan, "Wearable IoT sensor based healthcare system for identifying and controlling Chikungunya virus", *Computers in Industry*, 2017, 91, 33–44.

[41] S. K. Sood and I. Mahajan, "A fog-based healthcare framework for Chikungunya", *IEEE Internet of Things Journal*, 2018, 5(2), 794–801, doi:10.1109/JIOT.2017.2768407.

[42] I. S. G. Brites, L. M. Silva, J. L. V. Barbosa, S. J. Rigo, S. D. Correia, and V. R. Q. Leithardt, *Machine Learning and IoT Applied to Cardiovascular Diseases Identification through Heart Sounds: A Literature Review*, 2021. Preprints, 2021100161, doi: 10.20944/preprints202110.0161.v1.

[43] P. Malarvizhi Kumar, S. Lokesh, R. Varatharajan, G. Chandra Babu, and P. Parthasarathy, "Cloud and IoT based disease prediction and diagnosis system for healthcare using Fuzzy neural classifier", *Future Generation Computer Systems*, 2018, 86, 527–534, ISSN 0167-739X.

[44] J.-C. Park, S. Kim, and J.-H. Lee, "Self-care IoT platform for diabetic mellitus, *Applied Science*, 2021, 11, 2006. https://doi.org/10.3390/app11052006.

Chapter 13

Comprehension of melody representation and speed-up approaches for query by humming system

C.N. Trisiladevi[1] and P. Mahesha[1]

Query by humming (QBH) music framework under music information retrieval (MIR) is one of the most vital research domains. Largely, it adopts eminent techniques to identify the songs from the music repository based on the user humming for music retrieval applications. The focus of the suggested research work presented in this chapter is to establish novel techniques for the QBH system to popularise it for MIR applications.

Establishing more generic techniques for music retrieval through the QBH framework is a trivial task because of the inherent challenges like an accurate projection of music melody content using discriminating feature set, design and development of prominent melody matching criteria, indigent humming queries and intricate musical structures.

In the present book chapter, novel techniques are recommended and discussed for representing melody with various melody extraction perspectives. Further, two-speed optimisation approaches are suggested and elaborated.

After analysing the different views of the recommended approaches, an attempt is made to find the proximity with preexisting work. Also, the comprehensive view of all the projected approaches along with preexisting work is illustrated. This effort has instituted many eminent specifics that were not taken into account so far.

In order to comprehend the performance of the approaches, empirical metrics like mean reciprocal rank, mean of accuracy, Top X hit rate and response time are employed. Further, exhaustive experimentation is carried out with the target music fragment datasets and humming query datasets [28].

The book chapter is prepared to present the regular and distinct features of recommended and preexisting techniques. Further, in the chapter detailed experimental analysis of the suggested approaches is illustrated.

[1]Department of Computer Science and Engineering, S.J. College of Engineering, JSS Science and Technology University, Mysore, India

13.1 Introduction

At present, query by humming (QBH) is a highly preferred mode of music retrieval among contemporary music information retrieval (MIR) systems. Also, because of the encouragement received from all the music stakeholders and lovers, researchers are motivated and novel algorithms for the QBH system were contributed [1]. Hence, a substantial amount of QBH music signal processing research is noticed in the literature [2–16].

The work discussed here incorporates two major interlinked parts. First, new techniques for extracting melody patterns for music retrieval are developed. Precise melody representation is essential as it depends on unique feature extraction and estimation. In the second part, progressive filtering (PF) through multiresolution histograms (MRH) and complex multiscale music entropy balanced (CMMEB) Kd tree indexing techniques are visualised for speed optimisation.

Exclusively, the research work attempts to explore the progression made towards melody extraction and speed optimisation so that new approaches are incorporated into the QBH music retrieval system. The existing models incorporate pitch interval, pitch contour, zero-crossing rate (ZCR), beat, rhythm and energy-based approaches.

Similarly, the proposed work facilitates melody representation techniques anchored in statistical description, acoustic features, perceptual parameters and modulation frequency (MF) information. Although diverse techniques for the QBH system are existing, relatively limited work is available for representing melody using statistical descriptors, acoustic features, perceptual parameters and MF information.

Besides, two speed-up approaches are accomplished for the QBH music retrieval framework. The research work presents four melody representations as well as two speed-up techniques to address the challenges of the QBH system.

The first component of the research framework establishes the reliability of the statistical descriptors [17,18] as music melody features through empirical mode decomposition (EMD). The potential advantages of EMD are exploited in most of the projected approaches to reduce dimension and estimate the structural information of music. In addition, the accuracy of the recommended methods is instituted with benchmark performance appraisal metrics such as mean reciprocal rank (MRR), mean of accuracy (MoA), Top X hit rate (TXHR) and response time (RT). The results are satisfactory and close to human expectations.

Subsequently, three more approaches were used to accomplish the objective of melody representation. The potentiality of the acoustical features [19,20] was investigated in the second approach for music retrieval using QBH. The acoustic music melody repository was constructed during the training phase. Then, the system performance appraisal is carried out using humming queries. The results analysis confirmed the efficiency of the technique for music retrieval purposes.

The third criterion of melody representation is accomplished through perceptual parameters [21,22]. The outcomes of the approach were encouraging and also provided an evidence for using the human hearing perception model for melody representation. Undeniably, the approach yielded better performance scores and they were significant compared to the performance of the preexisting approaches.

Later after validating the perceptual features, in the subsequent approach, MF information [23] is used for the implementation of the QBH framework. Here, nine descriptors were used to portray tempo as well as rhythmic rationales. Promising outcomes were obtained with sort merge (SM) modality. The method demonstrated remarkable performance in comparison with statistical descriptors, acoustic features and perceptual parameters. Also, indirectly it established its suitability for melody representation and music retrieval.

The dominant performance is observed in the case of MF features over statistical descriptors, acoustic features and perceptual parameters. It is quite obvious that the results are predisposed by melody features, representation, indigent humming queries and other relevant parameters which are vital for MIR applications.

Consequently, in the latter part of the experimentation, the eminence of PF with MRH [24,25] and CMMEB Kd tree [26,27] based indexing speed optimisation criteria is observed with reference to the QBH music retrieval system. Enhanced performance with respect to the CMMEB Kd tree-based indexing technique is noted over PF with the MRH technique.

The chapter illustrates the performance of the recommended melody representation and speed optimisation approaches. Further, they are compared with existing techniques as well as between them on the basis of performance appraisal metrics like MRR, MoA, TXHR and RT.

This chapter is structured as follows. The performance comparison with existing approaches is illustrated in Section 13.2. The experimental analysis of the proposed work is discussed in Section 13.3. Approximation and envisioning of the relations among performance metrics are presented in Section 13.4. Finally, the chapter is concluded with future work in Section 13.5.

13.2 Comparison with existing approaches

The preexisting research works have reasonably addressed the challenges like poor humming query, appropriate melody representation, retrieval speed optimisation and matching the similarities of melodies.

In the current work, new methods are suggested for the QBH system for automatically retrieving the preferred song using humming query and melody patterns. In the present section, outcomes of suggested melody representation as well as speed-up approaches are compared with the preexisting techniques separately. Primarily, projected melody representation and then speed-up approaches are compared with the currently existing modern techniques.

Hence, in this research framework, the melody representation and speed-up perspectives are the basis. Largely, the aim of the proposed work is to assimilate a variety of invariant features which can accurately characterise the music melody and reduce the retrieval speed.

The exhaustive comparative study is carried out for analysing techniques for melody representation as well as for reducing the speed with preexisting techniques. Generally, in the literature, the duration, rhythm, notes, interval, fundamental and maximum energy frequency, pitch ternary contour, major and minor pitch, pitch and duration change contours are employed to represent music melody [2–16].

In addition, boundary location via Boundary Location (BALS) as well as multiple indices [7], frequency spectral modified discrete cosine transform coefficients used to extract pitches adaptively [13], pitch tracking with quotient space [5] and pitch vectors represented using Hidden Markov Models (HMMs) [14] too are used.

The proposed research framework projected four melody representation techniques, such as enhanced statistical descriptors, acoustic features, perceptual parameters and MF features. Here, an attempt is made to extract features considering human auditory perception, vocal tract shape, MF and probability parameters to suitably characterise and represent music melody.

Additionally, the latest enthralled techniques such as edit distance and hierarchical [13], instance string searching [2], top-down recursive alignment, DDP [4,7], linear scaling (LS) [12] and local sensitive hashing (LSH) [14] are realised by research community.

Four empirical performance appraisal metrics are used to exploit and analyse the potential of all the approaches. In Table 13.1, the values are tabulated after the comparative study between recommended and existing approaches with respect to researchers, datasets, melody features, classification techniques and retrieval performance. Keen observation of the tabulated values hints at the projected QBH melody representation techniques and ascertains superior retrieval performance with respect to MRR, MoA, TXHR and RT yardsticks in comparison with the preexisting QBH approach [2–16].

Also, PF and CMMEB Kd tree-based indexing approaches subsidise retrieval speed optimisation. Both approaches have exhibited significant enhancement in their performance in comparison with contemporary [2,4,7,12–14] QBH speed-up techniques.

13.3 Experimental analysis of the proposed work

A broader all-inclusive experimental analysis framework of projected melody representation and speed-up techniques is illustrated using relevance and exhaustive experiments. The research-oriented study has explored new approaches for melody representation. Also, it facilitates identifying the speed-up approach that helps to retrieve the desired song quickly.

13.3.1 Mean reciprocal rank

At the outset, the potential of statistical, acoustic, perceptual and MF information for music melody pattern recognition and retrieval is exploited. The MRR and average MRR values generated for all the techniques are tabulated in Table 13.2. Also, the same values are depicted in Figure 13.1.

Table 13.1 Comparison of projected methods with existing approaches

Author name	Datasets	Melody features	Classification techniques	Retrieval performance
Ghias *et al.* [2]	Dataset with 183 music songs in MIDI format	Pitch, notes and melody contour, and fundamental frequency	Based on searching using strings	90% Accuracy, 0.34–0.75 min RT
Lie *et al.* [8]	Songs 1000 in MIDI format and queries 42	Pitch duration, interval and contour	Algorithm to match using hierarchy	88% Accuracy, 59.52% Top 1, 74% Top 3 and 88% Top 10
Raju *et al.* [6]	Regional and Hindi film songs 20, queries from 5 users 100	Autocorrelation three-level pitch contour, pitch contour with note duration and absolute pitch	Dynamic programming (DP) and longest common subsequence	95% Accuracy
Yunyue and Dennis [9]	1000 fragments with 50 songs	Pitches and envelope transform time series	Sub and whole sequence matching, local dynamic time warping indexes	90% Accuracy
Unal *et al.* [15]	MIDI files 200, hummings 250	Major and minor pitch, pitch and duration change contours	Similarity score	Accuracy 94% with and 72% without musical training
Fu and Xue [4]	MIDI 1904 pieces, humming queries 38	Rhythm, pitch interval and pitch contour	Dual dynamic programming (DDP)	1st Approach: Top 1: 55.3%, Top 10: 84.2%, RT: 0.0745 min 2nd Approach: Top 1: 63.2%, Top 10: 89.5%, RT: 0.150 min
Dannenberg and Hu [10]	MIDI 8926	Tempo-invariant and transposition presentations using pitch and inter-note onset intervals	Two-stage search with fast n-grams and DP	80% Accuracy, 0.375 MRR

(Continues)

Table 13.1 (Continued)

Author name	Datasets	Melody features	Classification techniques	Retrieval performance
Wang and Zhang [13]	MIDI (Chinese folk) 1498 and queries 48	Pitch tracking and quotient space ([X],[f]) representation	Approximate string matching using edit distance & hierarchical	50% Accuracy, faster than the frame-based approach
Wu and Li [12]	MIDI 1180 and humming queries 15	Pitch sequence to semitone scale	LS and top-down recursive alignment	90.29% Accuracy, 83.77% Top 1, 90.29% Top 10, 0.055 min RT
Fu and Xue [7]	MP3 songs 300 (pop solo, male and female vocals 150 each), humming queries 20	Modified discrete cosine transform coefficients on frequency spectral to extract pitches adaptively	DDP	Accuracy: 70% Top 1, 85% Top 10, and 0.208 min RT
Ryynanen and Klapuri [14]	MIDI 48 and melodies 6030 Dataset 1, 427 files and melodies 159 Dataset 2	Hidden Markov models used to represent notes and pitch vectors	LSH	0.885 MRR,92% TXHR,4-20 Speed factor Dataset1,0.578 MRR,69% TXHR,4-20 Speed factor Dataset2
Amiya Kumar *et al.* [3]	MIDI	Pitch contour of three levels	DP-based edit distance	Accuracy close to 100%
Jing *et al.* [11]	Songs 100 and fragments 40	Pitch and notes vector	LSH and GA	85% Accuracy, 10% Top 1,50% Top 4–10, 65% Top 10–20, 85% Top 20–30
Zhiyuan *et al.* [5]	–	Key transposition recursive alignment, boundary location using BALS and multiple indices	Notes Local Sensitive Hashing (NLSH) and Pitch Local Sensitive Hashing (PLSH)	MRR:any hum 0.873, beginning hum 0.912

Proposed method 1: Enhanced statistical descriptors	Kannada devotional songs with instrumentals, Kannada devotional songs without instrumentals and Hindi film songs without instrumentals datasets; 1200, 500 and 500 MP3 songs with 1495 fragments; Kannada devotional humming queries and Hindi film humming queries, humming query datasets	First moment (mean), second moment (SD), third moment (skewness), fourth moment (kurtosis), three percentiles, Shannon entropy and ZCR	EMD and sliding window	80.67% Accuracy, 40–59% MRR, 36.34–51.34% MoA, 61.34–80.67% TXHR, 0.51–1.34 min RT
Proposed method 2: Acoustic feature	do	Mel frequency cepstral coefficients	DTW and EMD	82% Accuracy, 46–63.34% MRR, 40.34–55% MoA, 65.67–82% TXHR, 0.58–1.34 min RT
	do	Linear predictive coefficients	do	84.34%Accuracy,48–65.67% MRR, 39–53.67% MoA, 68.34–84.34% TXHR, 0.54–1.30 RT min
	do	Linear predictive cepstral coefficients	do	85.34% Accuracy, 49–67% MRR, 36.67–51.67% MoA, 69.34–85.34% TXHR, 0.51–1.27 RT min
Proposed Method 3: Perceptual features	do	Spectrum centroid, spectrum spread, sone, pitch class profile or chroma, harmonic spectral centroid, harmonic spectral spread, harmonic spectral variation and harmonic spectral deviation	EMD, context drift information through perceptual transfer entropy and new similarity measure	88.67% Accuracy, 50.67–69% MRR, 42.67–57% MoA, 71–88.67% TXHR, 0.44–1.24 RT min

(Continues)

Table 13.1 (Continued)

Author name	Datasets	Melody features	Classification techniques	Retrieval performance
Proposed Method 4: MF features and SM speed-up	do	Rhythm type: Rhythm patterns, statistical spectrum descriptor, rhythm histogram, pulse metric, band periodicity; Tempo type: Beat spectrum, cyclic beat spectrum, beat tracker, beat histogram	EMD, clustering and SM	90% Accuracy, 52.67–70.34% MRR, 44–58% MoA, 72.34–90% TXHR, 0.34–1.06 min RT
Proposed method 5: PF speed-up	do	MRH	EMD, PF uses multiresolution histograms distance and pruning	82.67% Accuracy, 43–62% MRR, 36.67–51% MoA, 65.34–82.67% TXHR, 0.37–1.18 min RT
Proposed method 6: CME, Shannon's information theory and speed-up	do	CMME	EMD, Indexing uses CMMEB Kd tree and K-NN	83.67% Accuracy, 47.34–66% MRR, 39.34–53% MoA, 68.67–83.67% TXHR, 0.34–1.08 RT min

Table 13.2 The tabulation of MRR and average MRR values in percentage

Projected techniques	Fragment subsets								
	100	**200**	**400**	**600**	**800**	**1000**	**1200**	**1350**	**1495**
Kannada devotional songs with instrumentals dataset									
1. Enhanced statistical	58	56	53	51	49	47	45	42	40
2. Acoustic	66	65	63	61	58	55	54	51	48
3. Perceptual	68	64	62	61	59	56	53	51	49
4. MF with SM	70	68	66	63	61	59	56	54	52
5. PF with MRH	61	59	57	55	53	51	49	45	43
6. CMME with Kd tree	65	63	61	59	56	54	52	50	47
Kannada devotional songs without instrumentals dataset									
1. Enhanced statistical	60	58	55	52	49	47	45	43	41
2. Acoustic	68	66	65	63	61	58	56	53	50
3. Perceptual	70	66	65	64	61	58	56	54	52
4. MF with SM	71	69	68	65	63	61	59	56	54
5. PF with MRH	63	61	59	57	54	52	51	47	44
6. CMME with Kd tree	67	65	63	61	59	57	54	51	48
Hindi film songs without instrumentals dataset									
1. Enhanced statistical	62	60	56	53	50	48	46	44	42
2. Acoustic	70	68	66	64	62	60	58	54	52
3. Perceptual	72	70	68	65	63	60	58	56	54
4. MF with SM	73	71	69	67	65	63	60	57	55
5. PF with MRH	65	63	61	59	56	54	52	48	45
6. CMME with Kd tree	69	67	65	63	60	58	56	53	50
Average									
1. Enhanced statistical	60.00	58.00	54.67	52.00	49.34	47.34	45.34	43.00	41.00
2. Acoustic	68.00	66.34	64.67	62.67	60.34	57.67	56.00	52.67	50.00
3. Perceptual	70.00	66.67	65.00	63.34	61.00	58.00	55.67	53.67	51.67
4. MF with SM	71.34	69.34	67.67	65.00	63.00	61.00	58.34	55.67	53.67
5. PF with MRH	63.00	61.00	59.00	57.00	54.34	52.34	50.67	46.67	44.00
6. CMME with Kd tree	67.00	65.00	63.00	61.00	58.34	56.34	54.00	51.34	48.34

The suggested MF with SM-based melody representation technique establishes the dominating MRR values than the other five approaches. It is possible because of the discriminating nature of the approach and positioning the desired song in top places as part of the resulting list. The lighter shaded bars in the bar charts of Figure 13.1 indicate the MRR values of better performing criteria.

13.3.2 Mean of average

Likewise, the MoA and average MoA values of all projected criteria are depicted in Figure 13.2 [28]. The PF with MRH method has yielded better MoA value during comparative analysis with reference to enhanced statistical, acoustic, perceptual, MF with SM and CMME with Kd tree techniques.

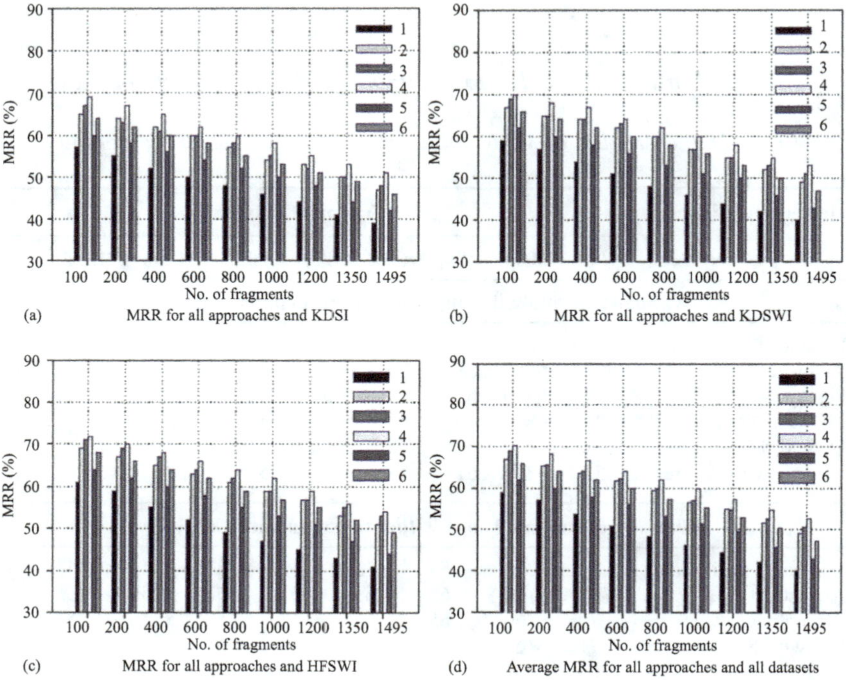

Note: Proposed approaches

1. Enhanced statistical; 2. Acoustic; 3. Perceptual; 4. MF with SM; 5. PF with MRH; 6. CMME with Kd tree

Figure 13.1 The analysis of MRR and average MRR

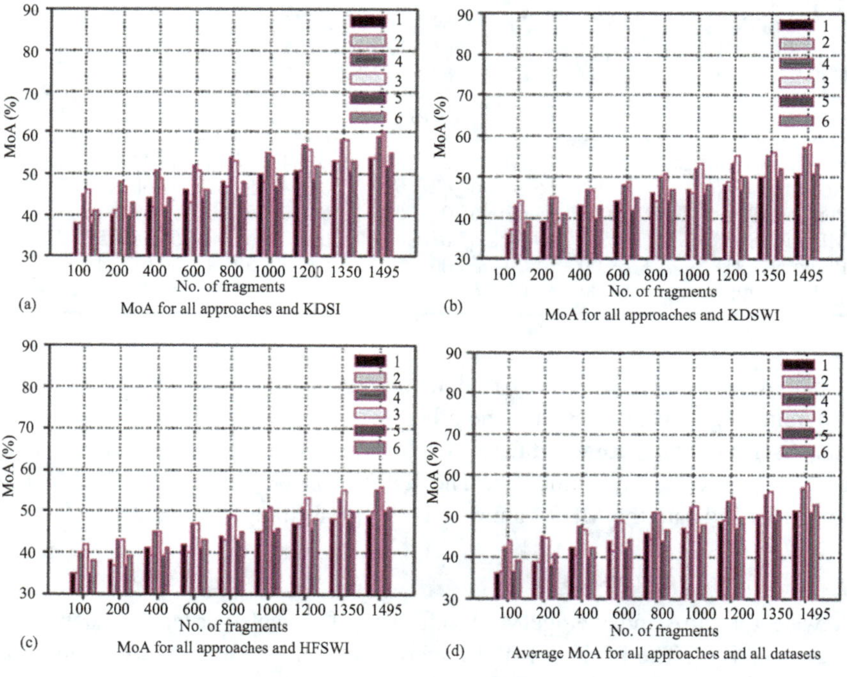

Note: Proposed approaches

1. Enhanced statistical; 2. Acoustic; 3. Perceptual; 4. MF with SM; 5. PF with MRH; 6. CMME with Kd tree

Figure 13.2 The analysis of MoA and average MoA

The PF with MRH approach facilitates obtaining a better MoA value for retrieving the preferred song because of its nature of efficiently locating the preferred song. The better performing techniques are shaded with medium darker bars in the bar chart portrayed in Figure 13.2.

13.3.3 Top X hit rate

The TXHR and average TXHR values are shown in Figure 13.3 [28]. The bars with a lighter shade in the bar charts of Figure 13.3 establish the dominating performance of MF with SM approach in comparison with enhanced statistical, acoustic, perceptual, PF with MRH and CMME with Kd tree approaches. The possibility of identifying queries in the music melody repository is more in MF with SM approach.

13.3.4 Retrieval time

The relative investigation of the experimental results in relevance to CMME with Kd tree method unveiled that the RT is good as compared to other projected criteria. The good performance is because of music melodies indexing with CMMEB Kd tree. Figure 13.4 [28] showcases the RT and average RT values with reference

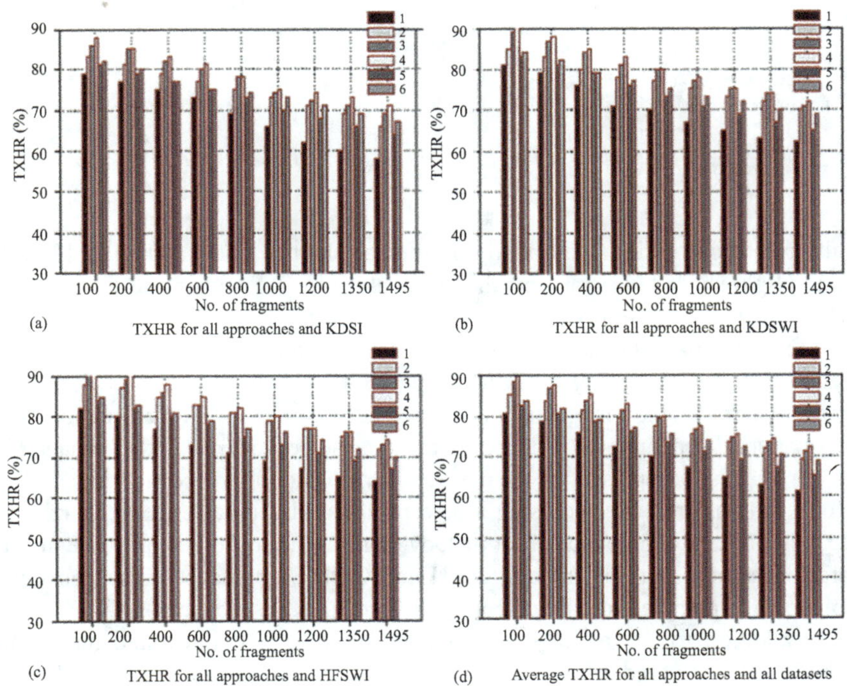

Note: Proposed approaches
1. Enhanced statistical; 2. Acoustic; 3. Perceptual; 4. MF with SM; 5. PF with MRH; 6. CMME with Kd tree

Figure 13.3 The analysis of TXHR and average TXHR

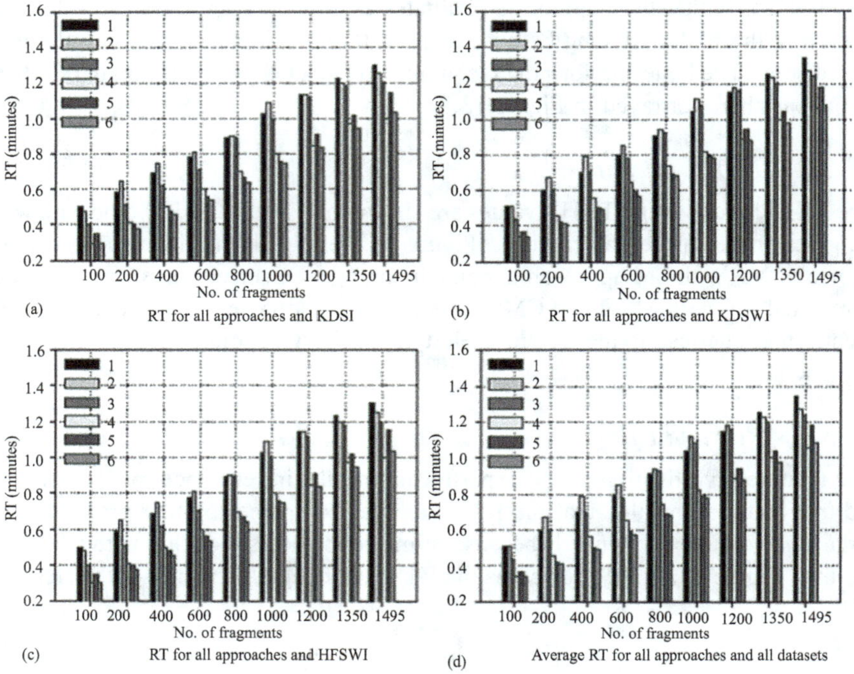

Note: Proposed approaches
1. Enhanced statistical; 2. Acoustic; 3. Perceptual; 4. MF with SM; 5. PF with MRH; 6. CMME with Kd tree

Figure 13.4 The analysis of RT and average RT

to all projected criteria. Additionally, in the bar chart depicted in Figure 13.4, the average light shaded bar confirms the least values for RT metric and distinguishes the criteria as the highest performing speed-up technique.

These criteria establish the probability, vocal tract, perceptual and spectral aspects of melody representation. In particular, the enhanced statistical descriptors approach optimises the performance with reference to the probability estimation of the query patterns. Similarly, the acoustic approach presumes to embody the human vocal tract parameters for representing the melody. Human auditory perception-based music melody patterns are extracted using a perceptual-based approach. Consequently, the spectral features are more crucial for melody representation. Additionally, the melody characterisation is realised through MRH and CMME approaches based on multi-resolution and complex multiscale entropy information, respectively.

In addition to the melody representation perspective illustrated above, the methods are compared in relevance to speed-up facet as well for SM, PF and CMMEB Kd tree indexing techniques. Approximately close to human expectation and optimal RT is obtained through the CMMEB Kd tree indexing approach.

The experimental outcomes presented showcase the enhanced MRR, MoA, TXHR and RT for subsets of various target music fragment datasets. Owing to the nature of enhancing the naturalness and quality of the musical song retrieved, the projected approaches are applied for music analysis. As a final point, musical signatures extracted from these approaches influence and enhance the music identification accuracy for various automatic music recognition applications.

13.4 Approximation and envisioning of relations among performance appraisal metrics

The work showcased in this paper is an in-depth study and analysis of melody representation and speed-up approaches for the QBH system. The experiments harness the enhanced statistical, acoustic, perceptual and MF features, in order to provide complete insight into the melody parameters. Besides, PF with MRH and CMME with Kd tree techniques seeks to epitomise speed optimisation.

The yardsticks used for forecasting the performance of the proposed criteria are MRR, MoA, TXHR and RT. Additionally, for estimating the potential and to get a comprehensive view of the attainment, approximation and envisioning technique is illustrated in this section.

The technique focuses on providing an estimation of the MRR and MoA relationship with reference to the region covered by MRR and MoA lines. Also, the subsequent association between the metrics TXHR and RT establishes the threshold which helps to find the trade-off between accuracy and speed in the form of intersecting point.

13.4.1 Relevance analysis of mean reciprocal and mean of average rank

The main aim of relevance analysis with reference to MRR and MoA is to institute the correlation between the possibility of getting the preferred song in the top place and the average rank of the preferred song.

The MRR and MoA rank distribution characterises the potential span of values and the possibility that the desired song will be listed in the region defined by MRR and MoA curves. Eventually, it establishes the possibility of the preferred song being listed in the topmost positions. A higher value of the possibility shows the encouraging lookouts and it is estimated using relevance probability percentage through a number of fragments. Each and every graph is generated using experimental values. They embody the region consisting of MRR and MOA curves for all projected methods and the same is shown in Figure 13.5(a–e and f) [28], respectively.

With reference to unveiling the potential of the projected methods, it is significant to generate the pragmatic supposition using these areas. Nevertheless, many-fold criteria can be harmonised using the research outcomes.

The region connecting MRR and MoA curve for every method is established through the following equation:

$$A = \frac{\sum_{i=1}^{N-1} \left[\left| \frac{\text{MRR}_1 - \text{MoA}_1}{100} \right| + \left| \frac{\text{MoA}_i + \text{MoA}_{i+1}}{2 \times 100} \right| \times \left| \frac{\text{MRR}_{i+1} - \text{MRR}_i}{100} \right| \right]}{10}$$

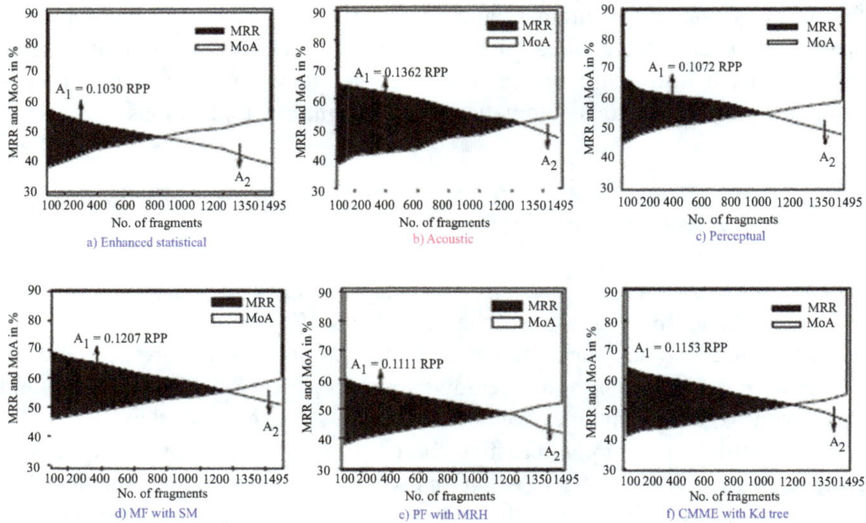

Figure 13.5 Analysis of the area for MRR against MoA curve

In every graph shown in the figure, there are two distinct regions indicated with labels A1 and A2. Further, region A1 is shaded in black colour. The shape and size of the region expand and shrink, respectively, as the crossover point moves from left to right and vice versa. The movement indicates the good performance of the proposed criteria with respect to MRR and MoA values. This hypothesis holds valid for the acoustic, perceptive, MF with SM, PF with MRH and CMME with Kd tree criteria as depicted in Figure 13.5(b–f) [28].

Assume the regions A1 and A2 are roughly of the same size; subsequently, the performance with respect to MRR and MoA values is considered average and the same is true with enhanced statistical technique as portrayed in Figure 13.5(a).

All the results documented indicate the varying level of performance of different approaches with respect to MRR and MoA ranks. Consequently, the area A1 value is high for an acoustic criterion which provides the evidence of good performance. Additionally, the area A1 value measurement for MF with SM, CMME with Kd tree, PF with MRH, perceptive and enhanced statistical techniques facilitate their placement in second, third, fourth, fifth and sixth positions, respectively, as shown in Figure 13.5(d, f, e, c and a) [28].

13.4.2 Synchronisation of accuracy and retrieval time with intersection point analysis

The presented analysis is aimed at the synchronisation of accuracy with RT using the localisation of intersecting points between TXHR and RT lines. The analysis institutes more perceptiveness and supports setting up the trade-off between accuracy and speed. Ultimately, it conveys the intersecting probability percentage which varies with reference to the synchronisation point.

Primarily, the graphs are drawn for TXHR and RT individually using the experimental values. Later, using the equation given below, the interesting point of TXHR and RT lines is calculated for each technique. Consequently, the curves portrayed in Figure 13.6(a–e and f) showcase TXHR and RT lines. Also, the intersecting point of both curves is shown for all instituted methods.

$$\mathrm{IP} = \frac{\sum_{i=1}^{N}\left[\left|\frac{\mathrm{TXHR}_1-\mathrm{RT}_1}{100}\right| + \left|\frac{1-\mathrm{TXHR}_1}{100}\times\frac{\mathrm{RT}_1}{100}\right|\right]}{10}$$

As indicated in the figure, the intersecting point in each graph is present at a different place in relevance to varying results of TXHR and RT parameters. Conversely, the point is located towards the left, middle or right side of the graph. It is indicating the low, medium as well as high trade-off in relevance with TXHR and RT lines.

Further, in consonance with this supposition, enhanced statistical, acoustic, perceptive, PF with MRH, CMME with Kd tree and the MF with SM technique's attainments confirm that the proposed approaches provide good performance. As a result, the proposed methods are assigned with ranks as first, second, third, fourth, fifth and sixth based on low to high performance. Also, the values of TXHR and RT metrics are considered for assigning ranks as portrayed in Figure 13.6(a, b, c, e, f and d) [28].

Collectively, the analyses of the proposed approaches provide evidence for different perspectives on the experimental results. Also, the in-depth patterns hidden in the relationships are visualised through outcomes.

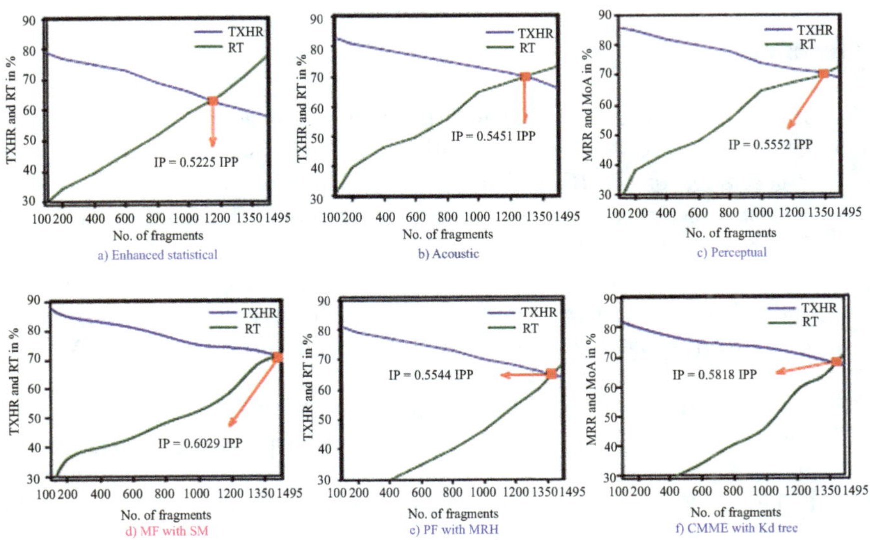

Figure 13.6 Analysis of intersecting point

13.5 Conclusion

In the current chapter, the complete analysis of the methods proposed for the QBH system is portrayed. In the beginning, the outcomes are compared with the pertinent pre-existing QBH research works; afterwards, they are compared with each other. It is instituted that the presented approaches step up the results of MRR, MoA, TXHR and RT in propinquity with preexisting methods.

Subsequently, the investigational outcomes of proposed music melody recognition and retrieval methods are analogised. It is established that the MF with SM method proficiently insinuates the preferred song and refines MRR and TXHR results. Additionally, MoA measure analysis with reference to proposed approaches designated that the PF with MRH is trivial when compared with the other methods. Also, the outcome analysis indicated that the CMME with Kd tree indexing approach is mainly the reason for the enhancement of retrieval speed.

Furthermore, much deliverance was garnered from the relative assay of the stated approaches by discerning the inclination of the graphs along with the cutoff points. The propinquity study of six methods reveals the connotation of each from a different aspect.

The research framework presented portrays a brewing QBH system with the melody representation and speed criteria that are crucial for the enactment. These sorts of research frameworks are obligatory to establish the comprehensive advantages of QBH as a service for passionate music enthusiasts.

Largely, the outcomes of the research framework unveiled, showcase evidence for the potential of melody representation features and speed-up approaches. Moreover, a thorough conception of results from all portrayed methods is endowed with comprehensive perception. Overall, the research framework showcases the pragmatic exploration of the contemporary vogue in the QBH music retrieval system.

References

[1] T. C. Nagavi and N. U. Bhajantri, Overview of automatic Indian music information recognition, classification and retrieval systems, In: *Proceedings of the IEEE International Conference on Recent Trends in Information Systems (ReTIS)*, Jadavpur University, Kolkata, India, 21–23 December 2011, pp. 111–116.

[2] A. Ghias, J. Logan, D. Chamberlin, and B. C. Smith, Query by humming musical information retrieval in an audio database, In: *Proceedings of the Third ACM International Conference on Multimedia*, Francisco, USA, 1995, pp. 231–236.

[3] A. K. Tripathy, N. Chhatre, N. Surendranath, and M. Kalsi, Query by humming system, *International Journal of Recent Trends in Engineering*, 2009, 2(5), 373–379.

[4] L. Fu and X. Y. Xue, A new efficient approach to query by humming, In: *Proceedings of the International Computer Music Conference, ICMC*, Miami, USA, 2004, pp. 1–6.

[5] Z. Guo, Q. Wang, G. Liu, and J. Guo, A query by humming system based on locality sensitive hashing indexes, *Elsevier Journal on Signal Processing*, 2013, 93(8), 2229–2243.

[6] G. Sasubilli and A. Kumar, Machine learning and big data implementation on health care data, In: *4th International Conference on Intelligent Computing and Control Systems (ICICCS)*, 2020, pp. 859–864, doi:10.1109/ICICCS48265.2020.9120906.

[7] S. M. Sasubilli, A. Kumar, and V. Dutt, Machine learning implementation on medical domain to identify disease insights using TMS, In: *2020 International Conference on Advances in Computing and Communication Engineering (ICACCE)*, 2020, pp. 1–4, doi:10.1109/ICACCE49060.2020.9154960.

[8] H. Y. Lie Lu and H.-J. Zhang, A new approach to query by humming in music retrieval, In: *Proceedings of the IEEE International Conference on Multimedia and Expo (ICME), Microsoft Research*, Tokyo, Japan, 2001, pp. 595–598.

[9] Y. Zhu and D. Shasha, Warping indexes with envelope transforms for query by humming, In: *Proceedings of the ACM SIGMOD International Conference on Management of Data*, San Diego, California, June 9–12, 2003, pp. 181–192.

[10] R. B. Dannenberg and N. Hu, Understanding search performance in query by humming system, In: *Proceedings of the 5th International Conference on MIR, Barcelona, Spain*, 2004.

[11] J. Qin, H. Lin, and X. Liu, Query by humming systems using melody matching model based on the genetic algorithm, *Journal of Software*, 2011, 6(12), 2416–2420.

[12] X. Wu and M. Li, A top-down approach to melody match in pitch contour for query by humming, In: *Proceedings of the 5th International Symposium on Chinese Spoken Language Processing*, Singapore, 2006, pp. 1–12.

[13] Z. Wang and B. Zhang, Quotient space model of hierarchical query-by-humming system, In: *Proceedings of the IEEE International Conference on Granular Computing*, 2005, vol. 2, pp. 671–674.

[14] M. Ryynanen and A. Klapuri, Query by humming of MIDI and audio using locality sensitive hashing, In: *Proceedings of the IEEE International Conference on Acoustics Speech and Signal Processing*, Las Vegas, Nevada, USA, 2008, pp. 2249–2252.

[15] E. Unal, S. S. Narayanan, and E. Chew, A statistical approach to retrieval under user dependent uncertainty in query by humming systems, In: *Proceedings of the 6th ACM SIGMM International Workshop on Multimedia Information Retrieval*, New York, USA, 2004, pp. 113–118.

[16] X. Jin, Computer music query by humming considering subsequence matching algorithm, *Journal of Physics Conference Series*, 2021, 2037(1), 12–28.

[17] T. C. Nagavi and N. U. Bhajantri, Effectiveness of empirical mode decomposition for query by humming systems, In: *Proceedings of the Eighth International Conference on Data Mining and Warehousing (ICDMW),*

Elsevier Science and Technology Publications, Bangalore, Karnataka, India, 2014, pp. 259–267.

[18] T. C. Nagavi and N. U. Bhajantri, Enhanced statistical descriptors of the intrinsic mode functions for query by humming systems, *International Journal of Information Processing (IJIP)*, 2014, 8(4), 45–55.

[19] T. C. Nagavi and N. U. Bhajantri, Perceptive analysis of query by singing system through query excerption, In: *Proceedings of the ACM Second International Conference on CCSEIT*, Coimbatore, India, 2012, pp. 580–586.

[20] T. C. Nagavi and N. U. Bhajantri, An extensive analysis of query by singing/humming system through query proportion, *The International Journal of Multimedia and Its Applications (IJMA)*, 2012, 4(6), 73–86.

[21] T. C. Nagavi and N. U. Bhajantri, Relating perceptual feature space and context drift information in query by humming system, In: *Proceedings of the ICCD, Vol. 2, Bhubaneswar, Odisha, India, Springer, Advances in Intelligent Systems and Computing, Intelligent Computing, Communication and Devices*, 2014, vol. 309, pp. 161–171.

[22] T. C. Nagavi and N. U. Bhajantri, Enhanced perceptual feature space with context drift information for query by humming system, *International Journal of Reasoning-based Intelligent Systems (IJRIS)*, 2015,7(3/4), 161–170.

[23] T. C. Nagavi and N. U. Bhajantri, A new approach to query by humming based on modulation frequency features, In: *Proceedings of the IEEE International Conference on Wireless Communications, Signal Processing and Networking (WiSPNET)*, Chennai, India, March 22–24, 2017, pp. 1675–1679.

[24] T. C. Nagavi and N. U. Bhajantri, Progressive filtering using multiresolution histograms for query by humming system, In: *Proceedings of the First ICMCCA, Springer LNEE, Vol. 213, Multimedia Processing, Communication and Computing Applications*, Bangalore, India, 2013, pp. 253–265.

[25] T. C. Nagavi and N. U. Bhajantri, Progressive filtering approach for query by humming system through empirical mode decomposition and multi-resolution histograms, *The Journal of Intelligent Systems (JISYS)*, 2015, 24 (2), 265–275.

[26] T. C. Nagavi and N. U. Bhajantri, Fast query by humming system based on complex multiscale music entropy and CMMEB Kd tree, *International Journal of Grid and Utility Computing (IJGUC)*, 2015, 6(3/4), 159–169.

[27] T. C. Nagavi and N. U. Bhajantri, Query by humming system through multiscale music entropy, In: *Proceedings of the International Conference on Intelligent Computing, Communication and Devices (ICCD), Vol. 2, Springer Advances in Intelligent Systems and Computing Series Intelligent Computing, Communication and Devices*, Bhubaneswar, India, 2014, vol. 309, pp. 139–150.

[28] T. C. Nagavi, *Retrieval of Music Through Melody: An Experiment with Humming (Doctoral thesis)*. Visvesvarya Technological University, Belgaum, India, 2017, pp. 1–203.

Chapter 14

Python for digital health solutions: elevated outcomes

Ayushmaan Khurana[1] and Mayank Aggarwal[2]

This chapter elaborates upon the current as well as potential future use-cases of Python-based solutions in various sectors of the healthcare industry such as research and diagnostics. The text discusses the improvements of such solutions made to the current healthcare system as well as the drawbacks of such implementations. Some of the implementations discussed are Python-based patient health portals, disease and hospital-workflow simulations, remote patient-diagnosis systems as well as medical image segmentation to aid in medical diagnosis. Wherever possible, examples of suitable real-world implementations have been included.

14.1 Introduction

Python[1] is an interpreter-based, object-oriented high-level programming language that was first released in the year 1991. Nearly 30 years later, Python is regarded as one of the most popular programming languages used nowadays and has been integrated into various processes in our everyday lives, such as transportation, finance and healthcare [1].

The general functioning of healthcare systems[2] around the globe is ever-evolving. From the discovery of herbal medicine in pre-modern societies to the exorbitant magnetic resonance imaging (MRI) machines that extravagant hospitals possess and use every day, the evolution of our healthcare systems is imperative for the efficient provision of medical aid to the public. And very recently, the wide-spread use of Python as a versatile programming language has opened the gates to a variety of technological advancements, and its implementation in the healthcare system presents a plethora of advantages to improve the functioning of the same.

It is imperative that one understands the significance of a powerful tool like Python, and the potential it holds in such a vast and diverse industry. The open-source, versatile nature of Python permits the existence of a variety of specialised solutions for specific use-cases, as well as incredible capabilities for real-time

[1]Global Indian International School, Singapore
[2]Gurukul Kangri University, India

interventions in terms of patient care. The plethora of libraries and frameworks available ready to be utilised in Python are responsible for its vast feature set.

Around the world, various researchers, scientists, and healthcare management officials are beginning to implement Python-based solutions to make workflows and tasks more efficient and accessible [2]. Ranging from genomics and bioinformatics research to patient data management and real-time healthcare intervention, Python's potential for improving the current state of the healthcare industry is enormous.

On the other hand, the disadvantages of such implementations must be noted as well. Switching healthcare processes to Python-based systems and automating aspects of healthcare sectors can have drawbacks in terms of ambiguous responsibilities as well as the costs involved in setting up such systems. The advantages and disadvantages of these systems must be weighed before large-scale implementation can be carried out.

Throughout this chapter, the current and potential future implementations of Python solutions in various aspects of the healthcare industry have been detailed, along with some appropriate real-life examples. Moreover, a basic overview of the evolution of the industry has also been discussed in the following section, along with a case study of the healthcare system of a well-developed modern nation, Singapore.

14.2 An overview of the evolution of the healthcare industry

The implementation of technology in the healthcare industry dates back to 1815 when the first basic stethoscope was used to detect a heartbeat. Since then, the integration of technology into medicine has been continuous and is still ongoing. Various notable advancements were made in the years that followed, such as the first use of the X-ray scan in 1896 and the ultrasound in 1956 [3].

Although healthcare information management as an industry was founded in the 1920s, only paper records were maintained in the decades that followed. It was only near the 1950s and 1960s that the technological boom led to the development of basic electronic medical databases by universities and researchers, which could then be purchased by hospitals. The first electronic health record (EHR)[3] systems were developed around the 1980s, and the use of EHR system spreads slowly over the two decades that followed. It was only near the early 2000s that the importance of properly maintained EHRs was realised, due to a significant number of deaths caused by overlapping paper records and consequent medical errors. EHR implementation soon increased exponentially [4].

14.2.1 A case study of Singapore

Singapore is a great example to portray the stages of development of the healthcare industry as a whole, and also to see the impacts of various changes that were implemented in the industry. This will allow us to better appreciate the extent to which Python can help further advancements in healthcare systems.

Singapore's primary means of healthcare remained to be traditional remedies till the nineteenth century, when it was first introduced to western medicine by the arrival of Thomas Prendergast, a sub-assistant surgeon, in the year 1819. He, along with an assistant surgeon William Montgomerie, oversaw the medical needs of the European community present in Singapore at that time. However, the Singaporean government states that there was no form of an organised public healthcare industry till the 1830s, with the establishment of the first Rural Health Officer [5].

Throughout the nineteenth century, Singapore's healthcare system slowly evolved, and various facilities were introduced in order to improve its reach and efficiency to the public. Administrative bodies such as the Municipal Health Department and the Rural and Urban Health Authorities were established in the years 1887 and 1896, respectively, to achieve organisational stability in the infant healthcare industry. The first form of organised, nationwide health policy came in the form of birth and death registrations along with public vaccination schemes, introduced by the British government in the middle of the nineteenth century [6].

The first few hospitals introduced in Singapore, such as the Singapore General Hospital (founded in 1821) and Tan Tock Seng Hospital (founded in 1884), all followed a paper-based record system, given the lack of technological facilities at that time. It was not until the Ministry of Health of Singapore (founded in 1955) implemented the National Electronic Health Record (NEHR) system in the year 2011, that Singapore finally saw an established database for patient data [7].

Although Singapore is yet to see any significant form of implementation of Python in its healthcare system, one of Singapore's objectives under its Smart Nation program is to implement artificial intelligence and analytics in fields such as healthcare. It seems as though the use of predictive analytics is not as far as some may have thought. However, it is imperative to discuss in depth the specific role Python can play in the healthcare industry before its viability can be judged [8].

14.3 Python's role in the healthcare industry

Python's current use in the healthcare industry is limited, due to its true potential in the system being discovered only very recently. However, experts project wide-spread implementation of the programming language due to its versatile set of libraries and frameworks, as well as the wide scope of implementation in the healthcare system due to ever-increasing needs for efficiency and automation.

Python is an open-source language that is easy to grasp because of its user-friendly syntax. The reduced volume and hence processing time of the source code make it efficient and reliable in use-cases such as healthcare technology. Owing to its versatility and interoperability with other widely used programming languages, it is easy to integrate Python-specific processes into an established workplace and industry systems with minimal time spent on testing and deployment. This integration has the potential of leading to increased efficiency as well as human-error reduction due to its rigid, robotic nature.

These features also allow Python to be an effective tool for implementations of artificial intelligence and machine learning, which is a helpful tool to tackle the enormous amounts of user data that such implementations may have to analyse and/ or sort. This also opens up the world of visual data processing, which is extremely beneficial in forms of image analysis including, but not limited to, setting optimal contrast levels for medical image data, as well as highlighting anomalies and abnormalities in the same [9].

The potential uses of Python in various sub-processes of a typical healthcare system have been discussed below.

14.3.1 Healthcare data management

It is estimated that nearly 2300 exabytes of patient data were generated in the year 2020, while the available storage capacity in the healthcare industry was projected to be close to 985 exabytes [10]. This exorbitant amount of data needs to be processed, sorted and maintained for easy access, and must also be analysed either manually or through the help of artificial intelligence in order to derive information and predictions about the patient's condition. As discussed before, EHRs are a vital tool in storing and maintaining patient data.

Singapore's EHR is called the National Electronic Health Record, or NEHR. The NEHR collects and summarises patient health records from a multitude of healthcare providers in a concise format allowing healthcare personnel to easily and quickly access a patient's medical history to provide appropriate medical care [11].

Data management and analysis is an integral part of hospital operations management, which has been digitised only in recent years with the implementation of EHRs. Thus, there is a growing need of optimising this sector of the healthcare industry to allow for a more efficient and streamlined process.

Python is an effective tool to tackle this problem. With its incredible data management and analytic capabilities fuelled by its wide variety of frameworks and libraries, Python is the ideal tool for this task. Around the globe, Python has started being implemented in order to develop patient health portals and databases of biomedical data.

One of the major factors that make Python such a remarkable tool for data management and analysis is its longevity. Python as a programming language is still being developed and amended to make it better and more efficient. This ongoing developer support allows for Python to be reliable as an implemented solution, not just specific to data management. Moreover, the various available methods of Python implementation offer industries flexibility in terms of satisfying their needs. Python as an object-oriented programming language can be utilised directly or via scripting, which makes it a versatile tool with a lot of different implementation options [12].

The wide variety of data analysis-oriented Python libraries allows Python to be extremely efficient when handling widely repetitive and robotic tasks which often accompany the management and analysis of large volumes of data. Similar to the concept of robotic process automation, Python implementation has the ability to free up the human workload and allow for the reduction of repetitive, redundant tasks, increasing workforce efficiency as well. The following case study displays the viability of such implementations in a real-world scenario.

14.3.1.1 Case study on Arateg

Arateg is a software development company situated in Minsk, Belarus founded in the year 2014, that specialises in the development of multi-user platforms and data-collection portals. When approached by a healthcare service provider in the UK, Arateg was tasked with the creation of a patient health portal to help the provider evolve from a paperwork-based, inefficient system that was getting in the way of efficient healthcare provision.

However, management of patient data is different from general big data management using Python. There are more hurdles to be crossed in terms of privacy and security, and the necessity of fault-tolerability. Patient data must be safe and secure and should be readily and rapidly accessible or stored with minimal chances of crashes and failures. Moreover, due to the amount of data involved, the system must have the ability to handle multiple data sources as well as to perform simultaneous data processing.

To achieve these goals, Arateg was able to take advantage of Python's interoperability with softwares such as RabbitMQ and Celery to develop a portal that could handle thousands of parallel data inputs. The utilisation of Proxy.Py allowed Arateg developers to ensure that their portals could handle data of large sizes easily, by providing a structural design pattern to ease the load on the system during simultaneous inputs. They were able to use security libraries and frameworks to implement data encryption and privacy.

All this culminated in Arateg successfully being able to provide their client with a patient health portal that was able to boost the productivity of their employees along with the overall efficiency of the provider by nearly 62 per cent as stated by their own metrics, and they are now reportedly able to service around three times the number of customers [13].

The implementation of Python allowed a paperwork-dependent healthcare service provider to gain the ability to provide online consultations and schedule online appointments, and heavily streamlined the process required to provide medical services to the public. Python's role in increasing the healthcare system's efficiency by being implemented as the primary tool to store and manage patient data can clearly be seen in this case study on Arateg.

Python's role in data management in the healthcare industry is not only limited to storing and managing patient data. Python's machine learning and artificial intelligence capabilities allow it to improve hospitals' workplace efficiency, and predictive analytics allow them to project patient inflow and estimate patient stay duration in the very early stages of the treatment, which is vital information.

These forms of predictive data analytics are one of the reasons Python is regarded as a strong programming language in terms of implementations in real-world processes. With the amount of data being collected by healthcare service providers nowadays, predictive analytics to gain crucial information about inpatient hospitalisation and treatment allow hospitals to plan for resource distribution and predict patient treatment plans and length of stay.

Many forms of such predictive data analytics have already been tested and implemented in the real world. A study by Cold Spring Harbor Laboratory and Yale University along with others attempted to map patient in-flow and hospitalisation patterns through predictive analytics. Data from over 60,000 hospitalisations in a large academic hospital was collected and scrutinised by machine learning algorithms. The goal of the study was to see whether the processed data could accurately predict patient data such as length of stay and allow the hospital to apply this information to improve workplace efficiency by having appropriate resource management.

The results displayed that predictive analytics were able to project the correct category of patients between short-term discharges and long-term patients as well as usage of the intensive care unit department nearly 80 per cent of the time. Although the number is not perfect, given that it is spanning over thousands of patients daily, access to such information could greatly boost workplace efficiency and manage the distribution of assets such as hospital beds and emergency room staff requirements [14].

Python is relevant in such use-cases as it allows users to implement big data analytics tools such as Swift and SQL Query Optimisation directly into a Python framework, thus being an independent tool. The variety and size of data taken in by healthcare service providers being processed with the use of purely Python scripts rather than the parallel use of multiple specialised softwares is a testament to Python's versatility and ingenuity. An experiment carried out by Vishal Tien, then a Master's student at the University of Pennsylvania, in which he processed the publicly available 'New York Hospital In-Patient Discharges in 2015' data set over multiple factors such as the severity of illness, income and race accurately predicted patient hospitalisation duration nearly 70 per cent of the time [15].

With this accuracy, hospitals in New York would be able to greatly reduce inefficiencies in care provision and significantly reduce patient stay duration, allowing for reduced healthcare costs and risk of infection and medication side-effects during the stay of the patient.

These examples accurately display the ability of Python to be a reliable tool in terms of healthcare data management. With the quantity of patient data being generated increasing over the years that pass, implementation of optimal data management and predictive analytics with the help of Python can prove to be a great advantage due to the added benefits of improved efficiency and resource management.

14.3.2 Healthcare simulations

The onset of the SARS-CoV-2 novel coronavirus, commonly known as COVID-19 in recent years, has greatly tested the strength and resilience of healthcare systems around the globe. Owing to factors such as a high basic reproduction number, COVID-19 was able to spread rapidly around the globe and increase healthcare expenditure. Singapore's Ministry of Health saw nearly a 45 per cent increase in healthcare costs between the fiscal year 2019 and 2020, the first year of the pandemic [16].

Owing to the number of epidemics and pandemics faced by the world in recent years, such as the COVID-19 pandemic, the Zika and Ebola outbreaks in the Americas and Africa, as well as H191 (Swine Flu) that broke out in the years 2009–2010, there is a demonstrated need for accurate predictions of the spread of communicable diseases and the implementation of such predictions in lockdown and quarantine strategies used by countries around the globe.

Python is a programming language that can be utilised for such uses, as various Python packages and frameworks have been developed with the aim of accurately modelling the spread of communicable diseases through the use of discrete-event simulations. These simulations take into account factors such as infectivity, population density and the effectiveness of implemented prevention methods to accurately map the spread of a communicable disease in a graphical format. These simulations can then be kept in mind while developing and implementing policies regarding controlling the spread of the disease.

To look at an implementation of Python in discrete-event simulations in healthcare, we take a look at SimPy, which is a Python framework built specifically for running simulations. SimPy allows the user to model simulations with custom variables that can be set using Python generator functions, allowing the user to vary things such as population density, population size and frequency of contact. There is also extensive documentation available to guide users through the process of developing a simulation.

How SimPy functions is by allowing the user to create an *environment* with specific variables that allow them to control various factors in the simulation. After the creation of such an *environment*, the user is now able to define various *events*, which are functions that simulate various circumstances and allow the user to return the required data or call further *events*. These events are dispatched asynchronously, depending on the *environment* that has been set up by the user [17].

Various pandemic simulations[4] have already been developed and run by experts to model the pandemic and project the extent of spread. One of the most important factors to keep in mind when it comes to modelling the spread of communicable diseases is the R0 value. The R0 number is used to estimate the number of people that an infected person can transmit the disease to, i.e. approximation of the disease's tendency to spread among the population. It is calculated using population-level models by running complex mathematical simulations, like the susceptible-infected individual level method. However, this information must be derived before accurate simulations can be modelled using SimPy. A simulation is only as accurate as the starting variables provided [18].

Apart from the R0 value, there are a lot of other variables that allow for a model to be more natural, and replicate the real-world events that we are trying to simulate. SimPy enables users to allow variable mutations due to the implementation of factors such as social distancing measures, natural immunity, etc. However, the accuracy of such simulations is often inversely proportional to the population size among which it is being simulated. The larger the population size, the more tiny variations in the starting variables affect the simulated outcomes.

This is the reason that although modelling communicable diseases has become common in recent times, we are yet to see widespread implementation of this methodology when coming up with public healthcare safety policy. However, the development of and research on stronger and more accurate pandemic and epidemic modelling methods, such as the susceptible exposed infected death (SEIR)[5] epidemic model, means that we are slowly moving towards a future where mathematical models will be soon an integral part of coming up with lockdown and quarantining policies when it comes to tackling pandemics and epidemics.

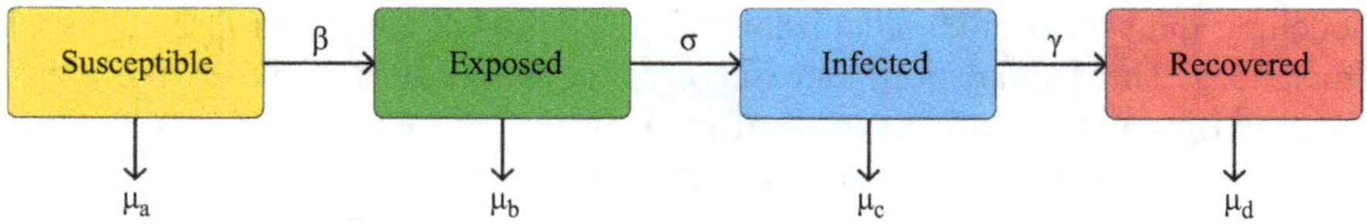

Figure 14.1 An overview of the SEIR model

The above flowchart gives a basic idea of the SEIR mathematical model and allows us to visualise the process to model the pandemic over a long period of time.

The flowchart depicts a flow of the affected population over four stages, namely susceptible, exposed, infectious and recovered. The population inflow in this flow is only under the susceptible stage, however, more complicated models can be used to include exposed and infected populations at later stages of the simulation as well. The population outflow in this flowchart can occur at any stage of the simulation, in terms of death. However, the outflow rate at different stages is affected by different variables that come into play when the population flows from one stage to the next.

Individuals progress from the susceptible stage to the exposed stage based on the variable denoted as β, which is the conversion rate of susceptible individuals to exposed individuals. This is different from the R0 value discussed above, which occurs during the jump from exposed to infected individuals, denoted by the variable σ in the flowchart. Similar to the definition of the R0 value, this variable takes into account the various factors that affect the probability of those exposed to a communicable disease and actually acquiring the disease. The rate at which the susceptible population progresses to the infected population is known as the infection or incident rate. The variable γ signifies the factors affecting the recovery rate, i.e. the rate at which the population infected with the disease is able to make a full recovery.

Throughout the flowchart, loss of life at various stages has been denoted by μ_a, μ_b, μ_c and μ_d. There are different variables as the mortality rate varies at different stages of the progression or spread of disease. The infected population is much more likely to meet a fatal outcome when modelling a dangerous communicable disease.

It should be noted that the SEIR model can have a plethora of variables and other processes that can provide a more accurate modelling of the disease. However, the given flowchart is able to summarise the process to a considerable extent [19].

More complex versions of the SEIR model have been used by various researchers to model the spread of the SARS-CoV-2 virus in Singapore. These models and simulations allow these researchers to estimate the number of infected, deaths and undetected infections after a certain time period, and changing the variables involved to match preventive government policies also give results that can test the efficacy of such measures [20].

It should be taken into account that modelling communicable diseases is not the only simulation that is beneficial to the healthcare industry when it comes to Python's applications. Building on the predictive data analysis discussed in the previous section, hospital simulations allow managerial staff to distribute resources even more efficiently and predict areas where excess healthcare staff may be necessary based on the starting factors.

There are various events and factors that can be predicted with the help of SimPy simulations that allow a healthcare service provider to improve its workings. One of the most important ones is patient flow. Patient flow is defined as the movement of patients through various departments of a healthcare system, be it the emergency room or a consulting doctor's office. Predicting and modelling patient flow allows the administration to ensure adequate resources and staff for specific departments, as well as predict how busy a certain sector of the healthcare system will be on a given day.

This comes under the broad umbrella of healthcare forecasting, which is defined as the act of using predictive analytics and simulations to project, to a considerable extent, the demand for future healthcare services and needs, as well as the extent of utilisation of current healthcare resources. It also considers the prediction of future disease outbreaks and spreads. It is an integral part of the future of predictive analytics in healthcare, as it allows for efficient preventative measures as well as accurate healthcare intervention policies both on a small and a large scale. In the year 2020, global healthcare industries saw the widespread use of healthcare forecasting to predict the need for personal protective equipment as well as ventilators and hospital beds amidst the COVID-19 pandemic, which shows the extent to which such forecasting can aid in healthcare resource management in situations of scarcity.

Researchers from ITMO University in Saint Petersburg, Russia, attempted to simulate patient flow across different departments, specifically for patients suffering from acute coronary syndrome (ACS). The simulation was modelled using a combination of data from patient records, and the use of machine learning and queuing theory to develop appropriate simulation variables. They specifically used Python simulation libraries and frameworks such as SimPy and SciPy and attempted to come up with more efficient clinical pathways for patients with ACS.

When the simulated results were compared with actual data, it was found that their proposed approach was able to provide a more realistic depiction of real-world events, and allowed for an accurate prediction of patient length of stay when data from the Almazov National Medical Research Centre was modelled. They concluded that such discrete-event simulations allow for better operational decision making and management optimisation among others.

Thus we have seen the potential Python-based healthcare simulations hold for real-world implementation in scenarios such as disease-spread modelling and improving efficiency of hospital and clinic workflow processes, again displaying the versatility of Python as a programming language.

14.3.3 Medical diagnosis, prognosis and treatment

Medical diagnosis is defined as the process of determining which ailment fits a patient's set of symptoms the best, i.e. establishing the disease from which the patient suffers. This is done by a medical evaluation of both qualitative and quantitative data collected from patients. However, when this task is done by humans, there are bound to be errors and mistakes. In the field of healthcare, diagnostic errors can be disastrous.

Table 14.1 displays the leading causes of death in the USA in 2019 according to the Center for Disease Control (CDC). Although the CDC does not officially recognise medical errors[6] as a distinct cause of death in its statistics, a study done by patient safety experts at Johns Hopkins extrapolated in-hospitalisation statistics in the USA and was able to estimate nearly 250,000 deaths annually because of medical and diagnostic errors. This estimation puts such errors as the third leading cause of death in the USA, behind only heart disease and cancer cases. This statistic displays the need for intervention in diagnostics and calls for solutions to reduce the number of diagnostic errors [21].

This is what makes the implementation of machine learning so tempting in the field of diagnostics. When machine learning frameworks were given access to patient data, they often evolved to prove to be superior to their human counterparts in terms of accurate diagnoses. But what role does machine learning actually play in this diagnostic process, and how can Python help in the same?

To gather an insight into the use of Python in medical diagnosis, the implementation of python in medical imagery analysis, often referring to scans such as computed tomography, X-ray and MRI scans can be taken into account. These scans often provide a useful insight to a patient's diagnostic team when they want

Table 14.1 Leading causes of death in the USA in 2019

Cause of death in the USA	**Estimated number of deaths**
Forms of heart disease	659,041
Cancer	599,601
Medical errors	250,000
Accidents and unintentional injuries	173,040
Chronic lower respiratory diseases	156,979
Stroke and other cerebrovascular diseases	150,005

to put focus on a specific region of the patient's body. However, it is possible to implement Python into the diagnostic process to narrow down options for the correct diagnosis.

Machine learning in Python can be applied to analyse medical imagery where the task is to classify the scan into categories based on physiological patterns displayed in the scan, with a minimal amount of subjectiveness and variance. Huge data sets derived from existing patient data and expert opinions are necessary to train the machine learning algorithm when it comes to classifying medical imagery. This is because analysis of scans is a subjective process rather than an objective one. Deep learning is able to extract this process of identifying patterns in images and is quickly able to grasp the classification process based on the visual data presented by the scan. Although not all ailments can be accurately diagnosed with this method, this does provide a more efficient and streamlined approach to a rather subjective process [22].

Along with the identification of patterns in medical imagery, machine learning is also a prospect for medical image segmentation. Image segmentation is defined as the process of compartmentalising images based on visual characteristics, specifically boundaries or changes in contrast. Medical image segmentation allows experts to localise objects of interest on scans, such as organs or tumours, allowing for greater focus to be put on the same.

There are specific Python packages developed to deal with medical image data specifically for image segmentation. For example, the open file format known as Neuroimaging Informatics Technology Initiative, i.e. NifTi, is used to store image data taken from MRI scans of brains and can be dealt with a Python package known as Nibabel or the Python framework MIScnn [23].

Medical image segmentation using machine learning consists of the Python framework analysing existing data to allow the neural network to learn generic segmentation techniques and patterns that are present in medical scans. After sufficient data has been processed, it is able to compartmentalise or segment parts of interest in the scans automatically, which may be taken as a conformational or preliminary step in the diagnostic process.

However, the use of Python is not just limited to visual analysis and processing. The study of EHRs by a machine learning algorithm can also be used to develop automated diagnostic aid assistants in the form of chatbots. These serve as remote diagnostic systems,[7] as they are void of human input on the diagnostic side. These systems are becoming increasingly popular due to them being an efficient, cost-effective measure that also frees up resources and improves patient satisfaction due to remote assistance [24].

Python machine learning frameworks can be paired with application programming interfaces for different messaging platforms to allow for the functioning of such a diagnostic system. These systems can also help the healthcare service provider map complex patient behaviour and collect patient data remotely, allowing for a streamlined process and better responses to medical emergencies due to pre-existing data on the patient's condition.

Python's versatility in medical diagnosis and prognosis can be credited, in part, due to its capability to produce deep neural networks. This comes under the process of deep learning and allows a neural network to 'learn' from pre-existing data in a supervised, semi-supervised or even unsupervised manner. The reason for neural networks being such an effective tool in the healthcare industry is the fact that this method of artificial intelligence implementation allows for Python implementations to be set up as a secondary part of the diagnostic process, i.e. it allows the neural network to learn from human diagnostic decisions, permitting it to potentially become an additional source of confirmation when it comes to patient diagnosis. The necessity for learning from human healthcare decisions has the potential of letting the neural network develop more organically and adopt a more critical decision-making process, making it more natural in real-world healthcare applications.

These advantages also have the potential of using Python in setting up patient treatment plans and implementing them. Some of the areas in which the use of Python is predicted in the near future are the administration of intravenous drugs as well as monitoring dosages depending on real-time analysis of changes in patient vitals. Such applications are currently merely in the testing stage. However, we have seen open-source development regarding the same. PiDose, an open-source Python library, has the primary application of oral drug administration to lab mice. Moreover, for more household-friendly applications, Python can be used to power automatic medicine dispensers. The number of potential applications of Python is endless.

14.3.4 Genomics and sequencing

Genomics is a medical research field focusing on the study of an organism's DNA as a whole. Analysis of genomics comes under the broader umbrella of bioinformatics, which is the utilisation of computational and analytical tools on biological data. The study of genomic data has expanded exponentially in recent years, generating large volumes of data annually. It is becoming increasingly popular as the scientific community begins to explore the science behind genetic sequences and inheritance among organisms in depth. Even for individual studies, millions or even billions of data points are generated while studying genomes. This data needs to be processed quickly and accurately in order to produce reliable results in a short period of time [25].

Python has been recently introduced into the field of bioinformatics and is projected to be a vital tool in the field. SciPy is an open-source ecosystem designed to integrate softwares necessary in the field of mathematics and science, including bioinformatics. A similar set of Python tools for biocomputation is BioPython. However, it is necessary to gain a basic understanding of genomics processes such as sequencing and annotations.

Sequencing DNA is a process by which the order of the four DNA chemical building blocks, or bases, namely adenine, guanine, thymine and cytosine, that make up the DNA double helix can be determined. The acquired sequence can help in the identification of the function of the DNA molecule being tested.

To prepare a DNA for sequencing, molecules of DNA are separated using a process known as electrophoresis, in which electric current forces DNA molecules to move through a gel-like medium. Different molecules move through the gel at different speeds due to varying sizes, resulting in successful separation. These separated molecules are then analysed to identify the constituting bases. This is done by introducing fluorescent dyes specific to each of the four bases, which fluoresce upon passing through a laser and are then analysed by the computer.

The result of DNA sequencing is a sequence of DNA with known bases but an unknown function. To specifically identify the function of a DNA molecule or strand, a process known as DNA annotation[8], specifically functional annotation must be carried out. This process can be carried out using specialised bioinformatics software, using tools such as signal and content sensors as well as similarity detection capabilities, i.e. identifying similarities with the help of various reference genomes [26].

DNA Features Viewer is one such tool. It is a Python library specifically designed by the Edinburgh Genome Foundry to visualise DNA features, capable of both basic and advanced plotting of DNA sequences while being compatible with BioPython [27].

However, given that automatic sequencing methods were only developed in the 1980s, large-scale implementation of Python in genomics and sequencing methods is yet to be seen.

The four subcategories discussed are just a fraction of the potential Python implementations hold in the healthcare industry. With everyday advancements in Python libraries and frameworks due to their open-source nature, Python processes such as deep learning and predictive analytics may be the future of the healthcare industry.

14.3.5 A double-edged sword: the disadvantages of Python's implementation

Although we have discussed various advantages of Python's implementation in both small and large-scale healthcare facilities, it is necessary to note that there are added disadvantages that make the prospect of Python's use in such cases grim.

A major factor posing as a hurdle to Python's implementation in the healthcare industry is the question of liability. In case of human medical errors, be it diagnostic or administrative, it is possible to find fault and thus label the person responsible liable. This is necessary as even minute errors can have disastrous consequences in a field like medicine.

However, when automation and artificial intelligence is implemented through the use of Python, the answer to this question of liability becomes blurred. If an incorrect diagnosis was given by a Python-powered remote diagnostic system, leading to consequences for the patient, who will be held liable? The answer is not clear, as machine learning and data analysis are unpredictable, just like real-world events.

It is necessary to realise that obscure events happen in the real world which cannot be foreseen. A data analytic program may project a relatively calmer day for the Emergency Department of a healthcare service provider, but an unforeseen pileup on a national highway may result in a shortage of resources and staff for the said department. This unpredictability necessitates an added safety network for systems with Python implementation.

It should also be noted that switching to an analytical, Python-based system requires a complete overhaul of the data management procedure of a healthcare system. It requires changes in both workplace responsibilities and the amendment of administrative barriers which may prolong the time taken to implement such measures. Moreover, the initial costs involved in setting up implementations such as patient data management systems, remote diagnostic portals and image analysis neural networks among others may deter healthcare service providers from completely or even partially switching to a Python-based healthcare framework.

We must also take into account the various legal and administrative standards of efficiency and accuracy that must be met by Python's implementations before they can be integrated into any public healthcare system. Any prospective Python-based solution would have to undergo rigorous testing as well as privacy and security checks before it could be approved for use in real-world scenarios. This also prolongs the integration of Python into today's healthcare industry.

14.4 Conclusion

The chapter has discussed in detail the potential of Python's implementation in various sectors of the healthcare industry, along with the various roles it can play. Ranging from simple data management for small-scale healthcare service providers, to large-scale communicable disease and pandemic modelling, there is no doubt that Python's prospects in this field are endless. However, we must take into account the various challenges that come with the complete overhaul of a functioning healthcare system, and weigh the advantages with the disadvantages. As of now, the use of Python specifically in the healthcare industry is minimal, and although it is constantly growing, it is unlikely that we see widespread, large-scale implementation of the language in the near future. For now, the various potential use-cases have been discussed, often with reference to real-world examples and implementations such as in countries like Singapore. This chapter has thus briefly described the efficacy of Python-based solutions in real-time interventions in the healthcare industry.

Glossary

1. **Python**: Python is an interpreter-based high-level programming language first showcased in 1991.
2. **Healthcare system/industry**: The industry whose various components serve to provide medical services including, but not limited to, diagnosis, treatment and rehabilitation.
3. **Electronic health records**: Electronic health records are a digitised collection of a patient's medical records compiled over time by medical service providers to gather a comprehensive medical history.

4. **Pandemic simulations**: An electronic simulation of the spread of a disease over a large region.
5. **SEIR model**: A method of modelling pandemics and epidemics based on the population of susceptible, exposed, infective and recovered individuals.
6. **Medical errors**: Medical error refers to preventable, adverse effects of medical care.
7. **Remote diagnostic systems**: A form of telemedicine service that provides medical information and monitors patients without the need for a physical examination.
8. **DNA annotation**: DNA annotation refers to the identification of the locations of various genes and coding regions in a DNA sequence and the determination of the function of such sequences.

References

[1] "*What Is Python? Executive Summary.*" Python.org, 2021, www.python.org/doc/essays/blurb/. Accessed 20 October 2021.

[2] "*Python for the Better Future of Healthcare – Healthcare Technology.*" Goodleven.health, 2021, www.goodleven.health/pythonhealthcare. Accessed 20 October 2021.

[3] "The history of X-rays: how an accidental discovery changed the world." *Radiology*, 18 February 2020, professionalradiologyep.com/the-history-of-x-rays-how-an-accidental-discovery-changed-the-world/. Accessed 20 October 2021.

[4] Net Health. "The history of electronic health Records (EHRs)." *Net Health*, 16 September 2021, www.nethealth.com/the-history-of-electronic-health-records-ehrs/#:~:text=The%20history%20of%20EHRs%20began,systems%20to%20adopt%20an%20EHR.&text=In%20the%201960s%2C%20EHRs%20were,in%20partnership%20with%20health%20organizations. Accessed 20 October 2021.

[5] "*Our History of Healthcare.*" Roots.gov.sg, 2015, www.roots.gov.sg/stories-landing/stories/history-of-healthcare-sg. Accessed 20 October 2021.

[6] Goh, S. "10 facts you should know about Singapore's 200-year healthcare history." *The Straits Times*, 6 January 2015, www.straitstimes.com/singapore/10-facts-you-should-know-about-singapores-200-year-healthcare-history. Accessed 20 October 2021.

[7] "Singapore's healthcare system through the years." *Global Is Asian*, 2018, https://lkyspp.nus.edu.sg/gia/article/singapore%27s-healthcare-system-through-the-years. Accessed 20 October 2021.

[8] "*Smart Nation Singapore.*" Smartnation.gov.sg, 2021, www.smartnation.gov.sg/. Accessed 20 October 2021.

[9] Copeland, B. *Advancing Opportunities in Healthcare with Python-Based Machine Learning.* Chief Healthcare Executive, May 2019, www.chief-healthcareexecutive.com/view/advancing-opportunities-in-healthcare-with-pythonbased-machine-learning. Accessed 20 October 2021.

[10] "Healthcare Data Storage Constraints Globally 2020 Forecast | Statista." *Statista*, 2020, www.statista.com/statistics/1038042/global-healthcare-data-storage-limitations/. Accessed 20 October 2021.

[11] "Home Page." Ihis.com.sg, 2019, www.ihis.com.sg/nehr/home. Accessed 20 October 2021.

[12] Zhidkov, R. "Why Python is essential for data analysis – RTInsights." *RTInsights*, 13 January 2020, www.rtinsights.com/why-python-is-essential-for-data-analysis/. Accessed 20 October 2021.

[13] Glebko, S., Co-founder and CTO at Arateg, "*Using Python to develop a patient health portal.*" Python.org, 2014, www.python.org/success-stories/using-python-to-develop-a-patient-health-portal/. Accessed 20 October 2021.

[14] Bertsimas, D., Pauphilet, J., Stevens, J., and Tandon, M. "*Predicting Inpatient Flow at a Major Hospital Using Interpretable Analytics.*" 18 May 2020, www.medrxiv.org/content/10.1101/2020.05.12.20098848v2.full, 10.1101/2020.05.12.20098848. Accessed 20 October 2021.

[15] Tien, V. "Predicting Inpatient Length of Stay at Hospitals Using Python + Big Data." *Medium, Towards Data Science*, June 2020, towardsdatascience.com/predicting-inpatient-length-of-stay-at-hospitals-using-python-big-data-304e79d8c008. Accessed 20 October 2021.

[16] "Budget 2021: Covid-19 measures meant Govt spent 25% more but raised 13% less revenue in FY2020." *TODAYonline*, 2021, www.todayonline.com/singapore/budget-2021-covid-19-measures-meant-govt-spent-25-more-2020-2019-raised-13-less-revenue. Accessed 20 October 2021.

[17] "*Overview – SimPy 4.0.2.Dev1+G2973dbe Documentation.*" Readthedocs.io, 2020, simpy.readthedocs.io/en/latest/. Accessed 20 October 2021.

[18] van den Driessche, P. "Reproduction numbers of infectious disease models." *Infectious Disease Modelling*, 2017, vol. 2(3), pp. 288–303, www.ncbi.nlm.nih.gov/pmc/articles/PMC6002118/, 10.1016/j.idm.2017.06.002. Accessed 20 Oct. 2021.

[19] "*SEIR and SEIRS Models – HIV Model Documentation.*" Idmod.org, 2021, docs.idmod.org/projects/emod-hiv/en/latest/model-seir.html. Accessed 20 October 2021.

[20] Chung, N. N. and Chew, L. Y. "Modelling Singapore COVID-19 pandemic with a SEIR multiplex network model." *Scientific Reports*, 2021, 11, 10122, https://doi.org/10.1038/s41598-021-89515-7

[21] Sasubilli, G. and Kumar A. "Machine Learning and Big Data Implementation on Health Care data." *4th International Conference on Intelligent Computing and Control Systems (ICICCS)*, 2020, pp. 859–864, doi:10.1109/ICICCS48265.2020.9120906.

[22] Sasubilli, S. M., Kumar, A., and Dutt, V. "Machine learning implementation on medical domain to identify disease insights using TMS," *2020 International Conference on Advances in Computing and Communication Engineering (ICACCE)*, 2020, pp. 1–4, doi:10.1109/ICACCE49060.2020.9154960.

[23] "Introduction to working with MRI data in Python." Github.io, *The Carpentries Incubator*, 2015, carpentries-incubator.github.io/SDC-BIDS-IntroMRI/aio/index.html. Accessed 20 October 2021.
[24] "*Machine Learning in Healthcare Communication*." Encyclopedia.pub, 2021, encyclopedia.pub/8078. Accessed 20 October 2021.
[25] Abdurakhmonov, I. Y. "Bioinformatics: basics, development, and future." *Bioinformatics – Updated Features and Applications*, 27 July 2016, www.intechopen.com/chapters/50934, 10.5772/63817. Accessed 20 October 2021.
[26] "*DNA Sequencing Determines the Order of DNA Building Blocks*." WhatisBiotechnology.org, 2011, www.whatisbiotechnology.org/index.php/science/summary/sequencing/dna-sequencing-determines-the-order-of-dna-building-blocks#:~:text=DNA%20sequencing%20is%20a%20method,to%20become%20(the%20phenotype). Accessed 20 October 2021.
[27] "*DNA Features Viewer*." Github.io, 2016, edinburgh-genome-foundry.github.io/DnaFeaturesViewer/. Accessed 20 October 2021.

Chapter 15

IoT-enabled healthcare – a paradigm shift

Parul Gandhi[1], *Preeti Chikara*[2] *and Meenakshi Yadav*[2]

There has been fast growth in the digitization of data, including health data which is termed as the "Internet of Things" or "IoT in Healthcare." In the wake of the COVID-19 outbreak, IoT in healthcare has grown exponentially. Having the correct medical equipment is a must for a successful medical procedure. The Internet of Things (IoT) is well-suited to both performing effective operations and assessing the results of such activities. Patient care during the COVID-19 Pandemic is improved because to the use of IoT. The IoT allows for real-time monitoring of a variety of health conditions. Using a smartphone, smart medical devices are able to transmit the necessary health data to the physician.

The future of IoT in healthcare is being shaped by a convergence of many fast growing technologies. Health care delivery and technology development face a number of obstacles. These include concerns relating to patient privacy and data security; the digital divide; governmental oversight; and the acceptance of new technologies by medical professionals and institutions. The chapter gives a worldwide overview of the IoT's impact on healthcare.

15.1 Introduction

As all the industries are growing up with the Internet of Things (IoT). It is the major area of concern nowadays. It plays a vital role in all the major sectors of the world like Economics, Medical Science, Defense Systems, Social Media, and other prospects to communicate with the people. The term "Internet of Things" may be defined as a combination of different types of technologies within a network, which works automatically without the help of human interactions. OR, IoT refers to the connection established by two internets of computing devices embedded in objects used in day-to-day life enabling them to send and receive data [1]. IoT has gained new limelight in the recent past and also it has gained much convincing research ground as a new topic to be researched in a huge list of academic, industrial, technological, economical, political, and other social aspects too. But in the field of

[1]Manav Rachna International Institute of Research and Studies, India
[2]Akhilesh Das Gupta Institute of Technology & Management College, India

medical sciences, the use of IoT has created a major change in the treatment of people. Diagnosis of every person has become easier than before because the conventional diagnosis systems have been replaced with more personalized systems which make monitoring patients and treatment better than before. IoT is now playing a more important role in medical science and healthcare due to lower expenses of IoT-based devices, much better services than the conventional systems, a very frequent and easily understandable user experience, and various advanced features too. In the case of IoT, there is no limit to the number of connected devices. It is all about exchanging data from one device to another or exchanging meaningful information between devices. However, the data have to be interpreted correctly and it needs many other technologies to come hand in hand to work side-by-side. IoT is not just a single type of technology; it is a collection of technologies and domain knowledge. In current situations which are being faced by the whole world due to the pandemic, IoT is playing a vital role. In the case of COVID-19, IoT-based technologies are being used for monitoring patients to maintain the protocols, different web applications are used for communicating with people and helping them to provide solutions for various problems being faced during quarantine times. Besides IoT, various web services also play a vital role in fighting these current situations and other future problems. Various web services are currently being used for the collection of data to find out specific information in a limited region or throughout the whole. These web services also work as a mediator between two different sources. It also helps to communicate with people in large numbers. IoT and web services are complementary to each other.

Science-based research institutions and companies of various sizes are pooling their resources to learn more about the virus, develop better testing methods, and ultimately produce an antibody as a result of their collaborative efforts. Many organizations and devices currently use cutting-edge compound science to measure smaller examples or achieve higher throughput. At each stage of our response to the virus, science is critical. In previous research, professionals are given the expert abilities expected to run tests, keep up the gear and oversee lab supplies – including the gifts of synthetics for hand sanitizer and security hardware that are going directly to specialists and attendants on the forefronts. We can protect ourselves using different chemical compounds like chlorine.

15.2 Architecture of IoT

Basically, IoT has three-layer architecture. But no specific architecture has been prescribed by an IoT-enabled system. Three-level architecture is the basic architecture used in IoT. IoT follows a layer-wise top-down approach in architecture and the flow of information also follows a top-down approach. Three-level architecture has a Perception Layer, a Network Layer, and an Application Layer. The perception layer is also known as the sensing layer. In the sensing layer, all sensors collect data from the user and send it to the IoT-enabled system. Network layer is used to collect data through Wi-Fi and process it for further tasks. Application layer delivers data for exact results according to the task.

Nowadays five-layer architecture is used instead three-layer IoT architecture. It is used to transmit a large amount of data and process it in a particular system. In the five-level architecture, there are the following layers: the sensing layer, the transport layer, the processing layer, the application layer, and the business layer [2].

(A) The perception layer:
The sensing layer is another name for the perception layer. This layer contains a plethora of sensors for gathering user data. Several sensors are employed to gather various kinds of data. Additionally, sensors for measuring cardiovascular and muscular activity are included in the body sensor network (BSN).

(B) The transport layer:
The network layer and the transport layer are both terms used to describe the transport layer. User data are sent from one layer to the next layer. It is used to move data up and down the stack from one layer to the next layer. NFC, LAN, Bluetooth, RFID, 3G, or any other wireless media can be used for transmission.

(C) The processing layer:
The processing layer is the middleware layer that is used to process data collected from the user. It is important in an IoT architecture because it removes the useless data which are not mandatory. This layer has innovative structures like storage policies, computation process, running process, and action performing abilities. It stores all the collected data and access it by their IP address. It is also used to take decisions based on different activities performed on data obtained from sensors.

(D) The application layer:
The application layer is used show result according to user data. The results are based on their usage like Alexa, smart watch, etc. The outcome is user-specific and may differ as per user data. For example, user gives command to Alexa by voice then it collect information regarding to the input and gives the result.

(E) The business layer:
With the appropriate application, the business layer participates in services and business operations. Managing the privacy of user data is a major concern and a fierce competition in this layer. Furthermore, it is used to share data for R&D and machine learning (ML).

When physical objects are connected to the Internet, information can be exchanged and exchanged through the Internet. Sensors, ML, real-time analysis, and embedded systems are all examples of IoT technologies that have been developed and changed over the course of the IoT. Fixed or wireless Internet is used to control various gadgets in the smart hospital. Everyday tasks can be completed with the help of mobile devices that can collect and share data. Figures 15.1 and 15.2 shows layer the architecture of IoT and IoT five layer architecture, respectively. Connected healthcare and smart cities are just two examples of how the IoT is expanding its reach. Medical IoT relies on a wide range of sensors, medical equipment, AI, diagnostics, and high-resolution imaging technologies. These gadgets boost efficiency and raise the standard of living for people in both

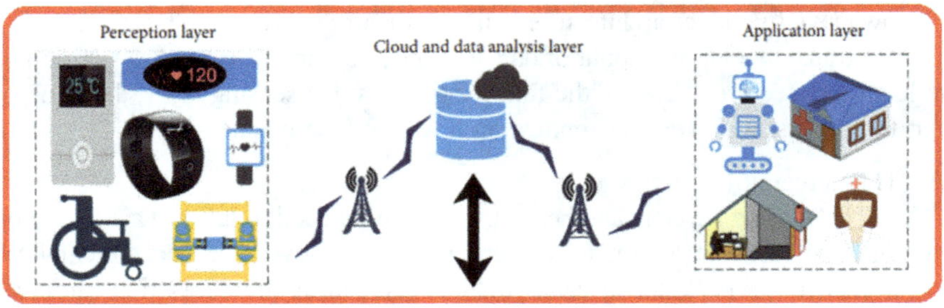

Figure 15.1 Layer architecture of IoT

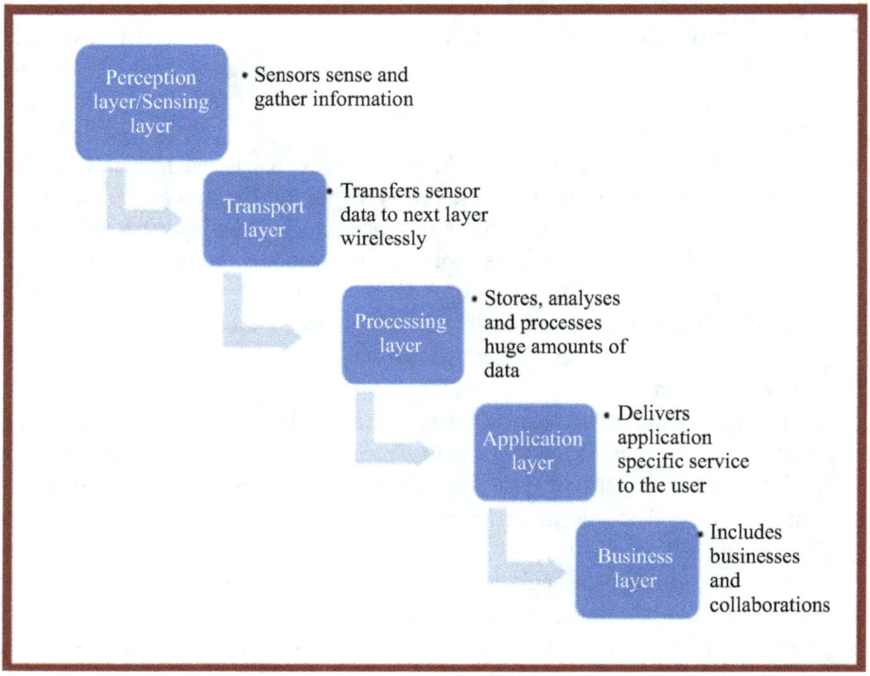

Figure 15.2 IoT five-layer architecture

traditional and modern sectors of the economy and society. There is no human intervention in the data transfer across the Internet as a result of the IoT. When it comes to monitoring healthcare in the wake of the COVID-19 Pandemic, this technology is thriving. A lot of people are dying today because they were given inaccurate and outdated information regarding their health. Through the use of sensors, this technology is capable of swiftly alerting users to potential health risks. There is a significant role for IoT in the medical industry [3].

Figure 15.3 shows the services provided by IoT in the medical field. It also shows applications of IoT in medical management, fitness, and so many different fields. These

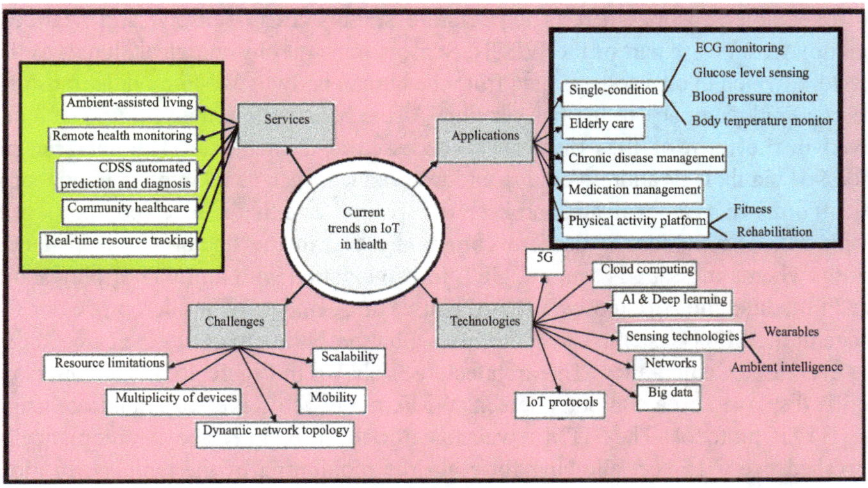

Figure 15.3 Role of IoT in the medical field

figures also show challenges faced by IoT in the medical field. There are many technologies used in IoT like cloud computing, Big data, and Artificial Intelligence (AI).

15.3 IoT implementation in medical field

The development of medical information systems relies heavily on the IoT. The tracking, charting, and monitoring of patients are critical for the advancement of the medical care system. Inadequate data in the healthcare system, on the other hand, leads to erroneous findings. An essential role for IoT in medicine has been examined, and the link between beneficial research and real-world applications has been highlighted in this chapter. The IoT has numerous applications in the medical field. IoT-based healthcare is examined for its innovative community structures, programs, and commercial characteristics, as well as its cutting-edge technology advancements. The unique protection issues are also discussed, and an innovative protection approach is proposed to minimize the security hazard. There are a variety of options for connecting medical devices to an open network, such as the Internet, on the marketplace. In this case, the issue is how to ensure the integrity of statistical data used in clinical trials. There are both limitations and potential for deploying IoT-based fully eHealth services.

15.3.1 Architecture of medical IoT (MIoT)

The main idea of the MIoT system is continuous care provided to the patient by the Internet. The framework for MIoT has three-tier architecture [4]:

(a) Wireless body server network (WBSN)
(b) Base station
(c) Graphical user interface (GUI)

Wearable sensors, such as temperature sensors, ECG sensors, and heartbeat monitor sensors, are part of the WBSN. Sensors transmit physiological signals to the network, which in turn collects data from the human body. A Wi-Fi module transmits the acquired data back to the base station. At the base station, the acquired data are saved in the form of files that can be accessed at a later date. The data can be accessed via their IP address on any end-user device, such as a laptop, a phone, or a PC. In order to save, evaluate, and keep track of the data that is received, the GUI is employed. The modem is used to communicate with the doctor or the patient's family when a message is sent via GUI. In combination with an LCD display and a Wi-Fi module, the Arduino Uno Board houses an ATmega 328p microprocessor for usage in the MIoT system. The doctor uses a buzzer button to notify the caretaker in the event of an emergency. The architecture is shown in Figure 15.4. This information is displayed on a computer screen. Wi-Fi and a web server communicate using the HTTP protocol. The HTML webpage for the user interface is automatically refreshed every 15 sec, allowing for real-time monitoring of the patients. In most cases, the FTP protocol is used to transfer files from a remote server to a recipient's email or mobile phone. Consequently, it is possible to create an environmentally friendly fitness monitoring device.

15.3.2 Types of sensors used in MIoT

Sensors

Analog sensors can translate a physiological signal into electrical impulses using digital components known as sensors. Numerous parameters can be detected with their help. Additionally, sensors are used extensively in IoT-based or IoT-enabled systems. The usage of particular sensors in BSNs and wireless sensor networks (WSNs) is critical in IoT-based medical care systems (WSNs). It is a tense situation.

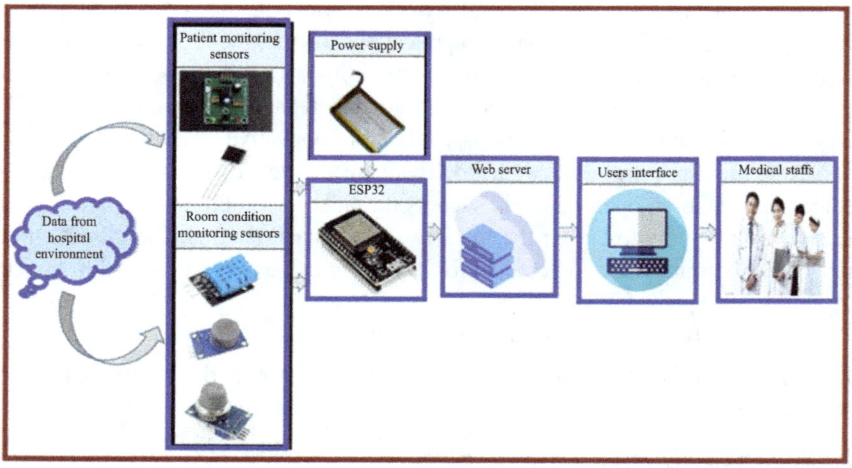

Figure 15.4 MIoT architecture

Some basic types of sensors are the following:

The sensors explained below have numerous medical applications in IoT-enabled medical care systems [5].

(A) *Biosensors*:

Biosensors are sensors that detect biological impulses and transform them into electrical signals. Human hormone, medication, and toxins levels can be measured non-invasively using biosensors. The authors, for example, propose an RF biosensor that may non-invasively test blood glucose levels using tear fluid.

(B) *Pressure sensors*:

An electronic pressure sensor translates the pressure in a liquid or gas into a form that can be read. Detecting the pressure of bodily fluids can be done by using pressure sensors. Researchers have proposed a capacitive intraocular pressure (IOP) sensor that may be used to identify and monitor glaucoma as part of their paper. Aqueous humor, the fluid that creates intraocular pressure in the eye, is the source of this pressure.

(C) *Image sensors*:

Sensors that convert optical signals into electrical signals are called image sensors. In healthcare, image sensors may be used to record neuronal communication, biological processes, etc. For example, the authors have developed a low-cost, lightweight, small size, and low-power sensor.

(D) *Accelerometers*:

Using an accelerometer, you can determine how quickly your speed changes. The rate at which velocity changes instantaneously can be determined using accelerometers. Accelerometers have a wide range of uses in IoT healthcare. An epilepsy monitoring system, for example, could employ accelerometers to track seizure attacks. Accuracy of physical activity can be measured by accelerometers, which are commonly used in fitness bands.

(E) *Temperature sensors*:

It is possible to utilize CMOS integrated temperature sensors in a variety of IoT applications because they can monitor the temperature of a body and transform it into an electrical signal. Researchers have developed a CMOS-integrated temperature sensor for usage in a wide range of IoT devices in this study.

15.3.3 Tools and technologies used to implement MIoT

The development of IoT-based systems can use various tools and technologies. These tools such as Arduino, Raspberry Pi, Intel's Galileo, and Beagle Bone are hardware platforms that provide the processing necessary to integrate sensors with IoT [6].

(I) *Arduino*:

Arduino is a widely used platform for the development of IoT systems. They have developed a system that uses pulse sensors and is based on Arduino to provide Ubiquitous Health Monitoring (U-Health).

(II) *Raspberry Pi*:

Due to its general purpose input/output (GPIO) pins, Raspberry Pis are quite popular because they can be used to control electronic components. An

IoT-based healthcare system developed using Raspberry Pi 3 can be used to monitor users' health and can also contact nearby hospitals and physicians if an emergency arises.

(III) *Intel's Galileo*:

Intel's Galileo development platform is a favorite among gadget makers and academics alike. This IoT-based healthcare kit uses an Intel Galileo second-generation board to collect data from patients, including blood pressure, ECG signals, and heart rate. An emergency alert can also be sent to a user's physician, if necessary.

(IV) *Beagle Bone*:

Using the Beagle Bone board, the authors have proposed a human activity recognition system. Human gestures and actions are detected by the system. Recognition of activity can aid in the early detection of certain diseases through symptom recognition and analysis.

15.3.4 Functioning of healthcare system

Sensors are connected to XBee modules using ADC pins. The temperature is sampled by 10-bit ADC according to the sample rate configured and the same is transmitted to the gateway every 5 sec (sample rate). To get the temperature value corresponding to the transmitted ADC value, the gateway collects these transmitted samples through UART and performs the required calibration [7]. These temperatures are then served up to the cloud via a web server running on the gateway. This is the experimental setup for the IoT device as shown in Figure 15.5. A XBee module and temperature sensor are used in this set up.

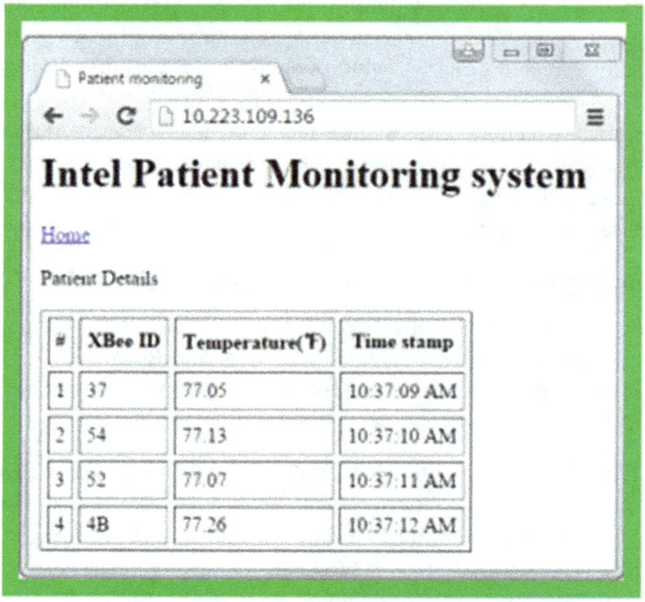

Figure 15.5 Health monitoring system using IoT

15.4 IoT-enabled devices in healthcare

For monitoring patients, health professionals have new scope by using IoT devices. These devices can be used by patients also for monitoring self-medical conditions. IoT provides various wearable devices for monitoring health. Following are the IoT monitoring devices.

1. *Monitoring for remote patients*: For healthcare, remote patient monitoring is a common application. This IoT device automatically collects the information of heart rate, BP, Temperature, O2 level, etc. when a patient is not physically present in the hospital. It reduces the traveling cost and time of remote area patients [8].
2. *Monitoring for glucose*: Around 30.3 million people are currently suffering from diabetes. The IoT-based glucose monitoring device can help diabetic patients to manage their glucose by providing their information to doctors time by time. IoT devices can provide the alarm in extreme conditions. It can also help to track the patient meals [12].
3. *Monitoring for heartbeat*: Plethysmography theory is used to develop a heartbeat sensor. Changes in blood volume have been measured. The critical part is to measure the timing of the pulses. Heartbeat has two parts: a transmitter and a receiver. A heartbeat is an electrical signal that is transmitted through the heart muscle. These electrical signals create an electrical activity and these activities are detected through the skin by the transmitter which is placed on the skin. After that transmitter transmits this information to the display receiver in the form of an electromagnetic signal [10].
4. *Monitoring for hand hygiene*: In the COVID-19 situation, hand hygiene is most important. Moreover, hand hygiene is good practice to minimize the risk of spreading infection. IoT-enabled devices give the reminder to wash their hands or sanitize their hands from time to time.
5. *Monitoring stress*: Using IoT-enabled devices, we can manage the stress. These devices suggest various ways to reduce the stress level of the patient on a mobile. Sentimental analysis is used for those suggestions [12].
6. *Monitoring for Parkinson's disease*: Parkinson's disease is characterized by movement disorders. Body wearable and IoT-enabled devices support the patient for movement. These devices also support in assessing disease status and efficacy of the medication. Parkinson's disease is currently considered incurable, so it is a strong requirement of sensor-based quantitative measurements and scales to identify the disease progression and treatment efficacy [9].
7. *Monitoring by ingestible sensors*: Using ingestible sensors, it is easy to collect information in terms of digestive, PH level, and internal bleeding. These sensors are very small can be swallowed easily. These sensors are connected with the center device which is placed on the human body.
8. *Monitoring of surgery by robot*: These small IoT-enabled robots perform the surgery on the human body. These robots can perform complex procedures

that would be difficult for a human. A small incision is required for faster healing for patients, using robotic surgery is possible.

9. *Patient body temperature monitoring*: Human body maintains a temperature that is 37°C. Increasing body temperature shows some infection in the human body, So measuring body temperature is necessary for healthcare. IoT-enabled body temperature sensors which are implanted or attached to the human body can sense the temperature and send the information to patients and doctors as well. RFID module is used for the transmission of data. There are various models available for measuring the temperature of the human body.
10. *O_2 monitoring*: Blood oxygen monitoring device includes wireless sensors, IoT, and Bluetooth technology. It measures the blood oxygen in time intervals. An oximeter is a small device that clips to a finger, toe. It sends infrared light into capillaries in the patient's finger and measures the amount of light is reflected off the gases. Medical professionals use the oxygen saturation level for various health conditions like chronic obstructive disease, asthma, lung cancer, anemia, heart failure, and heart disease. Using this measurement, doctors can evaluate medication is working or not. IoT-based O_2 monitoring device is helpful to watch the patient's condition without going specific place.
11. *Personal health monitoring for aged person*: After an age, routine checkup is required. For every routine checkup, they need to go to the hospital, and sometimes it is tough for aged people. An IoT-based personal health monitoring system will work for aged people. These systems have some wearable devices which are easy to wear in daily life. These devices have sensors and transmitters and receivers. They used protocol for transmitting the information [11].
12. *Activity monitoring*: Using IoT activity monitoring is possible for disabled or aged persons. This system uses computer vision, sensors, RFID, Wi-Fi, radar for detecting abnormal activities. Using IoT-enabled devices, it is easy to watch fall detection, fitness tracking, location detection, posture recognition, etc. [12].
13. *Stroke rehabilitation*: Using medical IoT, doctors can take help to identify the strength of muscles based on kinematics, provide the virtual environment for rehabilitation, and provide feedback on the patient's recovery process [12].
14. *Respiration monitoring*: These monitoring devices areas are either contact or non-contact. Using these devices, doctors can see the effects of medication on the respiration system and monitor the condition of asthma patients [9].
15. *Sleep monitoring*: Monitoring the patients during sleep is possible by using IoT enabled sleep monitoring system. It can also be used to observe sleep stages and disorders [15].

15.5 IoT technologies in medical field

1. *Big Data*: Traditionally medical data were stored in the hard form, which required extra time and cost. Using Big Data technology, it is easy to store the data in the digital form. Using Big Data is also get easy to capture, store, and

analyze the data. In a smart healthcare system, it is easy to maintain patient data, billing information, and clinical data using Big Data technology. Big Data technology stores the data in a systematic manner so it is easy to retrieve the data. IoT devices are used in various businesses, which mean that they have lots of data. Most of the IoT devices are installed in environment from where they collect the data [12]. These data need to be stores for later analyses. In IoT, they is no way of handling IoT data. Big Data helps to analyze and understand the data collected by the IoT devices [15].

2. *5G technology*: With this technology like SDN and NFV, network partitioning is used to provide medical treatment. Networks have their requirements like energy utilization, resource allocation security, and privacy. These requirements can be achieved using virtualization in this technology. Network partitioning can divide the main physical layer into an unattached logical network to provide the function for wearable gadgets. SDN architecture is used to reduce the complexity of the traditional network. NFV is used to allocate resources to the virtual network. 5G network provides lowest latencies with ultra-high speed. Frequency range of 5G network is divided into two groups FR1 called sub-6 GHz range, FR2 between 24 and 52 GHz.

   This technology is faster, stable, and providing more secure connectivity's. In AI-enabled robots, self driving vehicles, and in renewable energy, this technology provides secure and fast connectivity. IoT applications require high level of security and reliability. These applications are industrial robots, factory vehicles, predictive maintenance, connected tools, and wearable technology [9].

3. *Smart sensors*: For smart healthcare systems, smart sensors are used under the Wireless Body Area Network (WBAM). WBAN is the elementary network type of IoT. Wireless BSNs are based on IEEE 802.15.6 standard and according to this communication, architecture is presented in Figure 15.6 [14]. Wearable sensor nodes are placed into the patient body to monitor physiological signals such as blood pressure, respiration rate, and body temperature. In this architecture, several signal sensing devices are used to sense the data. These sensing devices

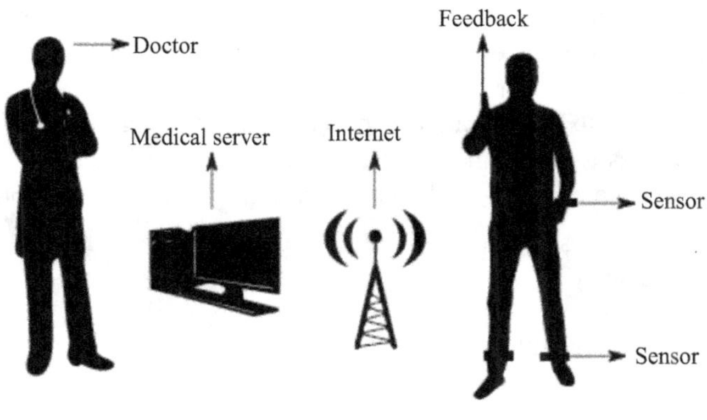

Figure 15.6 Wearable communication architecture for BSN

(wearable) collect the data from the source and send it to the base station (medical server). According to that data, right decision will be made using various techniques like ML, deep learning, and neural network [15]. ZigBee, WLAN, and Bluetooth are the wireless communication technologies used in WBAN.

4. *ML & AI*: Detection and prediction of issues in clinical tests are done by AI. ML is used to find the problems and provides the solutions continuous manner. Because of a large amount of data, it is tough for the medical professional to investigate every patient's report for analyzing the illness. Using AI and ML, it is possible to analyze the large patient's data and do the learning calculations. Using IoT, these analyzed data can be shared with the medical professional easily. Using IoT gadgets, we can collect the patient's data and an AL algorithm can deal with this large amount of data. There are some important benefits of ML technology for IoT like simplify ML training model, flexibility to use your data science library according to choice, and rapid model deployment to operational ML quickly. For operations on previous data stores prebuilt connectors and notebook integration. Notebook integration. For connecting IoT and ML, we have some steps. Those are as follows:
   - Sensors of IoT are added into the machinery in the form of discrete variables like noise, heat, and temperature.
   - Now ML model feeding on incoming data.
   - ML model now divide the incoming data into two parts: data for training and data for verification.
   - Once the validation is completed, executable endpoints are displayed.

5. *Fog computing*: In Fog computing, sensors processed the data to the nearby sensor or cloud servers. These cloud servers are connected with the remote healthcare system. Using Fog computing, it is easy for the organization to convey the information to the cloud in low latency. Fog nodes convey the information in low-energy consumption. From one IoT node to another IoT node, data can be communicated with low latency, but processing medical data is still a challenging task. For the processing of medical data, AI, ML, and deep learning methods are used. Using Fog computing, an IoT network can gather and distributed the stored data, compute the data, provide the network connectivity services, reduce the energy consumption, and also reduce the complexity. For example, smart cities and utilities.

   The value of IoT solutions can be enhanced by Fog computing. We can directly process a large amount of data on the cloud. There are some major benefits of Fog computing with respect to IoT app development process:

   - Greater business agility
   - Better security
   - Low latency
   - Network bandwidth efficiency
   - Uninterrupted services
   - Improved user experience

6. *Blockchain technology*: For smart healthcare systems, blockchain technology plays an important role. Privacy and integrity are major concerns in medical data. IoT-based healthcare systems transmit patient data so it rehires the encryption for privacy. Decentralization and centralization are also required in IoT-based healthcare systems, so Blockchain is used. Blockchain provides decentralized capacity, security, the association between patients and medical suppliers. There are some applications of IoT and Block chain like
   - Freight transportation
   - Tracking and compliance of component
   - Log operational maintenance data
   - ERP manufacturing and security and autonomous decisions
   - For energy industry smart buildings

   We have some benefits of IoT and Blockchain like

   - Building trust in IoT data
   - Added security
   - Flexibility
   - Efficiencies
7. *Virtual reality/augmented reality* [*main*]: Real-time information is required to add humans into the electronic systems. Virtual reality techniques help in planning, safe inpatient treatment. Augmented reality provides digital information in the form of imagery or sound. For interacting with IoT objects and their data, AR provides a convenient and spontaneous way to user. Users wearing a mobile or helmet, the device is capable of instantly connecting to an IOT product and ready to receive the specific data. AR data sets are associated with the AR user and understand the state with the current data sets which are received by the connecting IoT device. There are three key components of AR-enabled IoT platform:
   - Distributed AR data management
   - Object-guided tracking
   - Context-based augmented reality

     IoT and AR technology have some features when integrated them. Using AR, we can visualize 3D virtual object data sets and AR contents and, with IoT, we can access the distributed sensor data. Using AR, we can register the objects and, with IoT, we can monitor the resource objects that are embedded with the sensors. For controlling objects, AR and IoT technology give the interface to the user.

15.6 Security challenges

The health professionals faced various challenges while working in the field of IoT in healthcare.

15.6.1 Privacy and security

Data privacy and security are two of the most difficult issues IoT must deal with. Most IoT-enabled mobile devices do not comply to data protocols and standards in real time. There is a lot of confusion about who owns the data and how it is regulated. IoT-enabled devices are therefore vulnerable to data thefts, which makes the data more vulnerable to hackers who can get access to the system and steal personal health information [13].

15.6.2 Data overloaded and accuracy

Since data and communication methods are not standardized, aggregating data for critical insights and analysis is tough. In order to get the best results from IoT data analysis, it needs to be broken down into manageable bits that are not overloaded with perfect precision. In the long run, the decision-making process in the hotel industry may be impacted by the sheer volume of data.

Fraudulent health claims and the development of phoney IDs for drug transactions are two examples of IoT device data exploitation.

15.6.3 Outdated infrastructure

If the infrastructure of a facility is old, it can be difficult to take full use of the IoT. The health care system's ageing infrastructure is well-documented. It is tough for hospitals to find qualified workers when its infrastructure is in desperate need of repair. There is a great demand for skilled tech workers. Candidates may not want to deal with outdated systems.

15.6.4 Cyber attack

Reading the error logs of an infected system could reveal information about how a linked medical device works to malicious parties. Hackers could use the information they gather to make it easier to breach into a hospital network or manipulate devices so that inaccurate readings are sent to doctors and nurses.

By enforcing standards and standardizing safe protocols, these joint initiatives could lower patient risks by bridging the gaps that can arise between the many levels of an IoT system.

Cybersecurity hazards for healthcare firms in 2019 are impossible to predict. Facilities wishing to use IoT technology must, however, take precautions to raise public awareness of current hazards and learn how to safeguard their networks and devices from intruders.

15.7 Conclusion

Using the IoT, healthcare providers may monitor patients in remote areas and give emergency care for cardiac patients. The primary goal of this review is to examine the various IoT-based healthcare system research initiatives. Patients can be monitored effectively and data can be sent to a monitoring center via existing research. The ECG monitoring system is being studied to see if it can use ML to forecast illness signs. Some studies use optimization algorithms to reduce power consumption, and it is required to build a low-power system to achieve significant efficiency. Due to the IoT's lack of storage space, privacy is a serious concern because of some encryption technologies. The IoT adds a level of complexity to the cloud storage system, making it more difficult to manage vast amounts of data. Scalability and dependability are two of the strengths of the current IoT system for patient monitoring. With the use of a camera, speaker, and sensors, this technology keeps tabs on elderly patients. Increasing security, adaptability, and power consumption are all ways to improve the IoT.

References

[1] Y.I.N. Yuehong, Y. Zeng, X. Chen, and Y. Fan, "The Internet of Things in healthcare: an overview", *Journal of Industrial Information Integration*, 1, 2016, 3–13, ISSN 2452-414X, https://doi.org/10.1016/j.jii.2016.03.004.

[2] H. Bhatia, S.N. Panda, and D. Nagpal, "Internet of Things and its applications in healthcare – a survey", in 2020 8th International Conference on Reliability, Infocom Technologies and Optimization (Trends and Future Directions) (ICRITO), 2020, pp. 305–310, doi: 10.1109/ICRITO48877.2020.9197816.

[3] S. Tian, W. Yang, J. Michael Le Grange, P. Wang, W. Huang, and Z. Ye, "Smart healthcare: making medical care more intelligent", *Global Health Journal*, 3(3), 2019, 62–65, ISSN 2414-6447, https://doi.org/10.1016/j.glohj.2019.07.001.

[4] H.N. Saha, D. Paul, S. Chaudhury, S. Haldar, and R. Mukherjee, "Internet of Thing based healthcare monitoring system", in 2017 8th IEEE Annual Information Technology, Electronics and Mobile Communication Conference (IEMCON), 2017, pp. 531–535, doi: 10.1109/IEMCON.2017.8117245.

[5] D. Sehrawat and N.S. Gill, "Smart sensors: analysis of different types of IoT sensors", in 2019 3rd International Conference on Trends in Electronics and Informatics (ICOEI), 2019, pp. 523–528, doi: 10.1109/ICOEI.2019.8862778.

[6] R.K. Kodali, G. Swamy, and B. Lakshmi, "An implementation of IoT for healthcare," in 2015 IEEE Recent Advances in Intelligent Computational Systems (RAICS), 2015, pp. 411–416, doi: 10.1109/RAICS.2015.7488451.

[7] M.M. Islam, A. Rahaman, and M.R. Islam, "Development of smart healthcare monitoring system in IoT environment", *SN Computer Science* 1, 2020, 185, https://doi.org/10.1007/s42979-020-00195-y.

[8] M. Javaid and I. Haleem Khan, "Internet of Things (IoT) enabled healthcare helps to take challenges of COVID-19 Pandemic", *Journal of Oral Biology and Craniofacial Research* 11, 2021, 209–214, doi: 10.1016/j.jobcr.2021.01.015.

[9] B. Pradhan, S. Bhattachryya, and K. Pal, "IoT-based applications in healthcare devices", *Journal of Healthcare Engineering*, 2021, 18, Article ID 6632599, doi: 10.1155/2021/6632599.

[10] M.M. Islam, A. Rahaman, and M.R. Islam, "Development of smart healthcare monitoring system in IoT environment", *Journal of SN computer Science* 1,2020, 185, doi: 10.1007/s42979-020-00195-y.
[11] A.T. Thakar and S. Pandya, "Survey of IoT enables healthcare devices", in IEEE Proceeding of International Conference on Computing Methodologies and Communication (ICCMC 2017). ISBN: 978-5090-4890-8/17.
[12] A.K. Dubey, S. Narang, A. Kumar, S.M. Sasubilli, and V. García-Díaz, "Performance estimation of machine learning algorithms in the factor analysis of covid-19 dataset", *Computers, Materials & Continua*, 66(2), 2021, 1921–1936.
[13] M. A. Tunc, E. Gures, and I. Shayea, "A survey on IoT smart healthcare: emerging technologies, applications, challenges, and future trends", arXiv:2109.02042, https://doi.org/10.48550/arXiv.2109.02042.
[14] M. Ghamari, B. Janko, R. Siman Sharratt, W. Harwain, R. Piechockic, and C. Soltanpur, "A survey on wireless body area networks for ehealthcare systems in residential environments", *Journal of Sensors* 16, 2021, 831, doi: 10.3390/s16060831.
[15] S. Movassaghi, M.A.J. Lipman, and D. Smith, "Wireless body area networks: a survey", *IEEE Communications Survey & Tutorials*, 16, 2013, 1658–1686, 1553-877x/13, doi: 10.1109/SURV.2013.121313.00064.

Chapter 16

IoT-based cardiovascular prediction framework using deep learning algorithms

A. Devi[1], S. Matilda[2], G. Kavya[3], T. Ananth Kumar[2] and G. Glorindal[4]

Cardiovascular diseases (CVDs) are the biggest cause of mortality worldwide. The prediction of cardiovascular problems is a significant problem in the field of clinical data analysis. Heart disease is associated with various risk factors, and it is imperative that accurate, reliable, and reasonable methods be developed for early detection in order to accomplish timely treatment of the disease. Different machine learning (ML) approaches are available to predict CVDs. But they are not enough to predict the diseases effectively and efficiently. Healthcare generates a large amount of data, which makes deep learning algorithms increasingly viable tools for helping with decision-making and forecasting. We have taken note of the many features that have been included in recent versions of the deep learning model. This research framework proposes a new modified ConvNet for predicting cardiovascular problems with precise accuracy and reliability. The dataset contains 14 attributes for analysis. Several promising findings are obtained and verified using an accuracy and confusion matrix. To improve outcomes, the dataset is normalized and certain unnecessary attributes are isolated. The proposed Internet of Things-based framework achieves 95.1 percent accuracy in real-time analysis.

16.1 Introduction

People all around the world are growing their need for high-quality healthcare solutions with each passing year. According to the World Health Organization, 23.6 million people would die due to cardiovascular diseases (CVDs) problem in

[1]Department of ECE, IFET College of Engineering, India
[2]Department of CSE, IFET College of Engineering, India
[3]Department of ECE, SA Engineering College, India
[4]DMI-St. John the Baptist University, Malawi

the year 2030 [1], with the majority of deaths attributable to stroke and heart disease. In 2019, an estimated 17.9 million individuals died from CVDs, accounting for 32 percent of all fatalities worldwide [1]. Generally, heart disease, known as CVDs, includes different conditions that affect the functioning of the heart and is the primary problem of death worldwide over several decades.

In the industrialized world, the average age of death from coronary artery disease (CAD) is about 80, while in underdeveloped countries it is around 68. When comparing men and women, the diagnosis of illness usually happens 7–10 years earlier in males [2]. The majority of cardiovascular illness occurs in elderly people. The CVDs affect 11 percent of individuals between the ages of 20 and 40 years in the United States, 37 percent between the ages of 40 and 60 years, 71 percent between the ages of 60 and 80 years, and 85 percent of those beyond the age of 80 years [3]. The majority of cardiovascular illnesses are preventable by addressing behavioral risk factors such as cigarette smoking, an unhealthy diet and obesity, inactivity, and excessive alcohol consumption.

To predict CVDs, the healthcare industry has begun to embrace cutting-edge technology. The healthcare sector, in particular, has embraced the Internet of Things (IoT). Patients' contacts with physicians were restricted before the IoT. Currently, existing methods are not effective for physicians to constantly monitor patients' health and make appropriate suggestions. Distinct monitoring in healthcare is now feasible by IoT-enabled devices [4], allowing doctors to provide better treatment to the patients and keep track of their medical records in a safe manner. It has also improved patient involvement and satisfaction by making doctor interactions simpler and faster. IoT also has a huge impact on reducing the cost spent on healthcare substantially and improving the results of treatment.

Doctors and other health care workers use big data from medical research to support predictive models of cardiac disease, which is used to inform about treatment plans. Despite the fact that these diseases continue to be common in worldwide mortality rates, individuals may begin to take measures to avoid them. Let us know the general information about CVDs first.

16.1.1 Different types of CVDs

Heart diseases and blood vessel abnormalities are together called CVDs [5]. It includes

- Coronary heart disease – disease of the blood arteries that pump the heart muscle
- Cerebrovascular disease – disorder of the blood arteries that pump the brain
- Peripheral arterial disease – disease of the blood arteries supplying the arms and legs
- Rheumatic heart disease – caused by rheumatic fever, which is caused by streptococcal bacteria and results in damage to the heart muscle and heart valves
- Congenital heart disease – birth abnormalities in which anomalies of the heart structure lead to abnormal development and heart function in the womb

- Deep vein thrombosis and pulmonary embolism – blood clots that form in the leg veins and may affect the heart and lungs if they get dislodged.

The most important cause of the problem is an accumulation of fat deposit on the inner walls of the blood arteries supplying the heart/brain. Strokes may be caused by blood clots or bleeding from a blood artery in the brain.

16.1.2 Intermediate risk factors of CVDs

All populations today have a significant risk of CVD. A third of people in developed countries are likely to have these CVD problems because

- people do not follow a balanced diet
- people are physically inactive
- smoking or drinking excessively has an increased risk of heart disease and stroke
- people suffering from behavior-related health risks are more likely to exhibit signs of increased blood pressure, elevated blood glucose, raised blood lipids, and weight gain.

In primary care facilities, these risk factors are measured and suggest an increased risk of heart attack, stroke, heart failure, and other problems. People who abstain from cigarettes, limit their salt intake, eat more fruits and vegetables, undertake regular exercise, and avoid excessive alcohol usage may lower their risk of CVD. Policies that provide friendly settings for choosing healthy behaviors that are both cheap and accessible are necessary to persuade individuals to start and keep doing the right thing. There are other important influences on CVDs that are yet unknown. These factors are evident in the social, economic, and cultural changes occurring throughout the world. Factors other than diabetes, hypertension, and hyperlipidemia are also influential in CVD. To decrease the risk of heart attacks and strokes, medication therapy for hypertension, diabetes, and excessive blood lipids is required.

16.1.3 Symptoms and prevention of CVDs

There are usually no signs of blood vessel illness. The underlying illness may manifest itself in the form of a heart attack or stroke. Heart attack symptoms are like

- center chest pain, discomfort, or both
- significant aches or pain in the arms left shoulder, elbows, jaw, or back.

To add to this, the individual may have problems with breathing, breathlessness, nausea, and lightheadedness. For some reason, women have a higher likelihood of shortness of breath, nausea, vomiting, and discomfort in the back or jaw. Weakness of the face, arm, or leg, most frequently on one side of the body [6] is the most prevalent sign of a stroke. Those who are experiencing these symptoms should seek medical attention as soon as possible. Many forms of CVDs are avoidable. It is important to address risk factors by implementing the actions outlined in this section:

- Reducing the use of alcoholic beverages and cigarettes
- Consuming plenty of fresh fruits and vegetables

- Reducing sodium, sugar, and saturated fat consumption, as well as adopting a more active lifestyle, especially for youngsters.

It is possible that adopting harmful lifestyle choices, such as eating a high-sugar diet and not receiving enough physical exercise, will not result in CVD while a person is still young, due to the fact that the consequences of the disease build up over time.

16.2 Related works

The heart is one of the most important organs in the human body; it is responsible for the pumping of blood throughout the body, which is as important as oxygen for the human body; therefore, it must be protected at all times, which is one of the primary reasons why researchers are focusing their efforts on this. However, as time progresses, an increasing amount of research data and hospital patient records become accessible. It is possible to obtain patient data via a variety of open sources, and research may be carried out to determine how different computer technologies can be utilized to correctly diagnose patients and identify this illness early enough to prevent it from becoming deadly. Some of the proposed methods and approaches are summarized in Table 16.1.

Dun *et al.* [7] proposed different ML and deep learning methods for determining the presence of cardiovascular problems, and also performed hyperparameter so as to improve accuracy of the findings. The suggested neural networks improved the highest accuracy of 78.3 percent when compared to other ML models such as logistic regression, random forest, and support vector machine (SVM).

Singh *et al.* [8] performed a generalized classification analysis to extract nonlinear characteristics to reduce cardiovascular features. An extreme learning machine was employed to avoid overfitting and enhance training speed, as well as to increase the quality of rankings. The total accuracy for identifying coronary heart disease was 100 percent.

Zhang *et al.* [9] developed a multi-class classifier based on the AdaBoost algorithm to predict clinical outcomes in patients with breast cancer. In addition, an extra classification model is established using the same classifier learning strategy as the proposed method. The only difference between these two models is that the training inputs were different. The accuracy of the prediction was improved.

Gárate-Escamila *et al.* [10] developed a strategy for reducing the number of dimensions in the data as well as identifying characteristics of heart disease using a feature selection approach. The dataset has 74 features and is validated by six ML classifiers. ChiSqSelector identified characteristics of anatomical and physiological significance from the study, including cholesterol levels, the greatest heart rate, chest discomfort, traits associated with ST depression, and the presence of cardiac arteries. The proposed study showed that chi-squared principal component analysis (PCA) is more effective than other classification techniques in most cases. The PCA calculated directly from the raw data and with a reduced dimensionality are both preferable to the one with higher dimensionality.

Table 16.1 Literature survey of various researchers

S. no.	Author(s)	Year	Algorithm(s) used	Findings
1	Dun *et al.* [7]	2016	Neural networks	They have used a range of deep learning methods and ensemble approaches and have also carried out tweaking hyperparameters to increase precision.
2	Singh *et al.* [8]	2018	A binary classifier	In order to highlight significant characteristics, generic discriminant analysis and binary classifiers were employed.
3	zhang *et al.* [9]	2018	AdaBoost classifier	The PCA feature extraction was enhanced using AdaBoost classifier, and the accuracy of the prediction was improved.
4	Gárate-Escamila [10]	2020	DNN and ANN with statistical model	Parameters from the clinical data were utilized to validate the predictions
5	Shashikant *et al.* [11]	2019	Logistic regression, random forest, and decision tree	The random forest method achieves an accuracy of 93.61 percent compared to other models
6	Mezzatestaet *et al.* [12]	2019	SVM	Achieved accuracy of 95.25 and 92.15 percent for Italian and American datasets
7	Latha *et al.*[13]	2019	Ensemble methods	Using these ensemble methods, the accuracy of weak classifiers is improved to 0.7
8	Beunza *et al.* [14]	2019	SVM	SVM achieves the highest AUC value of 0.75 when compared to all other learning methods
9	Haq *et al.* [15]	2018	SVM, ANN, decision tree, naïve Bayes, and KNN	By comparing different methods, they concluded that SVM achieves better accuracy than that of decision tree and ANN
10	Ramprakash *et al.* [16]	2020	DNN and ANN	The prediction accuracy level is 84 percent
11	Buchan *et al.* [17]	2017	Naïve Bayes, MaxEnt, and SVM	SVM achieves 77.4 percent accuracy compared to all the classifiers and produced the best results
12	Sowmiya *et al.* [18]	2017	ANN, KNN, SVM, Decision tree, and Naïve Bayesian	The classification methods are evaluated to the pre-existing framework and discovered that the classification methodology provides the greatest level of accuracy
13	Karayilan and Kilic [19]	2017	ANN, backpropagation algorithm	The prediction system is capable of achieving a 95 percent accuracy rate when these algorithms are used
14	Kumar *et al.* [20]		Different logical regression	Attempted to determine which factors are most predictive of heart issues, and using the ROC test, they discovered that both fasting blood sugar (BS) and post-meal BS levels had a strong connection with heart problems

Shashikant *et al.* [11] suggested an approach for collecting datasets from MituSkillogies analysis team. The system's objective is to determine a patient's heart failure. It is composed of three statistical models, each of which has many inputs, specifically 19, and two output groups. The random forest method achieves an accuracy of 93.61 percent concerning accuracy and precision.

Mezzatesta *et al.* [12] proposed a method for predicting cardiovascular disorders in patients. The authors used ML techniques to analyze two datasets, one Italian and one American. Various techniques are utilized throughout all applicable methodologies, with the greatest results being obtained by SVM. For Italian datasets, it achieves 95.25 percent accuracy, whereas for American datasets it is 92.15 percent.

Latha *et al.* [13] used ensemble classification methods to create a predictive system for evaluating risk factors for heart disease. The authors are concerned about the accuracy of the dataset. To improve accuracy, they used techniques called bagging and boosting, which are ensemble techniques. These techniques improved the accuracy of weak classifiers by 70 percent.

Beunza *et al.* [14] conducted a comparison of methods for forecasting clinical occurrences that use classification or regression. The technique is analyzed using data from the Framingham heart study. Using random forests, SVM, and logistic regression, the authors achieved the greatest area under the curve (AUC) score of 0.75 of all the techniques.

Haq *et al.* [15] used all prediction methods to create a system that makes use of the Cleveland dataset. The authors suggested various ML methods in order to determine the most effective strategy. The SVM has the greatest accuracy of 86 percent compared to the artificial neural network (ANN) and decision tree.

Ramprakash *et al.* [16] suggested deep neural networks (DNNs) and a 2D statistical approach-based model. During the training and testing stages, the suggested model attains better results. The approaches were used to evaluate the model's effectiveness in correctly predicting the presence or absence of heart disease.

Buchan *et al.* [17] parsed data using an Apache method, which is a natural language processing-based system for extracting information from free medical text. The authors compare a functional removal puzzle to PCA and mutual information using naïve Bayes and SVM classifiers. Random-based testing process is used for comparative purposes. The authors obtained the best F1 score for the SVM technique out of all the classifiers.

Sowmiya *et al.* [18] compared prediction methods to the existing system for predicting cardiac issues. Among the applicable methods, such as the naïve Bayesian neural network, ANN, SVM, KNN, and decision tree, the naïve Bayes method produced the best results.

Karayilan and Kilic [19] created a framework based on ML for the purpose of predicting cardiac problems that provide a much more effective prediction than the conventional method. The author utilized the aerobic disease prediction technique in this work, which is based on a specialized neural network-based backpropagation approach. A total of 13 clinical skills are used as answers for the neural telephone system, and the neural network is taught through the backpropagation approach to predict the presence of CVDs with 95 percent of accuracy. Kumar *et al.* [20] developed a scalable three-layer structure for the store and process of a massive amount of wearable sensor data. Layer 1 focused on plethora data provided by a sensor based on IoT. Layer 2 makes use of Apache HBase to manage the enormous amount of data generated by wearing IoT sensors in cloud computing. Additionally, layer 3 makes use of Apache Mahout to improve the prediction of aerobic disease using logistic regression. Finally, receiver operating characteristics (ROC) analysis is used to determine one of the most critical health-related variables that contribute to amazing problems.

There are many approaches that have been used for performing testing operations on the different types of datasets. However, the results may be produced using any approach requiring a large amount of computation time. Some of the methods cannot be effectively applied to predict the problem. As a result, we need an effective and efficient method for predicting cardiac disease.

16.3 Introduction to deep learning

Deep learning is based on ML, which is a subfield of AI. As neural networks mimic the human brain, they will also be doing deep learning. The learning algorithms of deep learning are entirely emergent. An ML class uses several non-linear processing units for feature extraction and modification. Each subsequent layer is given input based on the previous layer's output. Deep learning models may be given a few pointers from the programmer to assist them to concentrate on the most relevant characteristics themselves, making them extremely useful in addressing the issue of dimensionality. These algorithms are utilized extensively when the inputs and outputs are large in number. The ML itself is part of deep learning. Deep learning uses an algorithm designed to imitate the human brain. Neural networks are used to execute deep learning and the inspiration for its design is biological neurons or brain cells. Basically, a deep learning network is a neural network that consists of several layers of hidden nodes.

Deep learning identifies the features and is capable of managing a vast volume of data that can be both structured and unstructured. On the other hand, deep learning may be overkill for certain tasks, particularly those involving complicated issues that require access to a large quantity of data. For example, **Imagenet** which is a popular deep learning tool identifies the images from its **14 million** datasets. It is an extremely excellent tool that has established new standard systems which use pictures as a dataset.

16.3.1 Deep learning vs. machine learning

A computer may learn without being explicitly programmed, yet it still thinks and acts like a machine. Their capacity to do complicated tasks like collecting data from images or videos still lags behind humans. Deep learning models, which are particularly designed after the human brain, are poised to handle these problems. Complex, multi-layered "DNNs" [21] enable data to flow between nodes (like

Table 16.2 ML method and deep learning method

Machine learning	Deep learning
Uses structured (like images or video) and labeled data to make predictions. So it goes through pre-processing	Process the unstructured data for prediction and avoid pre-processing.
Tasks are divided into smaller subtasks, solved separately, and then combined	It uses end-to-end approach to achieve the final result
Takes increased time for testing the data	Take less time to test the data
Takes less time for training the data	Takes larger time for training the data
Its operation depends on a low-end machine	Its operation depends on a high-end machine
Better results are obtained if the number of datasets is smaller	Better results are obtained if the number of datasets is larger

neurons). The outcome is a non-linear, abstract modification of data. However, once the programs are in place, there is minimal need for human involvement. Table 16.2 shows the difference between machine and deep learning methods.

16.3.2 Workflow of deep learning

With the use of a mix of data inputs, weights, and biases, deep learning neural networks (DLNNs) try to imitate the human brain. These components operate in concert to correctly identify, categorize, and characterize items included within the dataset in question. Unlike traditional neural networks, DNNs are composed of several layers of linked nodes, each of which builds on the preceding layer in order to improve and optimize the prediction or classification. Forward propagation is the term used to describe the flow of calculations across a network. The layers of a DNN are visibly seen at the top and bottom of the network. Deep learning models ingest and analyze data at the input layer; at the output layer, predictions and classification are produced based on the final prediction or classification.

The backpropagation method is used to compute errors in predictions by using gradient descent. Then, by going backward through the layers, it changes the weights and biases of the function, in an attempt to train the model. Forward propagation and backpropagation work together to enable a neural network to generate predictions and correct any mistakes that may occur in the process. The program improves in accuracy over time as more data is collected. The step-by-step analysis procedure of the deep learning working process is illustrated in Figure 16.1.

16.3.3 Type of deep learning networks or algorithms

The neural network models include many layers of neurons. In the next layer, each of these values reflects the data in the previous layer. Because unstructured datasets like picture data may contain millions of characteristics, using an ML method becomes impractical. While deep learning algorithms may learn more and more about certain kinds of data as they go through the neural network, this is not the

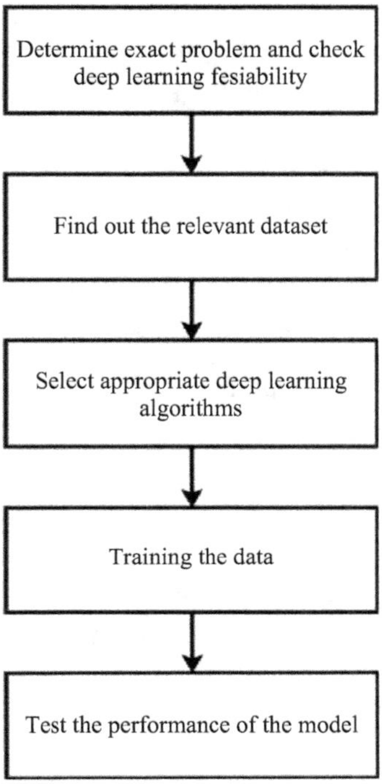

Figure 16.1 Analysis procedure of deep learning process

case with ML algorithms. Some of the lists of the best deep learning algorithms are discussed below to tackle difficult real-world issues:

- Multilayer perceptrons – for image and speech recognition systems.
- Recurrent neural networks – recognizing handwritten information, putting a caption on the picture, and time series prediction.
- Convolutional neural networks – identify the images of the satellite, medical images, and anomaly detection. Otherwise called as **ConvNets.**
- Long short-term memory networks – used for time-series prediction, speech recognizer, and music loop composition.
- Restricted Boltzmann machines – used for the reduction of dimensions of data, regression, and classification.
- Radial basis function networks – time series prediction, regression testing, and classification.
- Self-organizing maps – data visualization through artificial and self-organizing neural networks.
- Generative adversarial networks (GANs) – used in video games, creating cartoon characters, and rendering human face and 3D object.
- Deep belief networks (DBNs) – subset of DNN.
- Autoencoders – used for image processing and Sensex prediction.

The various deep learning techniques have been explored so as to predict the goal of the proposed research work by analyzing the dataset. So out of all these deep learning algorithms, the modified ConvNet is adopted for the proposed research work for predicting the cardiovascular problem in a patient effectively.

16.4 Proposed framework

16.4.1 Objectives of the proposed framework

The main aims of the proposed framework are as follows:

- To predict the cardiovascular problem in a patient when they are in a remote area because it is costly and time consuming to keep track of a patient's vital signs on a regular basis.
- To achieve the most effective and efficient prediction of cardiovascular problem in a patient by adopting appropriate features that are most important for a more accurate prognosis of CVD.
- Reducing several expensive tests of the patient since not all the features may be relevant in predicting cardiovascular problem outcome.

16.4.2 Proposed framework

The proposed research work intends to forecast the likelihood of developing heart problems as a potential factor of automatic heart disease prediction, which will be beneficial to doctors and patients in the medical profession.

The proposed framework consists of IoT with sensor systems and prediction/classification system for effective prediction of CVD. The proposed IoT-based cardiovascular prediction framework enables remote monitoring of cardiovascular patients in an effective manner and is shown in Figure 16.1. In IoT with a sensor system, different sensors are used for continuous measuring of pulse rate, blood pleasure, and heart rate of the patients. The measured values are stored in the patient database. If the sensor data from the IoT is not verified quickly, then it will take time to identify the sickness, and there is a potential that the diagnosis will be inaccurate. As a result, the suggested system is trained and tested in order to prevent the problem. In the proposed prediction and classification system, modified ConvNets are suggested for effective prediction of cardiac problems in the patient.

16.4.3 Methodologies

There are three stages followed by the proposed framework for training. They are as follows:

- Pre-processing
- Feature selection
- Classification/prediction and evaluation

The result of the training may be divided into two categories: (a) normal condition (i.e., persons with the non-heart diseased condition) and (b) abnormal (i.e., persons with heart diseases). Following the training stage, the testing stage is carried out. The sensor attached to the patient transmits information in a continual manner. This information is categorized based on training stage results, i.e., IoT sensor values compared with sensor values in the training stage. Based on this, the results are classified accordingly. The proposed framework compares the data and provides classification results. The following sections outline the processes involved in the training stage.

16.4.3.1 Pre-processing stage

- The first stage is diagnosing procedure. It includes missing values replacement, eliminating duplication, and data separation.
- After analyzing the complete history of the patient (age, cholesterol levels, and blood pressure), the missing value of a particular characteristic is updated.
- If the majority of the attribute values of the patients are matched, then the value is placed in the same location. The duplicate values are eliminated which tends to reduce dataset size.
- The patients are classified based on the chest pain they have (classic anginal pain, atypical anginal pain, and non-anginal pain).

16.4.3.2 Feature selection stage

- The goal of feature selection is to find the most important characteristics of heart disorders. Additionally, feature selection aids in the model development with accuracy by removing less relevant information, reducing training time, and improving the performance of learning.
- There are several feature selection algorithms available for effective feature predictions. The proposed framework uses a meta-heuristic algorithm, i.e., the cuttlefish algorithm.
- Changing their coloration allows cuttlefish to become either invisible or visible depending on the lighting conditions. Different cell layers produce different colors and patterns when light reflects off them.
- The most important processes performed by the cuttlefish algorithm are visibility and reflection. The cell layer of the cuttlefish adopts the light reflection mechanism simulated like a reflection process. The matching patterns visibility is simulated like a visibility process.
- By using these reflection and visibility processes, an optimal solution is obtained globally. The new solution obtained using this algorithm is represented in an equation form:

  $$NS = VI + RE \tag{16.1}$$

  where NS = New solution, VI = Visibility, and RE = Reflection.
- The algorithm divides the entire population into four different types. They are as follows:

Type 1: The algorithm obtains a new solution by combing the class 1 and class 2 (iridophores interaction with chromatophores). These two classes provide a complete search by employing each point value to create a new region around the best result within a specified period. These cases can be evaluated mathematically by using the following equations:

$$\mathrm{RE}(x) = A \times T1(c) \times k(x) \tag{16.2}$$

$$\mathrm{VI}(x) = B \times ((b(x) - T1(p)k(x)) \tag{16.3}$$

where c is the cth cell in type 1, $k(x)$ is the xth point of c. b is the best solution point. A and B are two random numbers.

Type 2: The algorithm makes use of class 3 (iridophores' reflection operator) and class 4 (iridophores' interaction with chromatophores) to provide a new solution by serving as a local search. It can be calculated using the following equations:

$$\mathrm{RE}(x) = A \times b(x) \tag{16.4}$$

$$\mathrm{VI}(x) = B \times ((b(x) - T2 \times k(x)) \tag{16.5}$$

Type 3: Class 5 (chromatophores interaction between leucophores) is used to generate better solutions than a local search solution. It can be calculated as follows:

$$\mathrm{RE}(x) = A \times T1(x) \times k(x) \tag{16.6}$$

$$\mathrm{VI}(x) = B \times (b(x) - \mathrm{avg}\ b) \tag{16.7}$$

where avg is average of b.

Type 4: All visible lights are reflected without any change so as to implement the leucophores' reflection operator as a global search. It can be calculated using the following equation:

$$F(i)k(x) = \mathrm{RN}(\mathrm{Upp}_l - \mathrm{Low}_l) + L_l \tag{16.8}$$

where $i,x = 1,2,3,\ldots,n$, and Upp_l and Low_l represent upper and lower limits.

16.4.3.3 Prediction/classification stage

In this proposed research work, ConvNets are modified by optimizing its weight values using a swarm-based meta-heuristic bee algorithm, hence it is called modified ConvNets. Once the features are selected, the selected features are categorized using the modified ConvNets. In the proposed framework, each chosen feature is used as input for the modified ConvNets. The weight values associated with each input are arbitrary values. The nodes present in the succeeding hidden layer execute the function of multiplying the input value by the weight vector of all of the input nodes related to that value. The weight vector of random values is utilized to improve results by using backpropagation technique, thus the optimization is carried out in this manner. After activation operation, the output is sent to the next layer. The classifier's output is heavily influenced by these weights. The classification/prediction algorithm steps are discussed below.

Step 1: The values of selected features and their equivalent weights are considered. They can be expressed as (16.9) and (16.10):

$$\mathrm{SF}_a = \{\mathrm{SF}_1, \mathrm{SF}_2, \mathrm{SF}_3, \ldots, \mathrm{SF}_i\} \tag{16.9}$$

$$W_a = \{W_1, W_2, W_3, \ldots, W_i\} \tag{16.10}$$

where SF_a represents the selected features input values, i selected features denoted as SF_1, SF_2, $\mathrm{SF}_3, \ldots, \mathrm{SF}_\mathrm{n}$, and W_a represents the equivalent weight values of SF_a.

Step 2: The input values of selected features are multiplied with the corresponding weight values in a random manner, and then added. It can be mathematically calculated by (16.11)

$$S = \sum_{a=1}^{i} \mathrm{SF}_a W_a \tag{16.11}$$

where S represents the addition value.

Step 3: Find out the activation function by using (16.12)

$$\mathrm{AF}_a = \mathrm{Ex}_a * \sum_{a=1}^{i} \mathrm{SF}_a W_a \tag{16.12}$$

$$\mathrm{Ex}_a = e^{-\mathrm{SF}_a{}^2} \tag{16.13}$$

where AF_i represents the sigmoid activation function, whereas Ex_a denotes the exponential of SF_a.

Step 4: Ascertain the output of the next hidden layer by using (16.14)

$$Y_a = \mathrm{BV}_a + \sum \mathrm{Ex}_a W_a \tag{16.14}$$

where BV_a represents bias value and W_a represents weight values of SF_a.

Step 5: Performs the previous three steps for every layer of the modified ConvNet. Finally, calculate the output by adding weights of all input signals together to get the value of the output layer. It can be expressed by (16.15)

$$U_a = \mathrm{BV}_a + \sum O_a W_b \tag{16.15}$$

where O_a denotes the preceding layer before the output layer, W_b denotes the weight of the hidden layer, and U_a denotes output part.

Step 6: Compare the network output with the target output. If there is any difference that occurs while comparing, then the difference is termed to be error value. This error value can be mathematically expressed by (16.16)

$$\mathrm{EV}_\mathrm{r} = T_a - U_a \tag{16.16}$$

where EV_r denotes error value and T_a the target output.

Step 7: Once the error value is determined after performing the comparison between the network output and the target value compute del value (δ) on this error. The del value is used to send the output part error feedback to all other parts in the network.

$$\delta_a = \mathrm{EV}_\mathrm{r}[f(U_a)] \tag{16.17}$$

Step 8: Finally, weight adjustment is accomplished by the use of the back-propagation approach.

$$\mathrm{WC}_a = \gamma\delta_a \mathrm{SF}_a \tag{16.18}$$

where WC_a denotes the correction in the weight and γ denotes the value of momentum. The weight values are optimized by the swarm-based meta-heuristic bee algorithm.

16.4.3.4 Swarm-based meta-heuristic bee algorithm

This algorithm is effectively used for optimizing numerical problems [22]. This algorithm works based on a Tereshko and Loengarov model for honey bee colonies foraging behavior. This model has important three key elements. They are as follows:

- Active foraging bees
- Inactive foraging bees consist of spectator bees and scouts
- Food resources

The active and inactive bees seek rich food sources near to their colony. The model outlines two popular behavior models such as rich food sources foragers recruitment (that provides positive feedback) and poor food sources foragers recruitment (that provides negative feedback). To apply the ABC algorithm for any optimization problem, follow the below steps:

- Step 1: The considered problem is converted into an optimum parameter finding problem so as to minimize the optimal solution.
- Step 2: Initial solution populations are identified by the bees and then enhance by adopting two techniques such as discarding the poor solution and swarming toward the best solution.

Pseudocode for ABC algorithm

```
Initialization stage
REDO
    Active bees stage
    Spectator bees stage
    Scout bees stage
  Store the archived best solution
UNTIL(Cycle = Maximum amount of time or maximum number of cycle)
```

Initiation stage: In this stage, all the food sources' populations are initialized by scout bees as vectors $\vec{a}_n$ (n=1,2,…,N; N is the population size). Then control parameters are set. The $\vec{a}_n$ is considered as the solution vector for each food source for the optimization problem. Each $\vec{a}_n$ vector has x variables ($\vec{a}_{nj}, j = 1,\ldots,x$) and it is optimized to reduce the target function. The initialization can be defined mathematically by

$$\vec{a_{nj}} = k_j + \text{r and}(0,1) * (\text{UPP}_j - \text{LOW}_j) \tag{16.19}$$

where UPP_j and LOW_j are the upper and lower limit of the parameter $\vec{a_{nj}}$.

Active foraging bee stage: Active bees lookout for new food resources ($\vec{S_n}$) which have more honey within the nearest food source region ($\vec{a_n}$) in their memory. Those bees search for the nearest food source region and then compute its fitness values. The nearest new food source can be determined by using the following equation:

$$S_{nj} = \vec{a_{nj}} + \phi_{nj}(a_{nj} - a_{kj}) \tag{16.20}$$

where $\vec{a_k}$ denotes the randomly chosen food resources, j denotes the randomly chosen parameter index, and ϕ_{nj} represents the random number ranging from $[-b, b]$.

The fitness value can be mathematically expressed by the following equation:

$$f_n\left(\vec{a_n}\right) = \begin{Bmatrix} \frac{1}{1 + o_n(\vec{a_n})} & \text{if } o_n\left(\vec{a_n}\right) \geq 0 \\ 1 + abs(o_n(\vec{a_n})) & \text{if } o_n(\vec{a_n}) < 0 \end{Bmatrix} \tag{16.21}$$

where $o_n(\vec{a_n})$ is the target function value of $\vec{a_n}$.

Spectator foraging bees'stage: In the swarm, the information regarding food source can be shared with spectators by active bees. Based on the received information, the spectators subsequently make their own food choices. The probability values are computed by utilizing fitness values provided by the active bees. Based on these probability values, the spectator bees choose their food resources. For this purpose, selection techniques are based on fitness value.

The probability value (prb_n) can be computed by the following equation:

$$\text{prb}_n = \frac{f_n(\vec{a_n})}{\sum_{n=1}^{N} f_n(\vec{a_n})} \tag{16.22}$$

When the food resource of the spectator bee is probabilistically chosen, the nearby food resources are determined and the fitness value is calculated using (16.19). Many spectator bees are selected to obtain rich food resources and to get positive feedback behavior by applying greedy selection between $\vec{S_n}$ and $\vec{a_n}$ in the active bee stage.

Scout foraging bees' stage: If the found solution of the active bees cannot be enhanced by predefined number of attempts, then it is called as limiting criteria or abandonment criteria. After then, the scouts begin to look for new ways to solve the problem. For example, consider if the $\vec{a_n}$ solution is abandoned, then the scout that is employed by the active bee finds the new solution defined by (16.19).

16.5 Discussion on experimental results

The results of the proposed IoT-based cardiovascular prediction framework are simulated using Python applications because Python is very powerful and flexible for IoT-based systems. Let us discuss the hardware description first followed by evaluation parameters and simulation results.

16.5.1 Hardware description

The prototypes of a proposed IoT-based framework have been developed by connecting the Raspberry pi module with different types of sensors to send the data to the cloud. The patient history can be generated by storing patients' age, gender, and required information related to them. There is no device commercially available to determine chest pain (cp). The cp parameter is computed using random number ranging from 1 to 3. The resting blood pressure of the patient (trestbps parameter) is measured using the HEM 6232T and measured values are sent to the cloud for storing the patient's data and alerting the patient by notification on their mobile. The continuous glucose monitor sensor measures the glucose level in the blood. Commercially, there is no device or sensor available for measuring serum cholesterol. So, it can be generated as a random number within a specified range. An electrocardiogram (ECG) sensor (AD8232 ECG sensor) measures the heart rate of the patient and generates the electrocardiographic date. Other parameters such as thalach, oldpeak, and slope are taken from the patient's history data. The Raspberry pi board performs the most important function for data gathering and processing.

16.5.2 Dataset description

The dataset utilized for training purposes is gathered from Kaggle (https://www.kaggle.com/ronitf/heart-disease-uci). There are 76 characteristics in this database; however, the vast majority of the published research only makes use of a subset of those 14. Many researchers have used the Cleveland database so far. Table 16.2 displays the first four rows and all of the dataset's characteristics without any preprocessing. Now, the attributes employed in this study are given in the table.

S. no.	Features name	Explanation	Range
1	age	Age of the patient (denoted in years)	29–77
2	sex	Gender	1: M, 0: F
3	cp	Different types of chest pain	1–3
4	trestbps	Rage of blood pressure (measured in mmHg)	94–200
5	Chol	Cholesterol (measured in mg/dl)	126–564
6	Fbs	Fasting blood sugar (measured in mg/dl)	0 or 1
7	RESTECG	Resting electrocardiographic	0–2
8	thalach	The heart rate reached its maximum.	71–202

(Continues)

(*Continued*)

S. no.	Features name	Explanation	Range
9	exang	Exercise-induced angina	0 or 1
10	oldpeak	Difference between depression caused by exercise and depression caused by inactivity	0–6.2
11	slope	Peak exercise	0,1,2
12	ca	Fluorescence-colored main vessels	0,1,2,3
13	thal	thalassemia (normal condition-1, fixed problem-2, 3- reversible problem)	1,2,3
14	target	Diseases diagnose	0 or 1

The dataset contains individuals aged 29–79. Periodic investigations have shown that adults beyond the age of 65 were hardly affected by heart disease. The patient's gender is determined by assigning a value of 1 to the male and 0 to the female. Male patients are more likely to get heart disease than female patients, according to a new study. According to recent research by the American Diabetes Association, women with diabetes problems are more likely to suffer from heart disease than males.

Chest pain is distinguished by its characteristics: classic anginal pain (assigned with the value of 1), atypical anginal pain (assigned with the value of 2), and non-anginal pain (assigned with the value of 3). The CAD is characterized by chest discomfort that occurs as a result of a lack of oxygen-rich blood that is given to the heart. The classical type of angina is caused by a decrease in blood flow to the heart muscle. A common cause of atypical angina is emotional or mental stress.

TRESTBPS is a measurement of an individual's resting blood pressure and is measured in the unit of millimeters of mercury. In the bloodstream, total cholesterol is called serum cholesterol. The arteries are narrowed when the "bad cholesterol," i.e., low-density lipoprotein, is high. Some experts believe that good cholesterol, i.e., high-density lipoprotein, reduces the risk of a heart attack.

When a person's fasting blood sugar is less than 120 mg/dl, the value is set to 1. When it is high, i.e., more than 120 mg/dl, the value is considered to be zero. The risk of heart disease and other health problems rises when the body's response to the insulin produced is less than the optimal level. The resting electrocardiographic is denoted as RESTECG. If the RESTECG is normal, then it is denoted as zero. If the RESTECG has ST wave abnormalities, then it is denoted as one. If the RESTECG has a left ventricular hypertrophy problem, then it is denoted as two.

The maximum heart rate (thalach) of the patients ranges from 71 to 202 in the given dataset. Cardiovascular mortality increases by at least 20 percent when the heart rate is increased by 10 percent or more. Exercise-induced angina (EIA) is assigned a value of zero when there is no pain. When there is pain, the EIA is assigned the value of one. Angina produces discomfort in the middle of the chest, such as pressure, fullness, squeezing, or pain. The most important issue to be considered is the length of the ST segment since recovery after high stress in the positive ECG stress phase.

The thalassemia is denoted by THAL, which stands for the length of an exercise test in minutes. When the downslope ST-segment depression is more than 1 mm at 60–80 ms during testing of ECG stress, it is considered to be an atypical condition. The class feature is represented as target. When the target=1, the person is suspected to have heart disease. When target=0, the person is normal (non-diseased) people.

16.5.3 Selected features and evaluation parameters

In the proposed framework, out of 14 features only 13 features are taken. The last feature called "target" is considered a predicted feature for determining whether the patient is having heart disease or not. The proposed modified ConvNets are compared with existing ANN, DLNN, random forest [23], and logistic regression models by considering the analyzing parameters such as precision, F1 score, sensitivity, specificity, accuracy, and error. For these analysis purposes, the general evaluation can be performed by using a confusion matrix. Table 16.3 shows the confusion matrix.

The evaluation parameters for simulation comparison can be discussed in the following:

(a) **Accuracy (A_{cc})**: In most cases, a comparison of many observations from the same or other sources is used to determine its quality, i.e., accuracy which depends on how the data is gathered. It can be expressed by (16.23)

$$A_{cc} = \frac{T_p + T_n}{T_p + T_n + F_p + F_n} \tag{16.23}$$

(b) **Sensitivity (Sen):** It is defined as the ability to determine whether the patient is having a heart disease problem or not. It is otherwise called recall. It can be represented by (16.24)

$$\text{Sen} = \frac{T_p}{T_p + F_n} \tag{16.24}$$

(c) **Specificity (S_p)**: This is a possible chance to diagnose a patient who is not having a heart disease problem. It can be expressed by (16.25)

$$S_p = \frac{T_n}{T_n + F_p} \tag{16.25}$$

Table 16.3 Confusion matrix

Test	**Results**	
	Disease	**No disease**
Positive	True positive (T_p)	False positive (F_p)
Negative	True negative (T_n)	False negative (F_n)

(d) **F1 score**: It is a classification parameter calculated by taking the average of the accuracy and sensitivity measurements. It can be presented by (16.26)

$$F_1 = \frac{2S \times P}{\text{Sen} + P} \tag{16.26}$$

The value of F_1 achieves the greatest value of 1 and the poorest value of 0.

(e) **Precision (*P*)**: It can be of two different types. They are as follows:

1. **Positive precision**: It can be defined as the ratio of number of people whose test results are positive to the total number of positive test people. This is the chance possibility that the patient has a heart disease problem if the diagnostic result is positive. It can be computed by (16.27) and (16.28)

$$P_{\text{pos}} = \frac{T_{\text{p}}}{T_{\text{p}} + F_{\text{p}}} \tag{16.27}$$

2. **Negative precision**: It can be defined as the ratio of number of people whose test results are negative to the total number of negative test people. It can be expressed by (16.28)

$$P_{\text{neg}} = \frac{T_{\text{n}}}{T_{\text{n}} + F_{\text{n}}} \tag{16.28}$$

16.5.4 Simulation results

Accuracy performance analysis: The accuracy analysis of the modified ConvNet compared to the existing models is shown in Figure 16.2. The performance evaluation depends on the total number of records.

From Figure 16.3, it can be inferred that the proposed modified ConvNet provides better accuracy than the existing models. The UCI dataset is trained by using modified ConvNet and then evaluated. The modified ConvNet achieves the highest accuracy (93.7 percent) for maximum records. But DLNN (85.2 percent), random forest (83.6 percent), logical regression (87.2 percent), and ANN (81.3 percent) [24] achieve only lesser accuracy.

Sensitivity performance analysis: The sensitivity performance of the proposed modified ConvNet is compared with the existing models and shown in Figure 16.4. For this comparison, the sensitivity of each and every set of data is evaluated separately.

The existing DLNN (83.33 percent), random forest (83.1 percent), logical regression (81.5 percent), and ANN (83 percent) gives lower than the modified ConvNet (93.1 percent). The highest level of sensitivity is achieved by the modified ConvNet. Better performance results are obtained by the proposed classifier.

Specificity performance analysis: The specificity performance value of the proposed classifier compared with the existing models is shown in Figure 16.5. For the first 200 datasets, the proposed classifier gives 92.7 percent, whereas existing

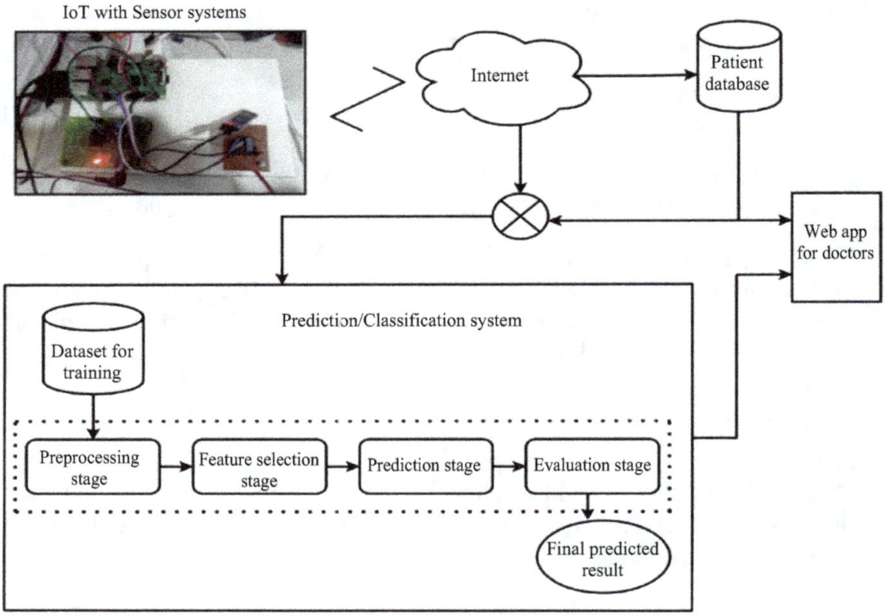

Figure 16.2 IoT-based cardiovascular prediction framework

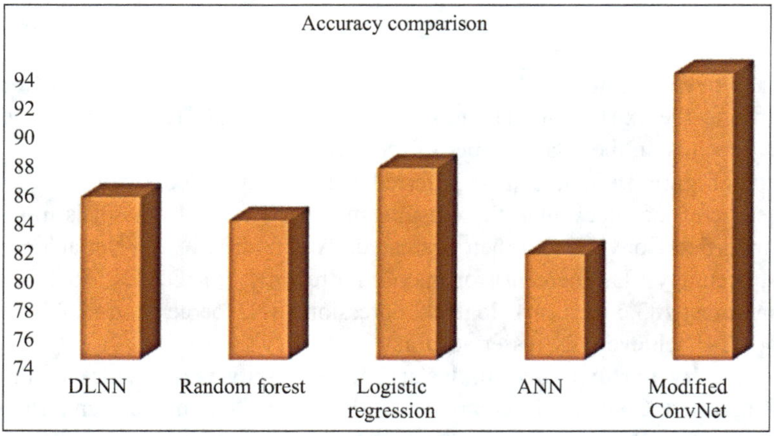

Figure 16.3 Accuracy comparison of proposed with existing model

DLNN, random forest, ANN, and logical regression give 91.5, 90.2, 91.0, and 91.4 percent, respectively. From this, it is concluded that the proposed classifier gives the highest specificity measurement value. So, from this comparison, it is understood that the proposed classifier renders excellent specificity performance.

F1 score analysis: The F1 score performance measurement of the modified ConvNet with the existing models is shown in Figure 16.6. The proposed classifier

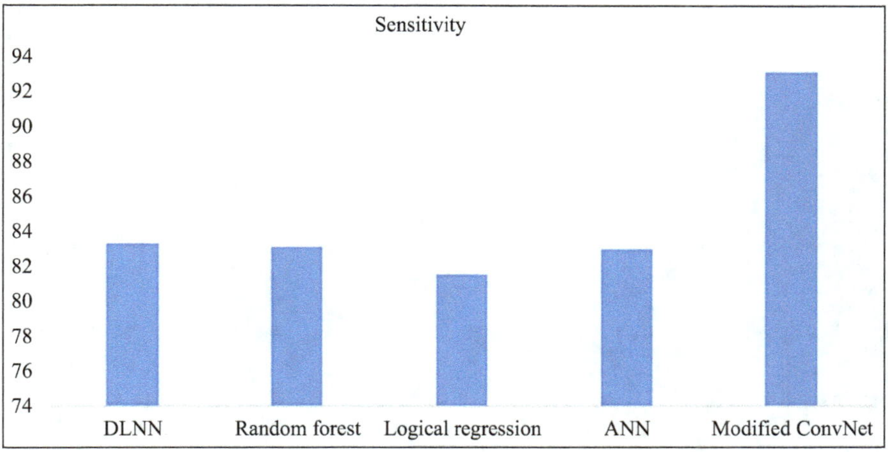

Figure 16.4 Sensitivity comparison

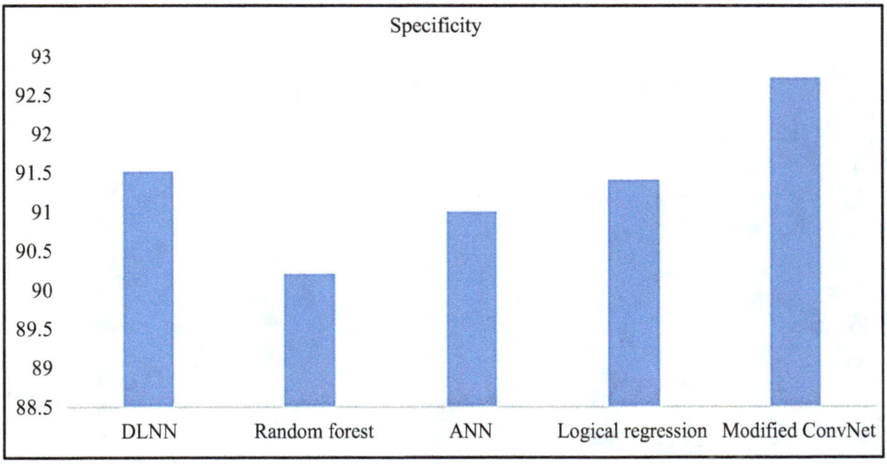

Figure 16.5 Specificity performance comparison

achieves 92.7 percent for the F1 score which is higher than existing models, namely, DLNN (91.5), logical regression (89.), random forest (89.0), and ANN (90.2).

Precision performance analysis: The proposed classifier precision performance compared with the existing model's performance by concerning the positive and negative precisions. The proposed framework is considered an excellent system because it has the best positive and negative precisions. The precision comparison of the proposed with the existing system is shown in Figure 16.7.

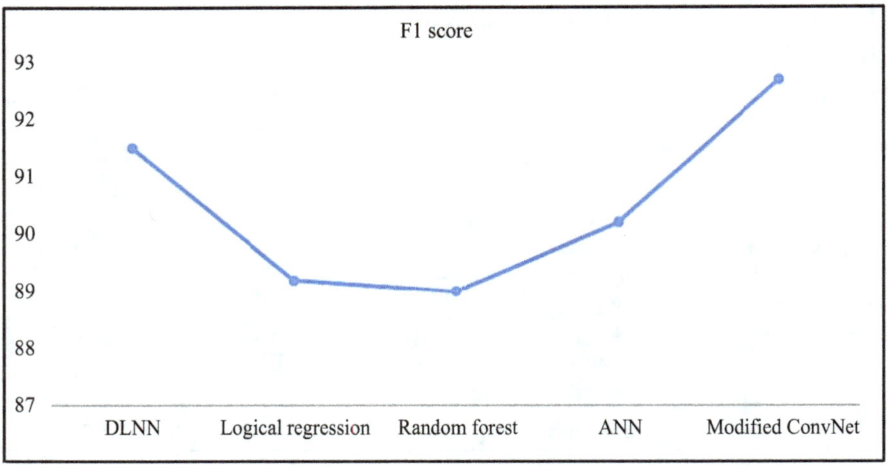

Figure 16.6 F1 score comparison

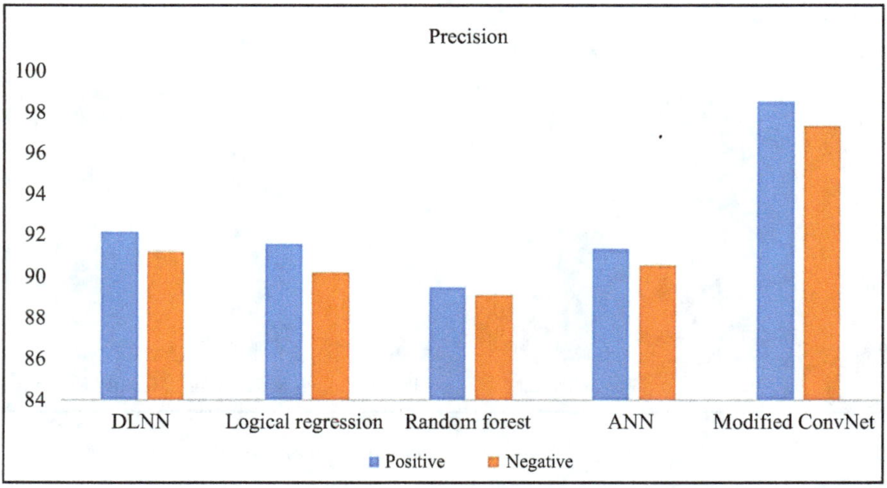

Figure 16.7 Precision comparison

Out of 300 records, the proposed classifier has 98.5 percent of positive precision value and 97.3 percent of negative precision value, whereas exiting models have 92.1 and 91.2 (DLNN), 91.5 and 90.2 (logical regression), 89.5 and 89.1 (random forest), and 91.3 and 90.5 (ANN). This shows that the suggested classifier outperforms the current approaches in this area. It is clear that heart disease prevalence increases by increasing positive precision value and vice versa.

16.6 Conclusion and future enhancement

The healthcare industry may be benefitted from the use of the proposed framework for effective prediction of cardiovascular prediction. When a patient is in a distant area, i.e., there are no medical services available near the area, the proposed framework performs monitoring and prediction operations with the help of IoT so as to save many lives by providing immediate care. Heart disease survival prediction is a difficult endeavor. Existing strategies for diagnosing heart disease were utilized by the existing systems for heart disease prediction. The modified ConvNet is used in the proposed framework to develop an IoT-based cardiovascular prediction system. The proposed framework performs these actions by following the pre-processing stage, feature selection stage, and prediction/classification stage. The suggested system is tested using the UCI dataset to see how well it performed. The suggested modified ConvNet classifier has a greater accuracy level when compared to existing methods. The results also show that the suggested technique is more accurate than the other methods. In the future, cardiovascular prediction can be improved by conducting many trials by adopting alternate feature selection and optimization methods so as to increase the prediction accuracy.

References

[1] CVDs (CVDs). (n.d.). Retrieved from https://www.who.int/news-room/fact-sheets/detail/cardiovascular-diseases-(cvds).

[2] Spence, J. D. (2013). Faculty opinions recommendation of heart disease and stroke statistics–2013 update: a report from the American Heart Association. *Faculty Opinions – Post-Publication Peer Review of the Biomedical Literature*, pp. 6–245. doi:10.3410/f.718113556.793483971.

[3] Sahni, G. and Narula, J. (2018). Global burden of cancer and cardiovascular disease. *Cancer and Cardiovascular Disease*, 1–2. doi:10.1007/978-3-319-62088-6_1.

[4] Devi, A., Therese, M. J., Dharanyadevi, P., and Pravinkumar, K. (2021). IoT based food grain wastage monitoring and controlling system for warehouse. *2021 International Conference on System, Computation, Automation and Networking (ICSCAN)*, pp. 1–5. doi:10.1109/icscan53069.2021.9526400.

[5] (n.d.). Retrieved from https://www.nhs.uk/conditions/cardiovascular-disease/.

[6] Cardiovascular Disease: Heart Disease Causes and Symptoms. (n.d.). Retrieved from https://my.clevelandclinic.org/health/diseases/21493-cardiovascular-disease.

[7] Dun, B., Wang, E., and Majumder, S. (2017). Heart disease diagnosis on medical data using ensemble learning. Available at: http://cs229.stanford.edu/proj2017/final-reports/5233515.pdf.

[8] Singh, R. S., Saini, B. S., and Sunkaria, R. K. (2018). Detection of coronary artery disease by reduced features and extreme learning machine. *Medicine and Pharmacy Reports*, 91(2), 166–175. doi:10.15386/cjmed-882.

[9] Zhang, D., Zou, L., Zhou, X., and He, F. (2018). Integrating feature selection and feature extraction methods with deep learning to predict clinical outcome of breast cancer. *IEEE Access*, 6, 28936–28944. doi:10.1109/access.2018.2837654.

[10] Gárate-Escamila, A. K., Hassani, A. H., and Andrès, E. (2020). Classification models for heart disease prediction using feature selection and PCA. *Informatics in Medicine Unlocked*, 19, 100330. doi:10.1016/j.imu.2020.100330.

[11] Shashikant, R., and Chetankumar, P. (2020). Predictive model of cardiac arrest in smokers using machine learning technique based on Heart Rate Variability parameter. *Applied Computing and Informatics*, Ahead-of-print. doi:10.1016/j.aci.2019.06.002.

[12] Mezzatesta, S., Torino, C., Meo, P. D., Fiumara, G., and Vilasi, A. (2019). A machine learning-based approach for predicting the outbreak of cardiovascular diseases in patients on dialysis. *Computer Methods and Programs in Biomedicine*, 177, 9–15. doi:10.1016/j.cmpb.2019.05.005.

[13] Latha, C. B. and Jeeva, S. C. (2019). Improving the accuracy of prediction of heart disease risk based on ensemble classification techniques. *Informatics in Medicine Unlocked*, 16, 100203. doi:10.1016/j.imu.2019.100203.

[14] Beunza, J., Puertas, E., García-Ovejero, E., *et al.* (2019). Comparison of machine learning algorithms for clinical event prediction (risk of coronary heart disease). *Journal of Biomedical Informatics*, 97, 103257. doi:10.1016/j.jbi.2019.103257.

[15] Haq, A. U., Li, J. P., Memon, M. H., Nazir, S., and Sun, R. (2018). A hybrid intelligent system framework for the prediction of heart disease using machine learning algorithms. *Mobile Information Systems*, 2018, 1–21. doi:10.1155/2018/3860146.

[16] Ramprakash, P., Sarumathi, R., Mowriya, R., and Nithyavishnupriya, S. (2020). Heart disease prediction using deep neural network. *2020 International Conference on Inventive Computation Technologies (ICICT)*, pp. 666–670. doi:10.1109/icict48043.2020.9112443.

[17] Buchan, K., Filannino, M., and Uzuner, Ö. (2017). Automatic prediction of coronary artery disease from clinical narratives. *Journal of Biomedical Informatics*, 72, 23–32. doi:10.1016/j.jbi.2017.06.01.

[18] Sowmiya, C. and Sumitra, P. (2017). Analytical study of heart disease diagnosis using classification techniques. *2017 IEEE International Conference on Intelligent Techniques in Control, Optimization and Signal Processing (INCOS)*, pp. 1–5. doi:10.1109/itcosp.2017.8303115.

[19] Karayilan, T. and Kilic, O. (2017). Prediction of heart disease using neural network. *2017 International Conference on Computer Science and Engineering (UBMK)*, pp. 719–723. doi:10.1109/ubmk.2017.8093512.

[20] Kumar, P. M. and Gandhi, U. D. (2018). A novel three-tier Internet of Things architecture with machine learning algorithm for early detection of heart diseases. *Computers & Electrical Engineering*, 65, 222–235. doi:10.1016/j.compeleceng.2017.09.001.

[21] Narmadha, S., Gokulan, S., Pavithra, M., Rajmohan, R., and Ananthkumar, T. (2020, July). Determination of various deep learning parameters to predict heart disease for diabetes patients. *2020 International Conference on System, Computation, Automation and Networking (ICSCAN)*, IEEE, pp. 1–6.

[22] Karaboga, D. (n.d.). Artificial bee colony algorithm. Retrieved from http://www.scholarpedia.org/article/Artificial_bee_colony_algorithm.

[23] Devi, A., Kavya, G., Therese, M. J., and Gayathri, R. (2021). Early diagnosing and identifying tool for specific learning disability using decision tree algorithm. *2021 Third International Conference on Inventive Research in Computing Applications (ICIRCA)*, pp. 1445–1450. doi:10.1109/icirca51532.2021.9544617.

[24] Jeyaranjani, J., Rajkumar, T. D., and Kumar, T. A. (2021). Coronary heart disease diagnosis using the efficient ANN model. *Materials Today: Proceedings*, pp. 1–12.

Chapter 17

An intelligent approach using convolutional neural network (CNN) for early detection of melanoma and other skin diseases

Sharmila Banu Sheik Imam[1] *and Preethika Immaculate Britto*[2]

Owing to medical technology's rapid growth, new techniques and algorithms appear every day to enhance our quality of life. Admittedly, skin diseases are more common than other diseases because the skin is easily affected by the environment. Some skin diseases may be caused with no symptoms or only mild skin problems that make people not consider it; however, they may result in severe conditions. The most common problem is the lack of dermatologists and delays in getting appointments. Inaccurate diagnosis can be fatal as most skin diseases appear similar. One of the real-life examples is the skin diagnosis as a sign of someone's severe disease that person had a mole in his leg, but he did not consider that as a symptom. That mole indicated cirrhosis of his liver, which led to death. So, artificial intelligence (AI) can help make the diagnosis automated, faster, and more efficient. The application is based on a pre-trained convolutional neural network (CNN) model to classify skin diseases as well as healthy skin. This work is based on an AI application for skin disease classification using machine learning. The deep learning algorithm in machine learning called CNN was used to build a model for the classification of skin disease. With the advancement of AI, we will build a mobile-based application that captures the patient's skin symptoms to provide the perfect diagnosis. In this application, the user will first sign in or register for the application. The user will then take a photo of the symptom on his/her skin. After that, the image will be processed and classified using the CNN model, where it will be previously trained to a dataset of skin lesions images. The model will recognize the images trained with whereafter the processing; the output will be whether a healthy skin or a disease diagnosis.

[1]College of Computing Sciences, King Faisal University, Saudi Arabia
[2]College of Engineering, King Faisal University, Saudi Arabia

17.1 Introduction

17.1.1 The skin

Skin is one of the largest organs in the human body, making up about 16% of the overall body weight [1]. Human skin has many functionalities such as protection against infection, chemicals, and ultraviolet rays, and maintaining body temperature [2]. Even though one of the skin functions is to protect the multiple tissues and organs from external invasions, the outer or the genetic factors can influence the skin producing skin disease. Having a skin disease is common in people's everyday life. Around 9 billion people, who nearly represent one-third of the world's population, are affected by skin diseases [3].

Having sparkling skin is one of the signs that indicate healthy skin. It is necessary to take care of the skin and make sure that it is in its best condition. The best way to do that is to check and examine the skin at least once a month for any suspicious moles. An early examination can minimize the damage and the evolvement of most skin diseases. The traditional method used to classify and diagnose skin diseases by dermatologists is using their knowledge base experience, where the symptoms on the skin will be subjected visually based on the knowledge the dermatologist already has. If the dermatologist recognized the symptoms, he\she would recommend the symptoms to some skin disease. Otherwise, he\she uses error and trial for the diagnosis by combining all the possible conditions and comparing them to reach the perfect judgment of the symptoms [4].

17.1.2 Anatomy of the skin

The skin is part of the integumentary system, which also includes the nails, hair, and exocrine glands (Figure 17.1). It is an incredibly large organ, accounting for 15% of an adult's total body weight. Total skin thickness varies depending on where it is found on the body. The thickest skin is found on the back, the palms of the hands, and the bottoms of the feet, where it can be up to 3 mm thick. The thinnest skin is found on the eyelid, where the epidermis measures just 0.05 mm with the little dermis and subcutaneous fat [5]. There are three major layers of the skin; each contains specialized cells, tissues, and appendages, and each performs unique functions for the body [6].

17.1.3 Problem statement

Many people around the world suffer from skin diseases. Owing to the lack of medical knowledge, most people with such disorders do not know their skin disease type. Sometimes people do not give enough attention to the symptoms showing on their skin that might indicate skin disease. Some of those diseases can be cured if detected at the initial stages only. It is necessary to identify those symptoms as soon as possible to prevent the development of the disease. However, a consultation with a dermatologist and skin disease diagnosis might be limited and costly. Dermatologists sometimes find it difficult to diagnose skin

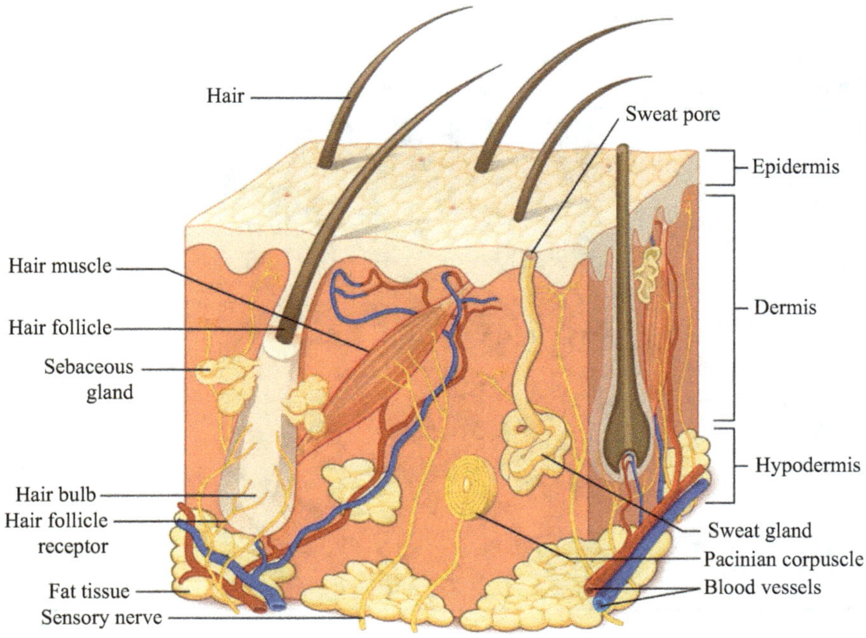

Figure 17.1 Cross-section of skin [6]

disease accurately without having laboratory tests. The diagnosis is often made visually by the dermatologist, and we have many visually similar skin diseases [6] as shown in Table 17.1. The treatment options for each skin disease differ depending on the type of the disease. So, having a mistake in defining the disease diagnosis will delay the therapy process and, in some cases, will make the patient's life endanger.

This paper proposes an artificial intelligence (AI) application for skin disease classification using machine learning. Machine learning is an application of AI that provides systems and applications the ability to learn and improve from experience without being explicitly programmed automatically. There is a deep learning algorithm in machine learning called convolution neural network (CNN) that can help us to build a model for the classification of skin disease. For that, our proposed application will be based on this model. With the advancement of AI, we will build a mobile-based application that captures the patient's skin symptoms to provide the perfect diagnosis.

In this application, the user will first sign in or register for the application. The user will then take a photo of the symptom on his/her skin. After that, the image will be processed and classified using the CNN model, where it will be previously trained to a dataset of skin lesions images [7]. The model will recognize the images trained with, whereafter the processing, the output will be whether a healthy skin or a disease diagnosis, as shown in Figure 17.2.

Table 17.1 Visually similar skin diseases

Skin disease name	Disease image	Disease stage
Melanoma	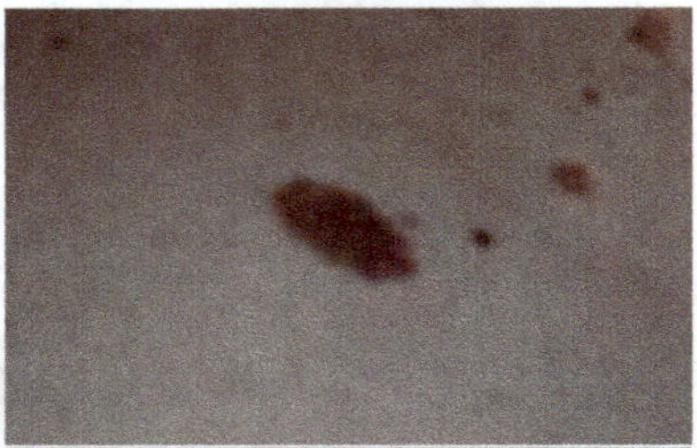	Malignant
Nevus	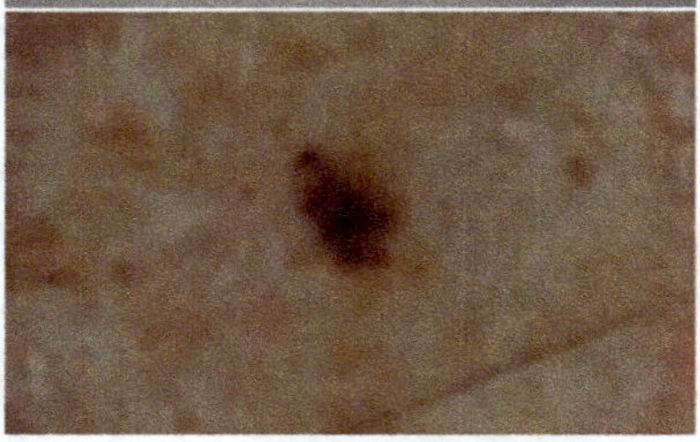	Benign

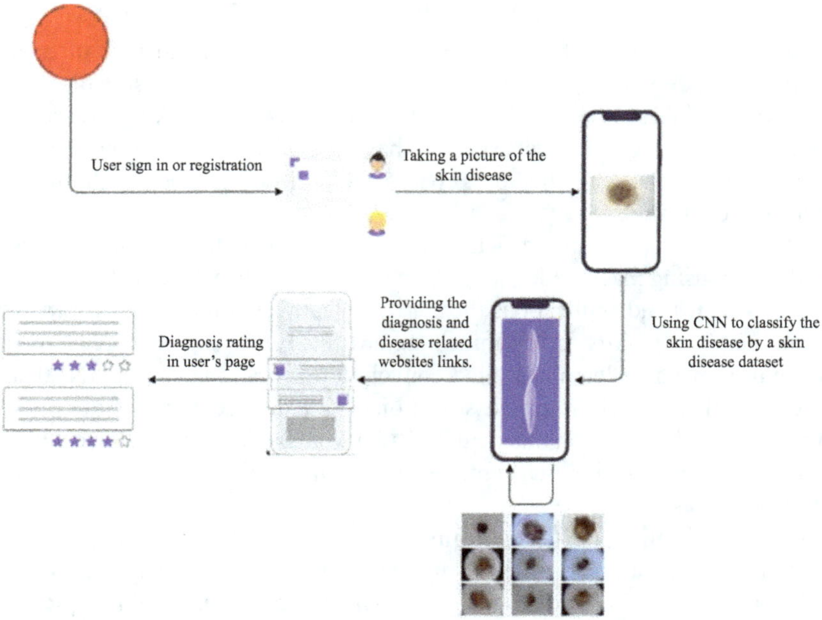

Figure 17.2 Application's architecture diagram

17.2 Scope of the project

The project's primary goal is to enable people to diagnose their skin diseases using their mobile phones. The application will be based on the CNN model. The application's main task will be pre-processing the image, extracting the features from the picture, and then classifying the type of skin disease. The working principle of the application will be as follows: foremost, the user will be asked about his/her username, email, birth date, gender, and password and then the user's information will be saved for creating the account. The user will then provide the skin disease image using the camera or the mobile phone image's gallery. After that, the image will be resized to unify the extracted features for all the pictures, increase the performance, and reduce the time consumed to process. Then, the CNN model will extract the image features, and after that at the last layer, the model will classify the skin whether it is healthy or has one of the skin diseases. Lastly, the result will be textual, including the disease's name, and a useful website link to assist the user knows more about the condition.

This application will help patients and dermatologists and could enable life-saving and fast diagnoses. Instead of visiting the dermatologists, the patient can install the app on mobile devices. However, the application will not prescribe the treatment; it only will give the diagnosis and a useful website link. The application will be able to classify only the trained skin diseases and the healthy skin. Moreover, the application recognizes the unknown diseases that belong to the unknown class while training. The application will be trained for four types of skin diseases, and it will classify the healthy skin and the diseases that belong to the unknown class.

17.2.1 Comprehensive analysis of related work

17.2.1.1 Early melanoma diagnosis with mobile imaging

This research aims to propose a detection system for the early diagnosis of melanoma using smartphone cameras [8]. Many melanoma diagnosis systems were proposed in several research pieces, but they were only focused on the images taken with dermoscopic that were taken under well-controlled conditions. In this research, they were offering a diagnostic imaging system with regular cameras equipped with smartphones. There are three main stages in the detection process for such systems: segmentation, feature extraction, and classification.

17.2.1.2 Segmentation

The segmentation stage points to determine the lesion region from the image. It will be hard to have an accurate segmentation due to capturing the photos from the smartphone cameras. For that reason, they have used two steps segmentation process. In step one, a mask of skin regions is generated using the skin detection method by discarding the pixels from the non-skin areas. To detect skin pixels, they have used a method based on the skin color model by converting the color space of the image from RGB into YCbCr. In step two, they have extracted the lesion by

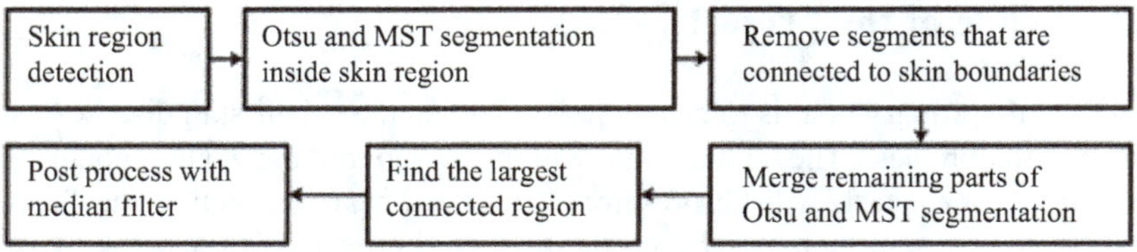

Figure 17.3 Flowchart of segmentation procedure [8]

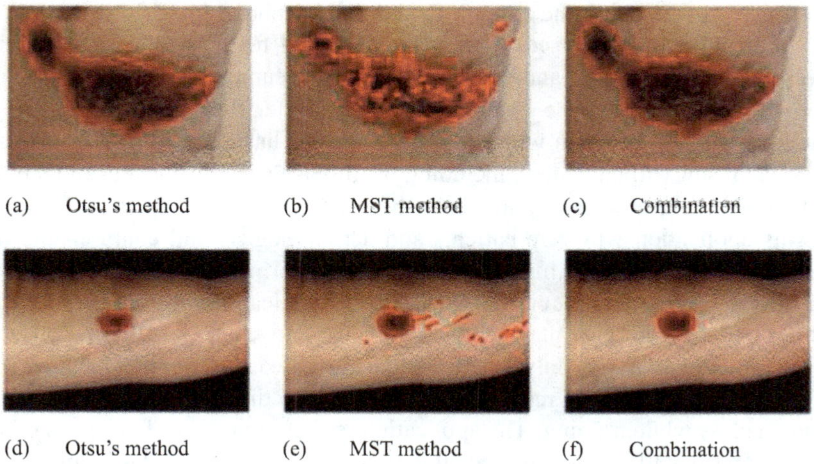

Figure 17.4 Results of the proposed segmentation scheme [8]

using two segmentation methods and then merging their results. The first method is a general histogram thresholding method that is called Otsu's method. This method uses color intensity to classify the image pixels, but it could not detect clear edges. The second method is a region-based called minimum spanning tree (MST); such a method could clear edges but may not detect smooth color intensity changes. The proposed segmentation procedure flowchart is shown in Figure 17.3. So, by merging both segmentation methods by first removing all the segments in both results and then having both the results' union is shown in Figure 17.4. They expected good segmentation results.

17.2.1.3 Feature extraction and selection

The usage of this stage is to extract the lesion features by examining 80 features that belong to four categories. The first category is the color feature; for the color lesion, they have used the mean and variance of the pixel values on the color channels: grayscale, red, green, and blue in RGB images, and hue and value in HSV images. In addition, they have used information from the histogram of pixel values to capture more color variation. In normal skin lesions, they have used the color triangle feature, which is used to capture the uniform variation of the color from the center to the border [8]. The second category is the border feature. By computing the shape features, like the compactness, solidity, and convexity of the distances from border points to the centroid of the lesion, they will describe the border [9]. They have also proposed a new method called border fitting to quantify the border's irregularity by approximating the boundary by the mean square error method with lines [8]. Then, the angles of every two adjacent lines are computed; after that, the angles' average and variance are used for the border irregularity description [8]. The third category is the asymmetry feature, which is used to compute the asymmetry of lesion shape.

17.2.1.4 Classification

The classification, which is the last stage, is used to classify the lesions' extracted features, whether they are for cancer with output (1) or non-cancer with the result (0). For each of the extracted feature categories – color, border, asymmetry, and texture – they have applied feature selection. They have then selected a subset of features for each feature category that provides the highest accuracy for classification. Later, a support vector machine (SVM) classifier is trained for each subset's feature category to make a prediction. The average accuracy of the proposed mobile application is 93.61%. For the cancer samples, it achieved the highest accuracy which is 96.67%, and for the non-cancer samples, it achieved 90.55%. For the mobile implementation, in the case of an immense size image, the image will be resized to 512 pixels to reduce processing time and memory to store the image.

17.2.2 Dermatological disease detection using image processing and artificial neural network

The dermatological disease detection system aims to detect dermatological skin diseases. The system uses image processing techniques to extract features of the skin image and an artificial neural network for the training and testing phase to detect skin disease. The system works as follows: first, we will need to collect all the necessary data from patients with such dermatological diseases. They took skin images from 128 patients with one of nine distinct types of dermatological diseases using a specialized camera. After that, a specialized doctor will collect all the necessary data from the patients such as medical history, feeling in the diseased part of the body, elevation of the diseased region, etc. Second, the system will take all the skin images taken from the patient's skin and pre-process them using image processing techniques.

In the pre-processing phase, eight distinct types of algorithms were used which are gray image, sharpening filter, median filter, smooth filter, binary mask, histogram, YCbCr, and Sobel operator. Figure 17.5 is the result of all the pre-processing algorithms. The system extracted features from the pre-processing results on the image and features from the user input. Features extracted from the image are the colors in the image, the size of the infected skin, and the edges of the infected skin which give us a shape. On the other hand, features extracted from the user input were if patient's history is liquid type, liquid color, elevation, duration, feeling,

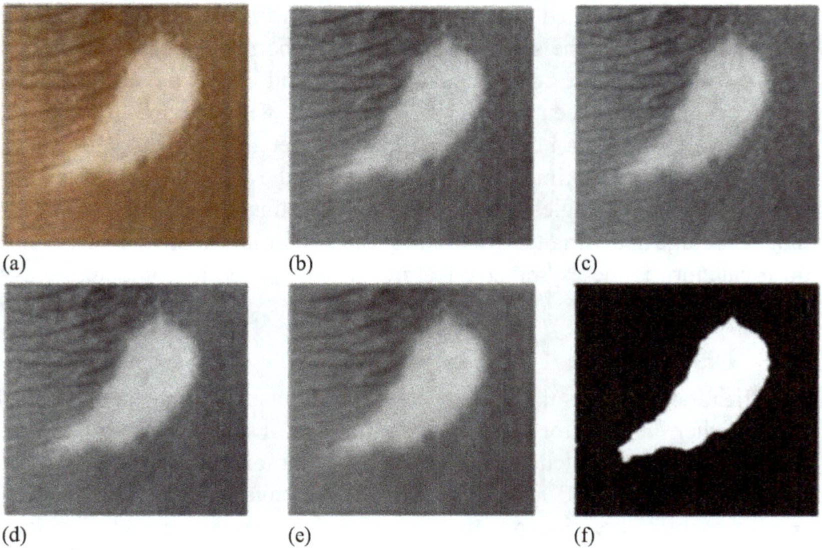

Figure 17.5 Result of pre-processing algorithms. (a) Original color skin image, (b) grayscale image conversion, (c) sharpening filter, (d) median filter, (e) smooth filter, (f) binary mask [10]

gender, and age. Lastly comes the training and testing phase with the feed-forward back-propagation neural network. The network has been trained with 10 distinct types of features, three of them from the image and the rest from the patient's input. To get the best result in detecting the skin disease, the system used 100 neurons in the hidden layer.

Below is a sample of how the feed-forward backpropagation neural network works with the ten feature values used in the input layer. Figure 17.6 shows a visualization of sample input from a patient to diagnose one of the nine skin diseases.

The system was able to successfully detect the nine different dermatological diseases achieving 90% correct detection accuracy. In Figure 17.6, we can see the detection rate of nine dermatological diseases. The system shows the best results for diseases that have low elevation in the infected area like foot ulcer, vitiligo, and psoriasis diseases with a 91–97% detection rate, while other diseases are between 85 and 88% detection rate.

17.2.3 Automatic detection and severity measurement of eczema using image processing

The eczema detection and severity measurement system use image processing to help the patient for early measurement of eczema severity and protect the patient from propagating eczema. This system uses several models, such as two segmentation algorithms, features, extraction, classification, and severity measurement—the two segmentation algorithms are working as: the first segmentation algorithm

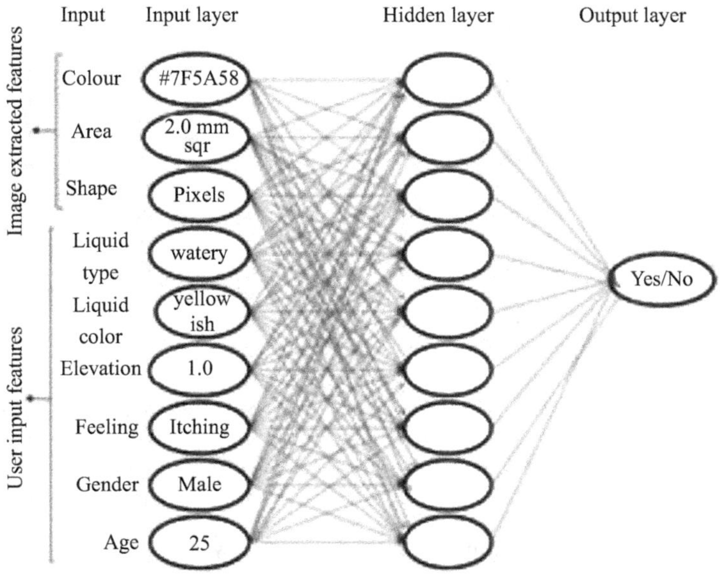

Figure 17.6 Feed-forward backpropagation artificial neural network [10]

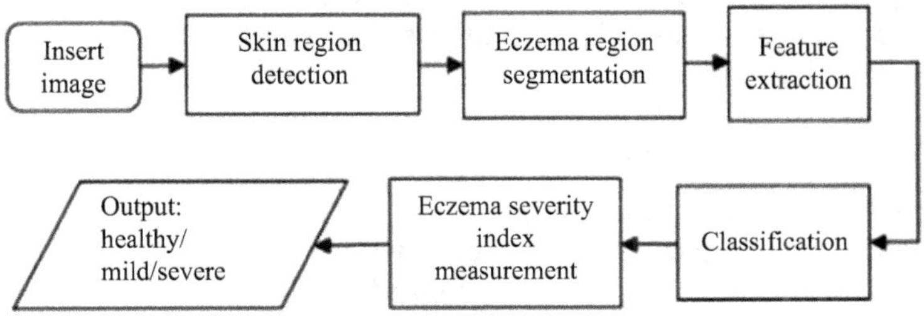

Figure 17.7 Brief flowchart to show how different sections are interconnected [10]

for separating the skin region from the image's remainder. If the skin region is separated, then the second segmentation can apply to find the skin image's eczema region.

Figure 17.7 shows the system's flowchart: in skin region detection, the skin color-based method is applied to discard the non-skin pixels and convert the images from RGB to the YCbCr color space. The purpose of segmentation of the eczema region is to identify the eczema region from the skin area using conventional methods employed in segmentation: thresholding, compression-based methods, clustering, region growing, and watershed transformation. The eczema region's biological features can be extracted in the feature extraction to help identify the disease state. Also, the extraction features contain color features, texture features, and border features.

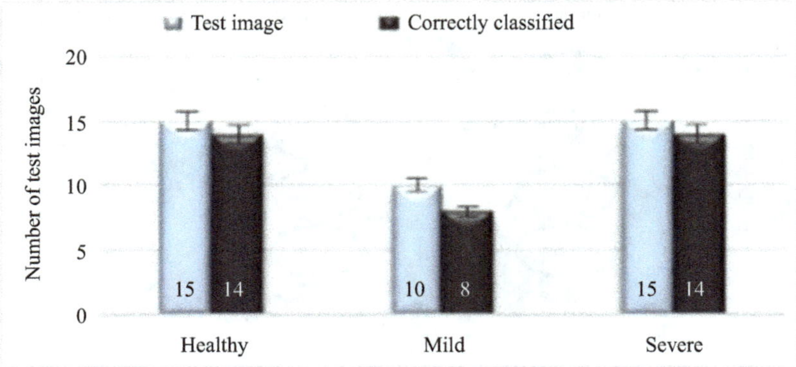

Figure 17.8 Classification result for healthy skin, mild eczema, and severe eczema test images [10]

The developer has done training, and test images were passed through the skin detection algorithm. Also, he tested the algorithm with 20 non-skin images collected from the internet. The algorithm was successfully worked with 100% accuracy to detect the skin pixels. The result is shown in Figure 17.8 for healthy skin, mild eczema, and severe eczema test images. Out of 40 test images, 36 were classified correctly, showing 90% overall classification accuracy.

17.2.4 Skin cancer classification using deep learning and transfer learning

In this paper, they are proposing a system for diagnosing three skin diseases as follows: melanoma, atypical nevus, and common nevus. They have applied a deep convolutional neural network (DCNN) to classify skin images. The main advantages of the proposed model are that it can work with any type of image and does not require any pre-processing technique. Also, a transfer learning method called AlexNet was used. They have compared their proposed method with other existing computer-aided methods. The existing methods have suffered from two major problems. First, the insufficient data in the dataset, and the second problem was with the image processing to extract the features from the images.

For the model results, they have performed two experiments with the same dataset. The first experiment evaluates the proposed method with the original dataset where they did not perform any augmentation in the images. And the second experiment evaluates the proposed model with the augmented images from the dataset. The first experiment achieved 80% correct accuracy. The second experiment achieved 98.61% correct accuracy. The performance of the proposed model with the augmented images has been compared with the other existing method using the same dataset. The results of the comparison are presented in Table 17.2, and it shows us that the proposed model achieved the highest classification rate in comparison with some other computer-aided methods using four quantitative

Table 17.2 Comparative results of the model performance

Model	Accuracy (training data), %	Accuracy (testing data), %
Alexnet-TL	84.94	79.29
FESVM	100	78.70
FESVM+PD (Type I)	100	80.16
FESVM+PD (Type II)	100	80.16
FESVM+PD (Type III)	100	80.39

measures: accuracy (ACC), sensitivity (TPR), specificity (TNR), and precision (PPV) [11].

17.2.5 Dermatological classification using deep learning of skin image and patient background knowledge

Feature extraction is added as a pre-processing step to speed up the model learning process. Lastly, feature extraction and SVM classification + patient data (FESVM+PD) is the extension of the feature extraction and SVM classification (FESVM) method by adding other patient data as background knowledge. Image data and background knowledge are then passed to the SVM modeling step. This research divided the dataset into two parts: one for training the model and the other for testing the model. The division ratio of training/testing in percentage is 70:30. Then, they applied the training dataset in three strategies for skin disease classification. In the first strategy, we create a CNN model with Alexnet using transfer learning (Alexnet-TL). It is the CNN model created by the pre-trained CNN model of Alexnet architecture using transfer learning by fine-tuning. The result of this model is called Alexnet-TL. The second strategy is to apply the Alexnet-TL for feature extraction. This strategy uses the previous Alexnet-TL model to extract the feature from the image data.

It then uses SVM to classify the data obtained from the CNN-based feature extractor. The model created by this strategy is named FESVM. In the last strategy, they create an SVM model from both feature extraction data extract from images and patient data that are used as background knowledge. This strategy employs an SVM classifier for the classification of the skin diseases of patients in the same way as the second strategy. But there are some differences in the data used for modeling. This model will also consider additional patient data. The model is then named FESVM + PD.

From Table 17.2, the best accuracy is from the FESVM+PD model. It can classify skin diseases more accurately than the FESVM model and the Alexnet-TL model. Among the three types of FESVM+PD method, the model from type III shows the best performance. This means that modeling with image data incorporating the patient's background knowledge for classifying skin disease, especially skin cancer, gets higher accuracy than simply using the images. They also observe that more details of background knowledge result in better classification results [12].

17.3 Project requirements

This section defines the main functional, non-functional, and software requirements of the application.

17.3.1 Functional requirements

17.3.1.1 Registration

The application allows the user to register. The user shall be asked about his/her full name, email, date of birth, gender, and password.

Response sequence: The first screen will be the splash screen. Then, the next screen that will appear is the log-in screen. If the user has an account, then he/she will be able to enter the application. Otherwise, the user shall sign up/register for the application. After that, the user will click the sign-up "button" to create the account.

Functional requirement: REQ#1: User registration.

17.3.1.2 Sign in

The application shall allow the user to log in. The user shall be asked about his/her email address and password.

Response sequence: The first screen will be the splash screen. Then, the next screen that will appear is the sign-in screen. If the user has an account, then he/she will be able to enter the application by providing his/her email and password and clicking the "login" button. If the user provides the wrong email or password it will display to him/her an error message.

Functional requirement: REQ#1: User login.

17.3.1.3 Reset password

The application shall allow the user to reset his/her password. The user will be asked about his/her email to send the new password.

Response sequence: In the sign-in screen, if the user forgot his/her password, he/she can click the "forgot password?" button. After that, it will go to another screen and ask about his/her email address and then the user should click the "Send password" button to send the new password to his/her email.

Functional requirement: REQ#1: User reset password.

17.3.1.4 View user account

The application shall allow the user to view his/her personal account information.

Response sequence: The user shall click the "profile" icon in the menu bar to view his/her personal account information

Functional requirement: REQ#1: View user account.

17.3.1.5 Update user account

The application shall allow the user to update his/her personal account information. The user shall be able to update his/her name, email, gender, birth date, and password.

Response sequence: The user will first click the "profile" icon in the menu bar to view his/her personal account information. Then the user will click the "update" button to reach the update screen. In the update screen, the user shall enter his new name, email, gender, birth date, or password based on his/her preferences. Then, click the "Done" button to confirm his/her update.

Functional requirement: REQ#1: Update user account.

17.3.1.6 Delete user account

The application shall allow the user to delete his/her account.

Response sequence: The user will first click the "profile" icon in the menu bar to view his/her personal account information. Then the user will click the "delete" button to reach the delete screen. In the delete screen, it shall view a confirm message to the user where the user can confirm his delete request by clicking the "Delete my account" button.

Functional requirement: REQ#1: Delete user account.

17.3.1.7 Skin diagnosis

The user can either take or choose a skin image from the gallery of his/her mobile. Then, the application will display the result of the diagnosis based on the trained model used of whether it is healthy skin, or one of the skins with the disease, or an unknown type of skin of the disease from an unknown class. In case of having a skin disease, a link will be provided to read about the disease.

Response sequence: The user can click the "Take a picture" or "Gallery" options after he/she logs in and wait for the result. Then, the diagnosis will be shown and a "Read about your disease" button can be clicked to open a website that provides some useful information about the disease.

Functional requirement:

REQ#1: Acquire skin disease image.

REQ#2: Display healthy skin diagnosis.

REQ#3: Display skin disease diagnosis.

REQ#4: Recognize the unknown skin disease.

REQ#5: Provide useful website links related to the disease.

17.3.1.8 View people's rate

The user shall be able to see other people's rates.

Response sequence: The user can click on the "Ratings" icon in the menu bar to see the previous rating of other people.

Functional requirement: REQ#1: View people's rate.

17.3.1.9 Rating a diagnosis

The user shall be able to rate the diagnosis to share his/her experience with others.

Response sequence: The user can click on the "Rate a diagnosis" icon in the menu bar to see his/her previous diagnosis and rate them.

Functional requirement:

REQ#1: Rate a diagnosis.

REQ#2: View previous diagnosis.

17.3.2 Non-functional requirements

- **Usability:** The application should satisfy the maximum number of skin diagnoses to be available to the users.
- **Privacy:** The skin disease images that are taken by the users shall not be shared with any party.
- **Availability:** The application shall be available 24 hours.
- **Accuracy:** The application shall not provide a wrong diagnosis.
- **Reliability:** The application shall run and execute all the functions without any failure.
- **Response time:** The time taken to classify and show the diagnosis of the skin condition shall be less than 4 s.
- **Maintainability:** The application shall be able to be modified and updated for further improvements.

17.3.3 Software requirements

- Android Studio to build the mobile application and run the application via its simulation.
- Python tool for image processing that contains libraries such as OpenCV.
- Keras API is a Google deep learning and built on top of TensorFlow.
- Scikit-learn is a free library for machine learning with various algorithms for classification.
- Database management tools such as MySQL.

Additionally, we will use skin diseases multiclass dataset that contains 23,906 images of various skin diseases to train the model. The problem is that the dataset does not contain healthy skin images, and it contains an unbalanced number of examples for each class. The solution found to overcome the first problem is to collect various healthy skin images from our families, friends, etc. Then add those images to the dataset and duplicate them using the augmentation method by rotating each image at 55 different angles to reach the number of examples in the other classes via an image data generator in Python [13]. The solution for the second problem is to oversample the images of the minority classes to have the same number of images as the majority class has [11].

17.4 Identification of alternative solutions and justification of selecting a solution

This section will introduce various solutions regarding each detail that could be used in our proposed project. Then, the selection will be made based on the most appropriate solution.

17.4.1 Acquisition of image

The skin disease images can be acquired from the mobile camera directly or from the mobile images' gallery. In both cases, the images can be easily obtained without making a problem with the classification. However, the input image needs to be clear so as not to affect the process.

17.4.2 Classification types

In our proposed application, we need to train a classification model to classify skin disease type or healthy skin. There are several models used for that. We are considering CNN and recurrent neural networks (RNNs) models. Table 17.3 lists the advantages and disadvantages of all of them. The chosen solution is to train the CNN model for classification since it is flexible and works well with image data. Moreover, it provides high accuracy and automatically detects the important features without any human supervision [14].

17.4.3 CNN pre-trained model

As the previous section's selected solution was the CNN model for the classification purpose, we need to choose the most appropriate CNN pre-trained model. CNN is a supervised learning method and is hence used to train data labeled with respective classes. There are various pre-trained models used for the CNN classification that is used for transfer learning. Transferred learning is a deep learning technique where we load another pre-trained deep neural network's weights to apply it in another domain [12]. We have compared two pre-trained models that are shown in Table 17.4. The selected solution due to the mobile usage in our proposed project is MobileNetV2 [15].

Table 17.3 Classification types of alternative solutions

Solution	Advantage	Disadvantage
CNN	Works well on image data There is no need for feature extraction Few training parameters	Requires a lot of training for the dataset
RNN	If it is used with convolutional layers It will operate effectively	It cannot process very long sequences It does not work with the RELU activation function It is not commonly used as an image classifier

Table 17.4 CNN pre-trained model

Solution	Advantage	Disadvantage
MobileNetV2	Fast in feature extraction Suitable for mobile applications	Lower accuracy as compared to the other CNN architectures
AlexNet	Easy implementation Provides high accuracy	Overfitting because it has 60 million parameters, so we need two methods to reduce this problem Only applicable for computer systems Not applicable for mobile application

17.4.4 Pre-processing of image

For high classification performance, the skin images should be as clear as possible. In some cases, the skin images may have skin hair that needs to be removed. Also, to reduce the processing time, the images have to be resized to a fixed size. For that reason, we will need two pre-processing techniques as follows: image resizing and image hair removal [16].

17.4.4.1 Image resizing

Different image sizes may increase processing time and cause the number of features not to be equalized in all images. The solution to this problem is to have image resizing. Each pre-trained network has a fixed input vector for it. As we have chosen the MobileNetV2 model, the input size shall be unified to 224×224 pixels.

17.4.4.2 Image hair removal

The existence of hair in the skin lesion may affect the classification of the disease type. Thus, we need to remove the hair from a skin lesion and preserve the lesion's features. The solution to this problem is to use image filters such as morphological closing or median filter. The advantages and disadvantages of these solutions are described in Table 17.4. The chosen solution is the median filter that even if we use it for hair removal, will preserve most of the image features.

17.5 Application analysis

This section is aimed to provide an analysis of the application. It contains the context, block, flowchart, and use-case diagrams.

17.5.1 Model block diagram

Table 17.5 shows the explanation of the modules of the training and testing diagram in terms of main tasks and the result (Figure 17.9).

17.5.2 Flowchart diagram

17.5.2.1 Model flowchart

Figure 17.10 shows the flowchart of the model used in the application. First, the application will let the user capture or select an image from the library. Then, the image will be resized into a fixed size. Next, the hair on the skin from the image will be removed. The CNN model will extract the features from the image. After that, the image will be classified based on these features, whether it is the healthy skin or the affected skin. If the skin is healthy or affected by the diseases that we specified, then it will show that in a textual form and display some useful links for more information about the skin disease.

Table 17.5 Description of the model block diagram

Image acquired	
Main tasks	*Result*
Capture/load an image	Acquired an image from a user
Image pre-processing	
Main tasks	*Result*
It is used to resize and remove the hair from the image	Fixed size and clear image
Feature extraction	
Main tasks	*Result*
Extracting the feature depends on the layers that the CNN has, so the convolutional and pooling layer to extract the features of the image has been used	Convoluted image
Classify the skin disease	
Main tasks	*Result*
Classify the skin if it is healthy or affected by using the fully CNN-connected layers.	From the labels and features that are extracted from previous layers, we will have the prediction results.

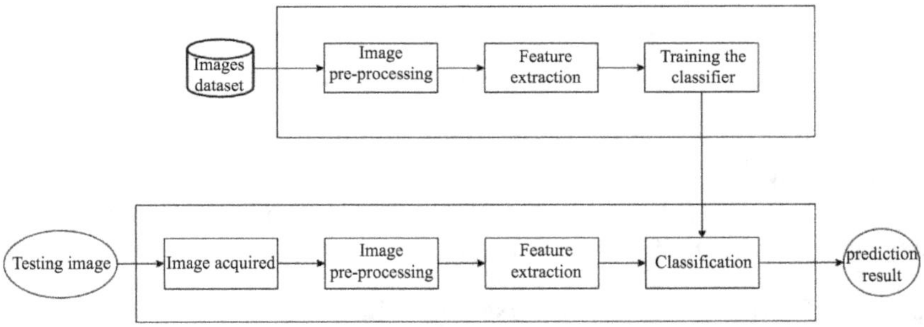

Figure 17.9 Model block diagram

17.5.2.2 Application usage flowchart

The flow of the application use is shown in Figure 17.11. First, the main welcoming splash screen will be shown to the user. After a few seconds, the sign-in page will be displayed to the user then he/she will be able to go to the forget password page or sign up if she/he does not have an account. If the user has an account, the verification process will be handled and if verified successfully, he/she will be able to use the application's functions and navigate them by using the menu bar. The user will be able to have a new diagnosis, rate the diagnosis, see people's ratings, and manage his/her profile.

17.5.3 Use-case diagram

In this section of the use-case diagram, there are two use-cases, which are the application us-case diagram and training and testing diagram.

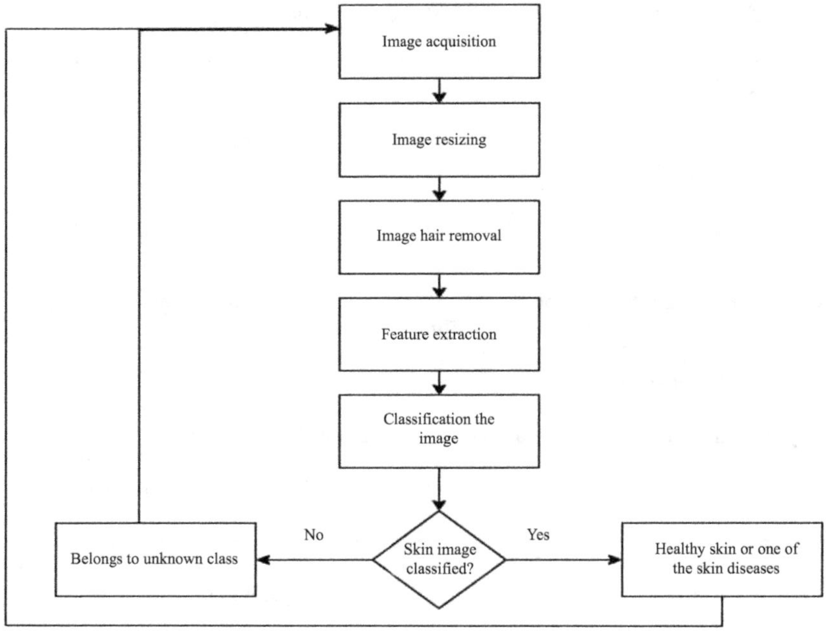

Figure 17.10 Flowchart for application during usage mode

17.5.3.1 Application of use-case diagram

Figure 17.12 shows the application of use-case diagram with six actions. Also, there is an explanation of these actions in Table 17.6.

17.5.3.2 Training and testing use-case diagram

Figure 17.13 shows the training and testing use-case diagram with its actions. Also, there is an explanation of these actions in Table 17.7.

17.6 Details of the project implementation conforming to the proposal phase

This section contains the complete implementation of the front-end and back-end parts of the mobile application development. Also, the data preparation, the classification model, and the process or method of removing the hair from the image's implementation.

17.6.1 Android mobile application front-end

To implement the proposed design, we have used XML for describing the user interface layouts, and to make them functional we have used Java programming language. The mobile application is supporting two languages which are English and Arabic, and the following are some of the figures of the user interface design in

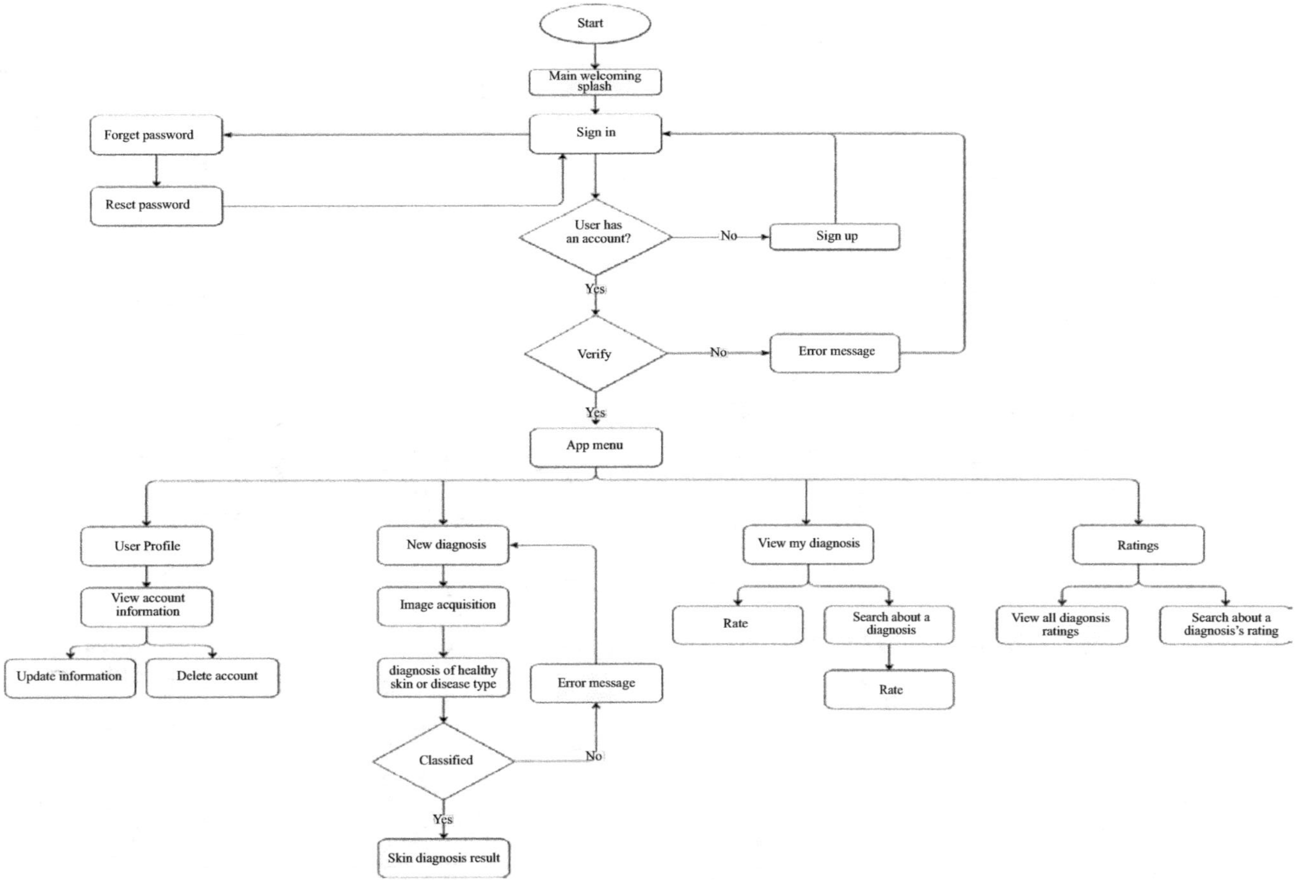

Figure 17.11 Application usage flowchart

Figure 17.12 Application usage use-case diagram

both languages, and the interfaces will be described in detail in the front-end part with the functionalities and the database design.

Here is the first interface which is the splash screen that will be shown to the user for a few seconds, and then the users will be able to go from one interface to another after logging in.

17.6.2 Mobile application back-end development

In this section, we will show the complete implementation of the back-end part of the app that includes the database connection to the application. We are using the Firebase database to handle adding, retrieving, updating, and deleting the data from the application. We have chosen the Firebase database because it allows the users to build a rich and collaborative application by providing secure and fast access to the database from the client-side as well as providing an unlimited number of users [16]. For the model deployment, the first step was to save the model in a TensorFlow Lite format. Then, we have integrated the model into the application locally because TensorFlow Lite lets us run deep and machine learning models quickly on mobile devices without using an external API or server. The first step

Table 17.6 Application usage use-case diagram description

Action	Description
Register	The patient/user will register in the application
Log in	The patient/user will log in to the application
Verify password	Verify the email and password of the patient/user in the login action from the database
Display error message	If the patient/user enters the wrong email or password in the login action
Skin diagnosis	Provide the diagnosis of the skin from the image taken from the patient/user
Maintain profile	Every patient/user in the application can access his profile information that contains the information entered when registering for the application
View profile	The patient/user can view his account
Update profile	The patient/user can update his account
Delete profile	The patient/user can delete his account
Show all previous diagnosis	The patient/user can view all his previous diagnoses in the application
Rate a previous diagnosis	After the user/patient gets his diagnosis he/she can rate it if he/she is satisfied with the diagnosis result

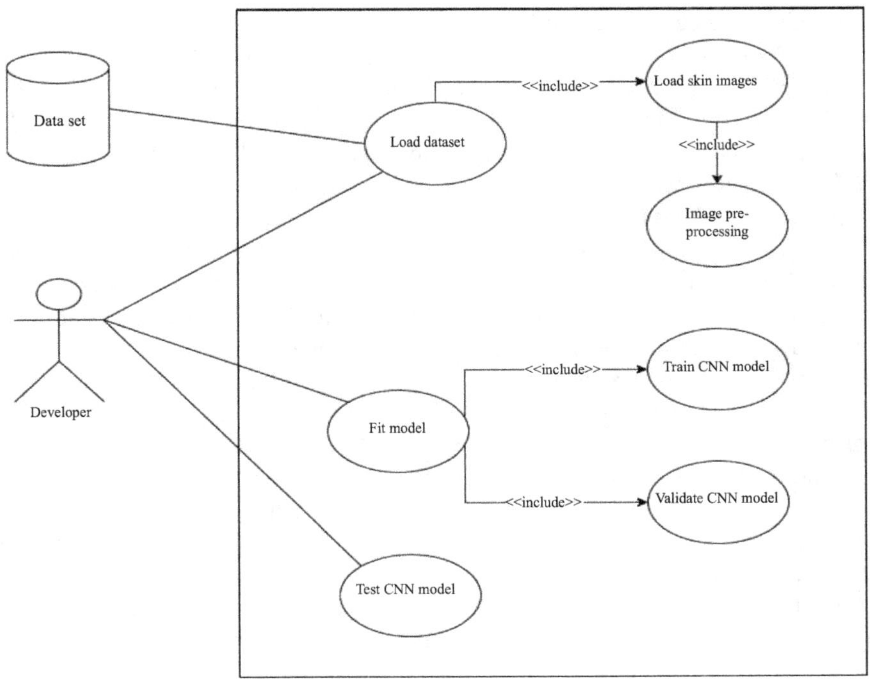

Figure 17.13 Training and testing use-case

Table 17.7 Training and testing use-case description

Action	Description
Load dataset	The training program loads the dataset specified by the developer inside a folder
Load skin images	The training program will load all the skin images it wants to train from the dataset
Image pre-processing	The developer will set the appropriate pre-processing techniques for training purposes
Feature extraction	Extract the features from the skin image
Classification	To classify the images based on the features that are extracted
Fit the model	The developer will use the model fitting to train and validate the model
Train CNN model	The developer chooses the appropriate algorithm for training the CNN model
Validate CNN model	The developer will validate the trained model during the training phase
Test CNN model	The developer will test the final trained model

needed to use the application is to sign up. From the "Login" page the user could click the "Join now" text to go to the "sign-in" page. After the user enters his/her information, he/she shall click the sign-in button to create an account. If the user forgot his/her account password, he/she will be able to reset it from the "Forget password" button that appears on the "Login" page (Figure 17.14).

Then the user will be asked to enter his/her email. After the user enters his/her email address and the email is found in the database, an email will be sent to the user containing a webpage link that shall be opened to set the new password. After the user signs up, the user will be navigated directly to the "login" page to insert his/her email and password. If the email and password are correct the user will be traversed to the "new diagnosis" page as it is the main functionality of the application. The user will be able to provide the skin image either by taking a picture on his/her phone or by providing a picture from the gallery. After the user provides the skin image, the application will display to the user the diagnosis of the skin and a button at the end of the page that will lead the user to a link about the skin disease the user was diagnosed with. The user's diagnosis information will be added to the real-time database and the Firebase Cloud.

The user will be able to check his/her information on the "Profile" page by clicking the profile icon in the bottom menu bar that will retrieve all the information entered on the "sign-in" page. Also, for updating the user's information account, the user will go to the same page, as it will view all of his/her personal information with an "Update" button at the end of the page. When the user wants to update his/her information, he/she will change the information required, then click the update button. The new information will automatically be updated on the Profile page and the real-time database. When the user wants to delete his/her account, he/she clicks the "profile" icon in the menu bar and will see a "Delete my

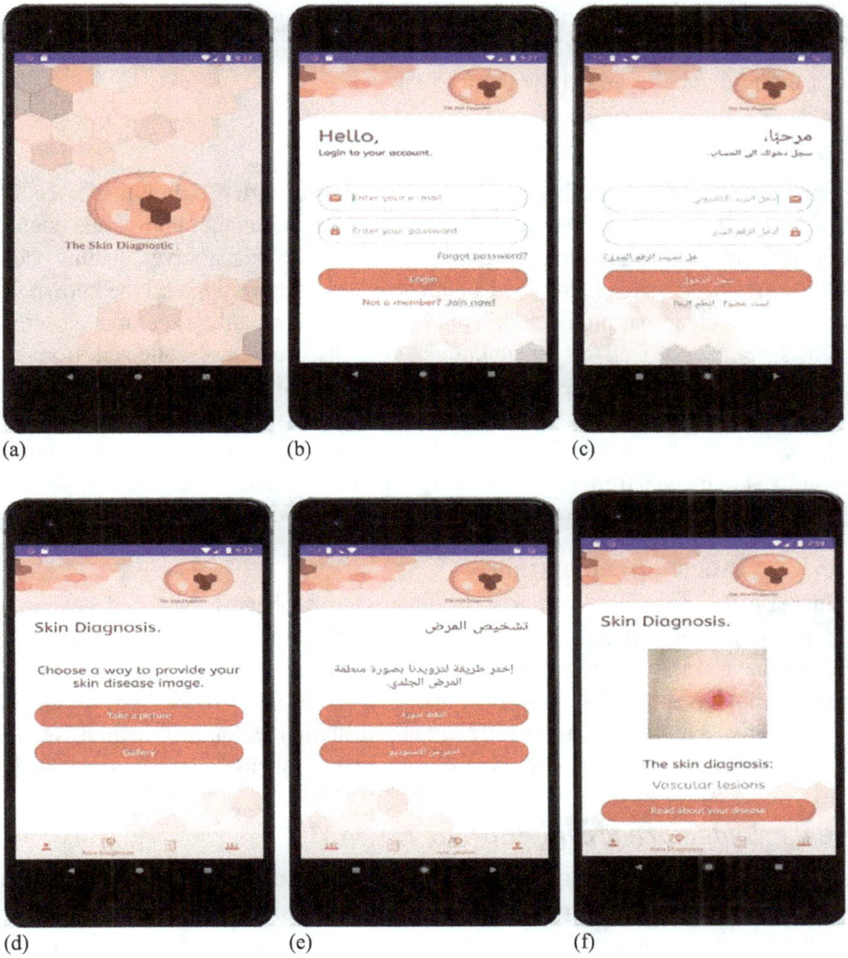

Figure 17.14 *(a) App splash screen. (b) Login screen in the English language. (c) Login screen in the Arabic language. (d) Skin diagnosis screen in English. (e) Skin diagnosis screen in Arabic. (f) Disease diagnosis screen in the English language*

account" text at the end of the page. After that, another page will appear where it will ask the user if he/she is sure about deleting their account.

If the user is sure, he/she will click the "Delete my account" button, and the user's account will be deleted. The user will be able to view all his/her previous diagnoses and rate them by clicking the "My diagnosis" icon in the menu. We will retrieve all the user's diagnoses from the "Diagnosis" table in firebase automatically by the user's special id number. If the user wants to rate his/her diagnosis, he/she will rate the diagnosis by the stars and then click the "Rate" button and the user's diagnosis will be saved. The user will be able to view other's people diagnosis experiences with the application by clicking the "Ratings" icon in the menu. We will retrieve all the user's diagnoses from the "Diagnosis" table in firebase automatically and show them to the user.

17.6.3 Data preparation

The images that we used in this project were integrated from different resources. The first resource is from ISIC Archive. Such an archive is collaborative work done by many international professional practitioners. The data in this archive contains photos and metadata, where the photos have a high standard of resolution. All the known four diseases are melanoma, pigmented benign keratosis, vascular lesion, and seborrheic keratosis classes as well as the unknown skin diseases class collected from this archive. As ISIC Archive does not contain healthy skin images, our next task was to collect them. So, we searched different websites for clear healthy skin images; however, the number of images was too few. For that reason, we collected a major number of healthy skin images from our community. Then, we selected the clear ones of them, and the images from these three resources were combined to form our dataset classes.

17.6.3.1 Image augmentation

To increase the diversity of the data available and to have a balanced dataset for training the model, without the need to collect additional data and consume more time, we have used image augmentation. The vascular lesion and the healthy skin classes do not have the same number of images as compared to the rest of the classes. So, we augmented them by using different augmentation techniques such as rotation, random cropping, mirroring, and color-shifting using principal component analysis.

17.6.4 Image processing for hair removal

Some of the images in the training, validating, and testing datasets may contain hair surrounding the skin lesion or placed over the skin lesion itself which will reflect the classification accuracy. For that reason, we came up with the idea of removing hair from the images using image processing. The methodology is to identify the dark hair in the images by first gray scaling the image, using morphological closing operation with back hat, applying thresholding, and lastly paint the thresholded image with the original image to get the final result. We applied such a method in two cases: if the image contains hair, or it does not contain hair.

First, let's consider the images with hair. Figure 17.15(a) shows an original melanoma skin disease image that will be processed to remove the hair from it. Figure 17.15(b)–(d) are all images that are pre-processed with different image sizes. Specifying the image size is an important factor in this step, as resizing the image will reflect the appearance of the hair. However, it affects the lesion as well. Whenever the image size is much smaller, the hair will be removed more but the lesion will be affected more as well. So, we tested the images with 1024, 512, and 224 image sizes. We preferred to preserve the lesion appearance and remove the hair but not totally, so we have chosen the 1024 image size as shown in Figure 17.15(d) to go with and be applied in the images for the classification model.

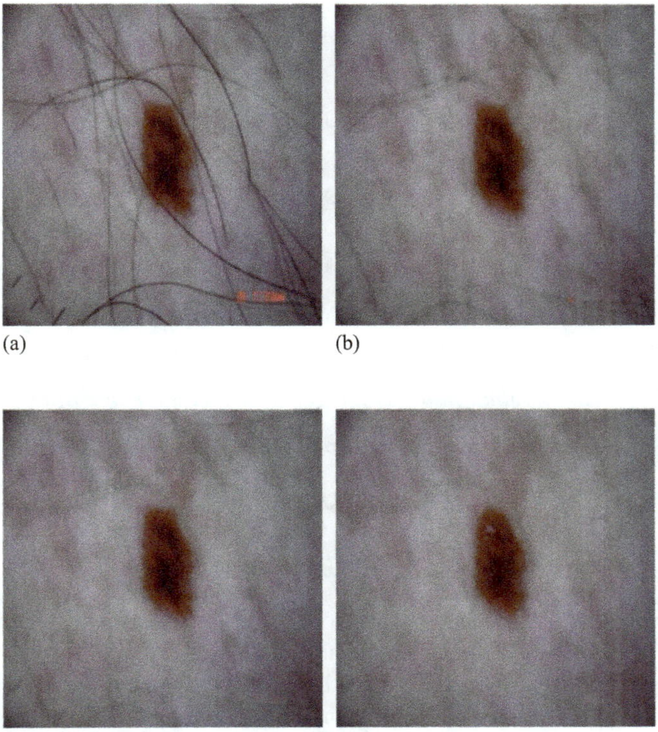

(a) (b) (c) (d)

Figure 17.15 (a) Original image with skin hair. (b) Image after removing the skin hair with size 1024. (c) Image after removing the skin hair with size 512. (d) Image after removing the skin hair with size 224

The same methodology was applied in images without skin hair, and the lesion appearance would not be reflected. Figure 17.16(a) shows an original image of skin with a vascular lesion before processing and Figure 17.16(b) shows the resulting image after the image processing.

17.6.5 Classification model building and training

After we finished setting up the dataset we started using and training the CNN model for the application to classify the four skin diseases, the unknown diseases, and the healthy skin classes. In this section, we will present the details of the model and the process of getting the best accuracy percentage of the classification. The model was built using CNN which is a mobile network version 2 as it is widely used for many real-world applications by using Python programming language. First, we loaded all the images from our dataset and resized them all as we are going to use the mobile network model which has a default input size which is 224 × 224 widths and height. After that, we split out the dataset into training validation, and testing sets randomly

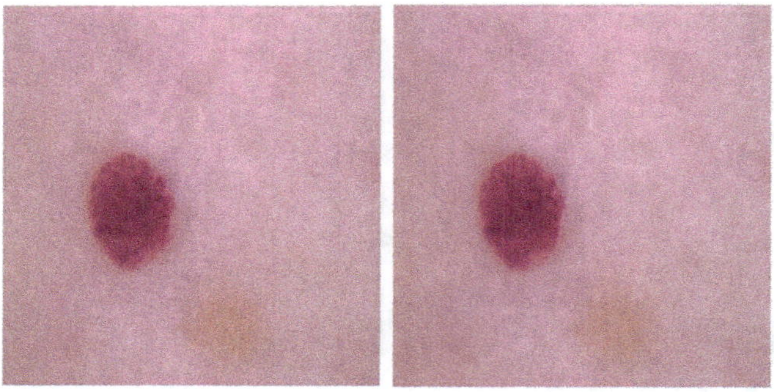

(a) (b)

Figure 17.16 (a) Original hairless image. (b) Hairless image after processing

```
baseModel = MobileNetV2(input_shape=(224,224,3),alpha=1.0, weights="imagenet",
                        include_top=False, input_tensor=Input(shape=(224, 224, 3)))
```

Figure 17.17 The base model weights

using the train_test_split function twice to get the three sets with a percentage of training 90%, and 10% for the validation and testing sets each.

To train our classifier, we used a deep learning API which is Keras. The first step was downloading a build-in model from tensorflow.keras.applications module. As we are using a pre-trained model from MobileNet there is some setup for the hyper-parameter in the pre-trained model. There are three arguments needed to be specified for the MobileNetV2 pre-trained model. One was pre-loading the weights of the model and we specified it to ImageNet which is the original dataset. The pre-trained model was trained. Two was specifying the include_top argument to false so that in the network, the classification layers do not be at the top, and this is ideal for feature extraction. The third was specifying the input shape and we used the default weights for the Mobile network which is (224, 224, 3) as shown in Figure 17.17.

For the model, we also freeze all the layers to prevent the weights in each layer from being updated during the training phase. Then we started constructing the model by using 159 layers, including average pooling for the mask size and dense layers to classify the six different classes. After that, by using Adam optimizer with MobileNetV2 we went through different experiments during the phase of training our model. For each experiment, we considered a different batch size and a different number of epochs.

Previously, with a batch size of 32 and 60 epochs, it gave us a training accuracy of around 96% and the validation accuracy of around 86%, but as the difference between the two accuracy values is around 10%, then that means we have an overfitting problem. So, the best result we came up with without having the overfitting problem is by having the batch size as 32 and the epochs as 30 which gave us a training accuracy of 92.78%, validation accuracy of 84.90%, and testing accuracy of 84.21%. Figure 17.18(a) shows the training and validation accuracy curves and (b) shows the confusion matrix on the training set.

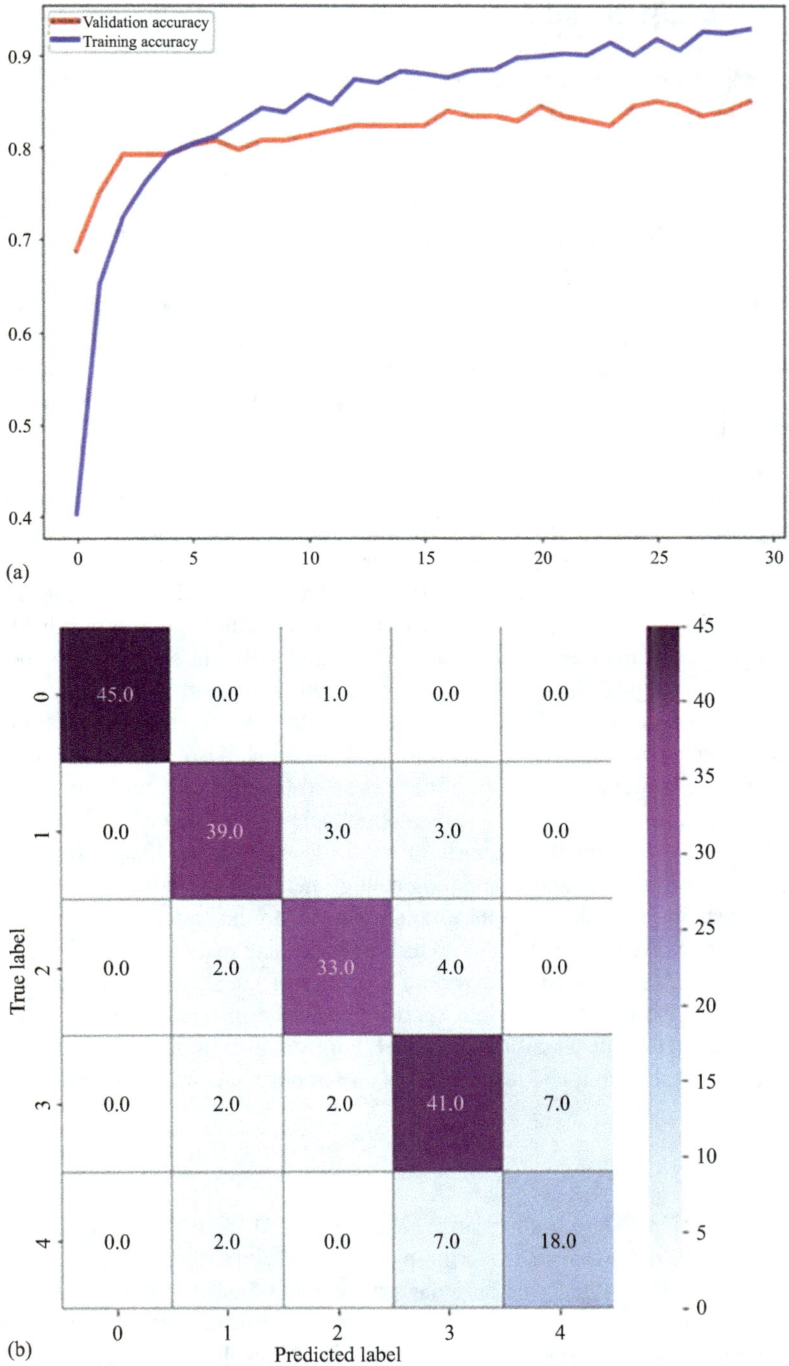

Figure 17.18 (a) Model training accuracy and validation accuracy curve. (b) Confusion matrix on the training set

17.7 Conclusion and future work

All the application functionalities related to a user account: registration, sign in, update account information, view account information, and delete account have been successfully implemented. Also, we have deployed the deep learning model into the mobile application to make disease predictions. So, all the diagnosis-related functionalities: having a new diagnosis, viewing all previous diagnoses, rating a diagnosis, and viewing all rated diagnoses by all users have been successfully finished. For the classification model, we have used a pre-trained CNN model which is MobileNetV2. The model was trained with six different classes each of which contains 500 images to have a balanced dataset. In addition, we have applied a hair removal method to all images that we have in our dataset to increase the model classification accuracy. Successfully the result of training the model achieved 92.78% training accuracy, 84.9% validation accuracy, and 84.21% for the testing accuracy. The classification of the symptoms showing on a person's skin is a significant step, where some of them might indicate severe skin disease. Diagnosing them early has a significant impact on reducing the development of the symptoms and the skin disease transmission. Despite other skin disease detection methods, they are still considered time-consuming and very expensive. This report has been presented for a smart application to perform the multi-class classification of skin diseases using deep a learning algorithm called CNN.

The proposed smart application can classify four different skin diseases, healthy skin, and can recognize unknown skin diseases. In this report, we investigated the features obtained by the DCNN for the multi-class classification task. The features are extracted using the MobileNET, a pre-trained CNN model. Also, we will use the CNN model for the classification purpose of the extracted features. In conclusion, with the help of machine learning algorithms and mobile technology, we were able to build an application that gets a quick diagnosis of a skin disease via an image. It can assist people with a real-time diagnosis and judgment of their skin condition. It can also help dermatologists in providing the appropriate treatment to the patients promptly.

For future work, we are planning to include more skin diseases to be diagnosed in our application. Also, instead of providing a picture of the skin disease area, we are planning to ask the user to provide a short video of the infected area from different angles to be treated to the classification model. For sure, increasing the model accuracy for all training, validation, and testing sets is one of our plans related to the project.

References

[1] Introduction to the human skin. (2010, April 4). Share4Rare. https://www.share4rare.org/library/xeroderma-pigmentosum/introduction-human-skin.

[2] Contributor NT. Understanding the structure and function of the skin. *Nursing Times*. 2019, August 1; https://www.nursingtimes.net/clinical- archive/dermatology/understanding-the-structure-and-function-of-the-skin-05-08-2003/.

[3] Hay RJ, Augustin M, Griffiths CEM, and Sterry W. The global challenge for skin health. *British Journal of Dermatology*. 2015;172, 6.

[4] Lawton S. Skin 1: the structure and functions of the skin. *Nursing Times [online]*.2019;115(12),30–33.
[5] https://www.nursingtimes.net/clinical-archive/dermatology/skin-1-the-structure-and-functions-of-the-skin-25-11-2019/.
[6] All Parliamentary Group on Skin. *An Investigation into the Adequacy of Service Provision and Treatments for Patients with Skin Diseases in the UK.* 1997, pp. 1–34.
[7] Liao H, Li Y, and Luo J. Skin disease classification versus skin lesion characterization: achieving robust diagnosis using multi-label deep neural networks. In: *2016 23rd International Conference on Pattern Recognition (ICPR)*. IEEE; 2016.
[8] Brinker TJ, Hekler A, Utikal JS, *et al.* Skin cancer classification using convolutional neural networks: systematic review. *Journal of Medical Internet Research.* 2018;20(10):e11936.
[9] Do TT, Zhou Y, Zheng H, Cheung NM, and Koh D.Early melanoma diagnosis with mobile imaging. In: *2014 36th Annual International Conference of the IEEE Engineering in Medicine and Biology Society*. IEEE; 2014.
[10] Rosebrock A. Keras image data generator and data augmentation. *PyImageSearch*.(2020, June 11); https://www.pyimagesearch.com/2019/07/08/kerasimagedatagenerator-and-data-augmentation/.
[11] Sasubilli SM, Kumar A, and Dutt V. Machine learning implementation on medical domain to identify disease insights using TMS. In: *2020 International Conference on Advances in Computing and Communication Engineering (ICACCE)*, 2020, pp. 1–4, doi:10.1109/ICACCE49060.2020.9154960.
[12] Sasubilli SM, Kumar A, and Dutt V. Improving health care by help of internet of things and bigdata analytics and cloud computing. In: *2020 International Conference on Advances in Computing and Communication Engineering (ICACCE)*, 2020, pp. 1–4, doi:10.1109/ICACCE49060.2020.9155042.
[13] Kotturu PK and Kumar A. Comparative study on machine learning models for early diagnose of Alzheimer's disease: multi correlation method. In: *2020 5th International Conference on Communication and Electronics Systems (ICCES)*, 2020, pp. 778–783, doi:10.1109/ICCES48766.2020.9137872.
[14] Bernard B and Britto PI. Design analysis and manufacturing of device for rehabilitation of hand fingers based on illusionary movements by tendon vibrations. In: Mandal DK and Syan CS (eds), *CAD/CAM, Robotics and Factories of the Future. Lecture Notes in Mechanical Engineering*. Springer, New Delhi, 2016. https://doi.org/10.1007/978-81-322-2740-3_22.
[15] Britto P, Vijayakumar R, and Sivarasu S. Design evaluation of REMAP exoskeleton. In: Goh J and Lim C (eds), *7th WACBE World Congress on Bioengineering 2015*. IFMBE Proceedings, vol 52. Springer, Cham, 2015. https://doi.org/10.1007/978-3-319-19452-3_30.
[16] Sankaran S, Murugan PR, Chandrasekaran D, *et al.* Design of IoT-based health care monitoring systems using Raspberry Pi: a review of the latest technologies and limitations. In: *2020 International Conference on Communication and Signal Processing (ICCSP)*. IEEE; 2020.

Chapter 18

Self-organizing deep learning approach for controlling movements of wheeled apparatus through corneal connotation

Shubhangi Rathkanthiwar[1]

This chapter presents a novel approach for facilitating the patients suffering from developmental coordination disorder (DCD). The motivation for this work is to develop a wheeled apparatus through corneal connotation related to sternly disabled patients facing difficulties in operating their wheelchair despite control operations provided on the handles or any other parts of the wheelchair. The main focus is on patients suffering from a lack of motor skills, e.g. "Dyspraxia." An exemplary implementation described in this chapter relates to "Design and Development of Applications using Human–Computer Interaction," a mechanical and electronic control system in a wheelchair whose operating movements are controlled by corneal association or eyeball movements. More specifically our work relates to the field of image acquisition. The related data are processed through self-organizing deep machine learning of artificial neural networks.

18.1 Introduction

Human eyes provide an instinctual way of collecting the information related to the surrounding observation, and the information is then processed in the brain and various parts of the body then respond accordingly. For a normal person, this is a normal routine sequence; however, for the patients suffering from DCDs, even if the instructions are received from the brain, some of the body parts do not respond as expected. As far as impairments in legs are concerned, a wheelchair is the best solution if the patient can operate the wheels of the chair. However, for those people, whose physiological capabilities are severely restricted to produce controlled movements, and also for the patients suffering from multiple sclerosis, total paralysis, locked-in syndrome, and muscular dystrophy, it becomes difficult to operate the wheelchair through fingers or hands. Some patients are not able to operate the wheelchair, due to various complicated technical features associated with various functions.

[1]Yeshwantrao Chavan College of Engineering, India

It is highly required to provide engineering solutions in the field of medical science correlated to patients with physical disabilities who lack motor skills. Motor skill disorder implicates a growing postponement of deportment causing a crusade drop in intelligence level which is substantially below that of the normal person. Such patients lose their confidence in responding to simple tasks. Other consequences could be down syndrome, muscular dystrophy, cerebral palsy, multiple sclerosis, etc. Dyspraxia refers to trouble with motor skills, balance, and coordination.

Our emphasis also includes patients suffering from Quadriplegia and Paraplegia. Quadriplegia is also termed as tetraplegia. Tetraplegia is a spinal cord injury that causes paralysis after an accident or resulting from injury to the spinal cord. Another point of focus is patients suffering from Dyspraxia, a DCD. Patients suffering from Dyspraxia face difficulty in important skills, including fine motor skills, which deals with the ability of a person in making actions through small muscles in the wrists and hands. Gross motor skills address tasks concerning large muscles, specifically, torso, arms, and legs. The term "Motor planning" is used as a skill for planning for all physical activities. We can remember all the steps to perform a set of actions. The last important skill is the "motor coordination" which critically involves complete integration of all the processes, including the interaction of muscles with the skeletal system as governed by the neural processes required for the actions in the spine and the brain. Thus, it involves the neurological pathways required to create the respective image and perform a desired predefined task.

We have developed a methodology for designing an apparatus for a disabled person having a lack of motor skills and difficulty in operating wheelchair functions. This would include patients suffering from Quadriplegia and Paraplegia. Quadriplegia (tetraplegia) is a spinal cord injury that causes paralysis after an accident or resulted from injury to the spinal cord, or disabled person, suffering from "Dyspraxia," or DCD. Dyspraxia refers to distress in movements. Patients suffering from Dyspraxia face difficulty in fine motor, gross motor planning, and coordination skills.

According to a survey carried out by World Health Organization (WHO), a high percentage of people are suffering from some form of disability, and the number of people with disability is growing dramatically. Because of diversified demographic trends, chronic health conditions, mental stress, old age, and several other causes, disability, either temporary or permanent, is increasing and is a major factor of concern. Disability is also related to mental health conditions and the problems related to cerebral palsy, down syndrome, and depression are also to be considered while dealing with solutions pertaining to the physical health.

Patients with disability resulting from a lack of motor skills frequently encounter a range of barriers when they try to attempt to live a normal life. If they wish to access healthcare services, in that case, affordability of health services and transportation are two main issues that prohibit them to use the services. People with disability are generally from a low-income group and it becomes difficult for them to afford healthcare as compared to people without disabilities. Second, due to limited or lack of appropriate services for people with disability, especially in rural and remote areas, and also due to inadequate skills and knowledge of health workers, the scope of addressing the problems becomes less.

To provide comfort in terms of movements, especially for those patients, suffering from a lack of motor skills, various classes of interfaces have been urbanized for powered wheelchair control. Joystick, sip-puff, and head control have been incorporated by other researchers. However, the major problem with people with disabilities caused due to lack of motor skills remains unresolved. They cannot operate the powered wheelchair through some of the interfaces and also this is so expensive that the common people cannot take its benefits.

Through this chapter, a novel technique for providing wheelchair movements through corneal connotation will be presented. Expected outcomes of our proposed product are the following:

- enhanced computational features, including feature extraction and tracking through self-organized deep learning technique;
- low-cost solution;
- ease in operation.

18.2 Previous works

To inspect the novelty of our work, related aspects are studied via a prior art search. Previous works are reconnoitred in this chapter to cover two contexts namely for existing intelligent models for wheelchair systems through corneal connotation, and methods for adding computational intelligence.

A prior art search gives an idea of various types of technologies used for developing the wheelchair so that the physically challenged person can easily move the wheelchair without depending on any other and the system will also provide maximum safety. Brain, eye, hand, head, and tongue movements are mainly used for developing a wheelchair. Arduino and Raspberry Pi have also been previously used by researchers for micro-processing.

A US patent application "US5360971A" describes an eye-tracking interface to generate communication as well as control functions defined by eye gestures. A perceiving device adapted herein detects bioelectromagnetic signals which are generated by movements of eyes. Out of the two processors used, while one receives the bioelectromagnetic signals, and generates tokens corresponding to eye gestures, the other one receives the keepsakes to generate grasp signals, based on the required protocol.

Another patent application "US9557812B2" describes an interface and calibration set up for eye gaze and for operating direction sensors, parameters including "angular accuracy" and "time averaging tradeoffs" are used. The method incorporated scales eye gaze target codes, the system transmits a large array of different ciphers with nominal training.

Literature [1] provides an overview of the methods used tracking the wheelchair. Information about voice operated wheel movements and in [2] wheelchair control operations are provided by user's facial expressions [3]. Primary difficulty in operating with voice control is because of various distortions, interference effects, and noises, the system responds to undesired commands. In the case of facial expression control, patients who have facial paralysis and other problems provide commands through facial expressions.

Some researchers have used infrared (IR) camera with light emitting diodes for adjusting automatic illumination [4]. Techniques for image processing were deployed for detecting the eye gaze [5], and the pupil knowledge was collected for detecting pupil location. Wheelchair commands were provided based on the threshold angle. The algorithm used in the patent application claims for fastest processing time. It also claims for high degree of robustness against the changing conditions. Limitation was that the threshold used was suitable only for precise illumination conditions and automatic illumination was enabled only through camera adjustments. The end result was the degradation in performance of the system for any strong illumination, e.g. sunlight. Another drawback of the system was in difficulties faced by the patients while operating the wheelchair through conceptual key control.

In the wheelchair proposed which is controlled by eyeball movements, the images captured by a camera (head mounted) are processed and a threshold is then applied for creating the binary images, erosion which is followed by distention is applied to reduce white Gaussian noise.

So as to move the chair to the left, right, and forward directions, some researchers have proposed the wheelchair model [6,7], where pupils of the eye are tracked using processing the images through open CV library. Raspberry-Pi plays an important role in providing the control operations. For the detection of movements of eyeball and also for recognition of facial expressions, HAAR cascade algorithm is used. Edge detection is done through Canny edge and circle detection using Hough Transform. Circle detection is basically required to recognize the eyeball circular borders. Two corner points are located using corner detection method, and based on the mean of these points, eyeball center is located. The distance between the mean points is measured and center point of the eyeball is determined. Min distance is then calculated which indicates direction in which eyeball is moving, and accordingly, commands are given for navigating the wheelchair in a specific direction. When the corneal is in the mid position, chair proceeds in the forward direction.

For tracking the eye gaze, a method, which can be operated independently of movements of camera, is the "EOG" i.e. Electro-oculography was also proposed by some researchers [8].

Due to low response time and low operating power, EOG is the preferred method than the other methods, as those are video based. EOG electrodes are positioned near the eyes. A ground electrode is positioned on the forehead. A human eye may be sculpted as an electric dipole. The retina can be considered as a negative pole and the cornea can be viewed as a positive pole. When there is no eye movement, or when the eyes are in original state, a steady electric potential is measured by the electrodes. The dipole's alignment changes for any movements in the eye.

18.3 Methodology

Our proposed model provides an apparatus for operating movements of wheelchair through movements of the eyeball. This apparatus includes a drive mechanism, which is configured on a chassis of wheels and the drive mechanism is provided to adapt movements of the apparatus, in the region of interest. Sensors are configured on the chassis to detect one or more physical attributes of the region of interest. Arm elements are adapted to collect information about corneal connotation and processed by image processing toolbox in the MATLAB® software, which is further processed through deep learning toolbox of MATLAB; and a microcontroller, Arduino, operatively coupled to the drive mechanism, the one or more sensors and the one or more arm elements, the microcontroller configured to operate and drive the mechanism to navigate the wheelchair. Data received from one or more sensors are analyzed to extract one or more physical attributes of the region. The wheelchair moves towards one or more physical attributes in the region of interest and receives information about corneal eyeball movements from the computer. It analyzes the received image data to extract one or more actions of the wheel movement according to the function approximation and pattern recognition properties of the artificial neural network. The comparison of the extracted one or more physical attributes of the corneal region of interest in the eye and the extracted one or more resource attributes of the neighborhood size, learning rate parameter, and the database with which the neural net is trained with the corresponding reference weight vectors.

18.4 Structural details

Image is captured using webcam and the image resolution is 640 $\times$ 480, skipping 30 frames to get an image. After the image is captured, the image is send to a PC/LAPTOP through Raspberry Pi. In image processing, captured image is read by the MATLAB software. After this, the image of the eye is segmented out. This segmented portion is turn into the black and white image (i.e. gray-scale image). A threshold is adjusted to recognize the corneal position, which is compared with the threshold value to know the eyeball position. This position of the eye is stored in the output variable as a result of image processing.

It involves the process of Data Transmission to Arduino, the data stored in the output the variable is transmitted to the Arduino for further processing. Here the Arduino uses the detected position of the eyeball. Accordingly, the processor takes the decision and converts the digital output signal to the electric signals which will direct the wheelchair to move in the desired direction. For the movement of the wheels, the H bridge principle is used. Arduino will generate the instructions, according to the digital output of the MATLAB, Arduino Uno kit will control the wheels of the wheelchair by actuating the L298D IC which is embedded in the H bridge module. The H bridge module will control the wheels of the wheelchair according to the instructions given by the Arduino kit. Transmitter unit comprises of Web-Camera, an image processing interface, a microcontroller, LED indicator, and an antenna to carry out transmission. Receiver unit comprises a receiving antenna, voltage regulators, a microcontroller, a motor drive, an emergency switch, a DC motor, and LCD display. The drive mechanism may include one or more servo motors, a direct current (DC) motor, an encoder, and a motor driver can be located in the one or more servo motors. RF module transmits or receives wireless radio signals. Transmitting unit is connected to wheelchair through the USB port.

One or more wheels provided on the wheelchair follow the commands required for the operation of the drive mechanism, adapted to move the wheelchair. Sensors are configured on the chassis of the chair to detect the physical attributes of the surrounding. The Arm elements are adapted to collect information about corneal connotation and processed by the image processing toolbox in MATLAB software, which is further processed through the deep learning toolbox of MATLAB, and a microcontroller, Arduino, operatively coupled to the drive mechanism, the one or more sensors and the one or more arm elements, the microcontroller configured to: operate, and the drive mechanism to move the apparatus in the region of interest. It analyses data received from one or more sensors and the region of interest to extract one or more physical attributes. The region of interest receives information about corneal eyeball movements from the computer. It also analyzes the received image data to extract one or more actions of the wheel movement according to the result of the function approximation and pattern recognition based on a comparison of the extracted one or more physical attributes of the corneal region of interest in the eye and the extracted one or more resource attributes of the neighborhood size, learning rate parameter and the data base with which neural net is trained with the corresponding reference weight vectors.

Image processing with intelligent features is an important aspect of this transceiver system. Such an intelligent system can be configured to enhance performance of the integrating system. Here ANN, i.e. artificial neural networks, can be configured to recognize recorded as well as extract patterns. ANN can also derive meanings from complex as well as imprecise data. When self-organized neural networks are trained, they act as an "expert system." The fact that these neural nets can learn to adapt and carry out responsibilities entirely based on experiences that they gain from previous experiences is a fantastic feature. Computational intelligence is added through self-organizing map (SOM) that works exceptionally when configured on the receiver side. SOM can be used to transform an incoming signal 1- or 2-dimensional distinct maps adaptively and that too in topologically ordered fashion. Such neurons can be positioned in 1- or 2-dimensional discrete map of the SOM.

The proposed model of wheelchair includes a method for generating a deep-learned unit, comprising several hidden units of SOM. Neural networks can transmit data through neurons and provide outputs values through the connections. Deep learning is related to the transformation and feature extraction and it establishes a relationship between neural responses associated with the brain and the stimuli.

The step of calculating radius of the neighborhood around the best matching unit (BMU) can be implemented such that neighborhood size decreases during every iteration. This lets SOM to make adjustments to respond to the signals, those are not mapped well, while not making changes in response to inputs it is well adjusted to. In an implementation, proposed architecture of SOM can be used for self-organizing the learning to adapt to discrete signal levels. Topology of the SOM network configured is a planar array of adaptive cells, where every cell corresponds to a particular lattice point in signal constellation. The hidden SOM network is a group of adaptive vectors beforehand adjusted neighborhood. Every node of the SOM has a time-varying weight vector that defines the BMU, or best matching labels, those can asylum all possible classification results. When the SOM is used in deep classification, distance between the sample vector and all other weight vectors consisting the nodes of the the map can be calculated. Classification result can then be evaluated as the label of a node with best response i.e. node having the nearest weight vector.

We have incorporated "Self-organized Deep learning," in which our multilayer model learns to perform classification tasks from the images captured by Webcam. Deep learning is implemented using multiple hidden layers of self-organizing neural network architecture. Deep mechanism enhances the classification task.

Multiple layers of SOM make adjustments in response to unfamiliar input signals. This feature makes the wheelchair used by any person, whose eye images were not used to train the neural network. Because it has been adequately adjusted to the data, Deep SOM does not require any changes to be made in reaction to the inputs. In an implementation, proposed architecture of SOM can be used to organize and learn to adapt to different signal levels. SOM topology can be configured as a planar array of adaptive cells. Learning network is a group of adaptive vectors that have a beforehand adjusted neighborhood. Every SOM node of the has a time-varying weight vector which defines the BMU. Best matching labels can cover all possible cataloging results. When the self-organizing neural network is used in a cataloging, the distance between sample and weight vectors can be calculated. Classification result can then be evaluated as the label of a node with best response i.e. node having nearest weight vector.

We have used ultrasonic sensors to stop the wheelchair in case of obstacles. Ultrasonic sensors with microcontroller unit detect the obstacle before it touch the chair. Microcontroller unit can then provide necessary series of actions for the movements of the wheels to proceed.

Figure 18.1 illustrates exemplary representation of the wheelchair operated by corneal connotations, in accordance with an embodiment of the present disclosure.

Exemplary representation of the wheelchair to be operated by corneal connotations is shown in Figure 18.1. It comprises a computing device or a computer with image processing and deep learning toolbox. Webcam, a transmitter, a receiver, and an ultrasonic sensor unit for obstacle detection are shown.

Referring to Figure 18.2, which pertains to the acquisition of images comprising an image of the eye with centrally placed eyeball/cornea, the image of the eye with left connotated eyeball/cornea, the image of the eye with right connotated eyeball/cornea.

The image acquisition block of the image processing tool of MATLAB software is an obstacle detecting circuit. A deep learning neural network block with deep layers is configured through a self-organizing feature map. In control block, microcontroller provides control actions to the wheels. These blocks constituting the transmit unit are shown in the exemplary representation. Receive unit, which include receiving antenna, voltage regulator, microcontroller to control wheel

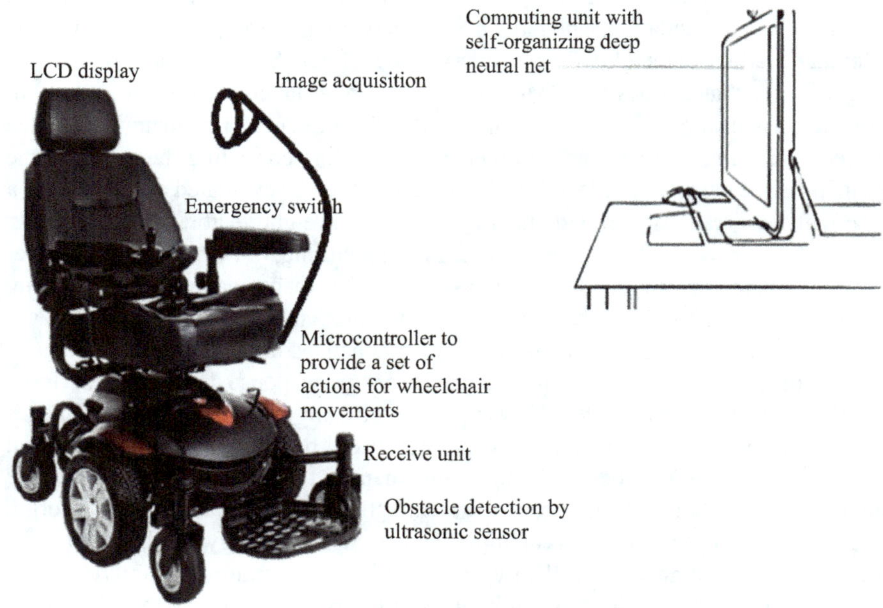

Figure 18.1 An exemplary representation of the corneal connotation controlled wheelchair with intelligent features

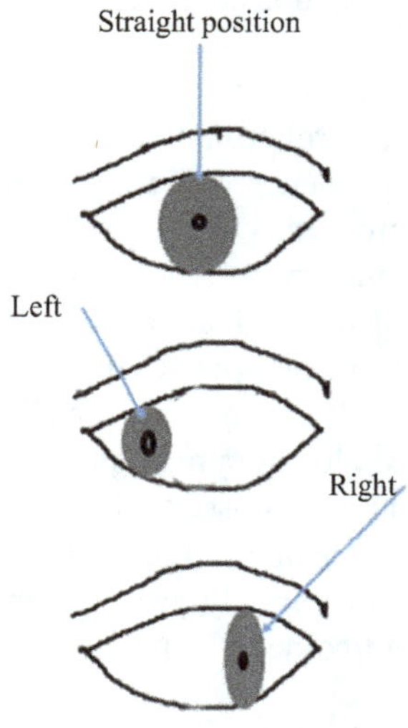

Figure 18.2 Images of eyeballs (centrally placed, left, and right connotated)

actions through DC drive control, LCD display, and emergency switch, is also shown.

This project facilitates the disabled user to have ease in mobility of the wheelchair. This mobility can be gained through four commands, namely, forward, right, left, and reverse. We provide two commands for starting and stopping the movements of the wheelchair. Triple eye winking starts the acquisition of the images and moving the wheelchair. Closed eye stops acquisition after double winking. Database was created for five users. Videos of five users making their corneal movements in three directions, i.e. right, left, and forward, reverse; triple eye winking and closing their eyes after double winking were captured under two different lighting conditions. The dataset consists of the six classes, namely, right, forward, left movements, reverse movement, and for starting and stopping the process of movements of the chair. For each handler, a set of 1,000 images were collected. The collected dataset includes images for in-door and outdoor lighting conditions. Besides, a dataset was collected for different 10 users, yet, only for indoors lighting condition. Then, 500 frames per each class for each user were taken to constitute a benchmark dataset of 10,000 frames. Next, the dataset was partitioned into training set and testing set, where 60% of the data were used for training. Consequently, the SOM algorithm was learnt from 5,000 frames and was tested on 4,000 unseen frames.

The system is especially designed for severely disabled patients having difficulties in operating powered wheelchair by hand; they lack motor skills and strength and, as such, they find it difficult to operate wheelchair functions. The context of development and implementation of a "Corneal Motion Controlled Wheelchair" is for disabled or paralyzed person.

Functionality of the system typically includes three steps:

- Detection of corneal movement of the eye
- Image processing
- Detection of the obstacle

Detection of corneal movement of the eye involves a process for checking the corneal movement of the eye using the camera that is provided in the transmitting unit. The installed camera will capture an image of eye of the person. In order to carry out the task for image processing, the camera will be continuously taking the image of the eye & checks the position of the corneal movement of the eye. The captured image will be compared with the pre-saved database of the images. By checking the position of the cornea, the wheelchair will move in any direction. A process is designed to detect the obstacles if any, so that wheelchair would get stopped and give alarm.

18.5 Conclusion

This chapter is based on the work that basically aims to help and build confidence in those patients who cannot move their legs and hands due to lack of motor skills, who may not give clear voice commands, or facial expressions, and those who can make movements in eyeball. Survey of existing wheelchairs clearly indicates that the available wheelchairs are very costly. Our project is a low-cost solution; however, additional computing device is to be connected to provide image processing and deep learning.

References

[1] Dahmani M., An intelligent and low-cost eye-tracking system for motorized wheelchair control, *Sensors* 2020, 20, 3936, doi:10.3390/s20143936.

[2] Naeem A., Qadar A., and Safdar W., Voice controlled intelligent wheelchair using Raspberry Pi, *International Journal of New Technology and Research* 2014, 2, 65.

[3] Rabhi Y., Mrabet M., and Fnaiech F., A facial expression controlled wheelchair for people with disabilities, *Computer Methods and Programs in Biomedicine* 2018, 165, 89–105.

[4] Arai K., Mardiyanto R., and Nopember K.I.T.S., A prototype of electric wheelchair controlled by eye-only for paralyzed user, *Journal of Robotics and Mechatronics.* 2011, 23, 66–74. [CrossRef]

[5] Mani N., Sebastian A., Paul A.M., Chacko A., and Raghunath A., Eye controlled electric wheel chair, *International Journal of Advanced Research in Electrical, Electronics and Instrumentation Engineering* 2015, 4, 4. [CrossRef]

[6] Gautam G., Sumanth G., Karthikeyan K., Sundar S., and Venkataraman D., Eye movement based electronic wheel chair for physically challenged persons, *International Journal of Scientific and Technology Research* 2014, 3, 206–212.

[7] Patel S.N., Prakash V., and Narayan P.S., Autonomous camera based eye controlled wheelchair system using raspberry-pi, in Proceedings of the 2015 International Conference on Innovations in Information, Embedded and Communication Systems (ICIIECS), Coimbatore, India, 19–20 March 2015.

[8] Al-Haddad A., Sudirman R., and Omar C., Guiding wheelchair motion based on EOG signals using tangent Bug algorithm, in Proceedings of the 2011 Third International Conference on Computational Intelligence, Modelling & Simulation, Langkawi, Malaysia, 20–22 September 2011.

Chapter 19

Prediction of breast tumour outcome to chemotherapy using statistical MR images through deep learning approaches

Md. Khaja Mohiddin[1], *Neha Mehta*[2], *V.B.S. Srilatha Indira Dutt*[3], *Seira Shinde*[4], *Amit Thakur*[4] and *G. Anand Kumar*[5]

Generally, breast cancer is generally of the most common cancers for women nationwide. Several diagnostic imaging tests have been performed recently. As a result, multiple image enhancement methods for diagnosing the progression of tumour tissues during chemotherapy were created. The aim of neoadjuvant chemotherapy (NAC) is to reduce cancerous cells before therapy. The ability to predict NAC behaviour may assist to decrease toxicity and time to psychological recovery. Deep convolution neural network (CNN) computational study of dynamic contrast-enhanced magnetic resonance imaging (DCE-MRI) seems to set a remarkable benefit in determining recipients and no recipient's patients. The objectives of this research are to introduce a new deep learning (DL) model that uses several MRI inputs to predict breast cancer reaction to NAC. For pre-processing and smoothing, an adaptive median filter has been preferred, whereas for segmentation K-means has been used. The best model for predicting pathological complete response (PCR) based on pre- and post-chemotherapy DCE-MRI was justified using a few external cases. The parameters such as accuracy, sensitivity, and specificity were used to evaluate the proposed model's efficiency. The graphical findings indicate that the peripheral area contains the much more significant extracted transforms from non-PCR tumours. Grey-level co-occurrence matrix and linear discriminant analysis have been suggested as feature extraction methods. Long short-term memory has also been suggested as the classification approach, which is more efficient than the other techniques. Use of the previous and first chemical pictures acquired with DCE-MR, the suggested and established CNN model was able to identify PCR and non-PCR patients with significant precision, even with a small

[1]Department of ET&T, BIT Raipur, India
[2]CSVTU Bhilai, India
[3]Department of ECE, GTIAM, India
[4]Department of CSE, BIT Raipur, India
[5]Department of ECE, GVPCOE, India

training dataset. After further assessment based on additional data, this model may be used in clinical research.

19.1 Introduction

The most often diagnosed form of breast cancer is seen all around the world. Many medical imaging examinations have been performed recently. As a result, numerous image processing approaches for diagnosing the progression of breast cancers during chemotherapy therapy were developed [1]. Neoadjuvant chemotherapy (NAC) will usually be used to treat patients with locally advanced prostate cancer before surgery. However, a significant number of patients did not respond completely to this treatment. To avoid needless chemotherapies, new approaches to forecast response during or earlier therapy are urgently needed. Furthermore, early detection of patients who do not respond to NAC would allow for treatment switching to other options [2–4].

Diffusion-weighted magnetic resonance imaging (DW-MRI), dynamic contrast-enhanced MRI (DCE-MRI), T1-weighted (T1W), and T2-weighted (T2W) pictures are among the various sub-sequences of MRI [5,6]. Each of these sequences may visualise and interpret intra-tumoural features in different ways. Quantitative MRI sequences are utilised to detect breast tumour response during NAC treatment (DCE-MRI and DW-MRI). Several studies have determined that statistical tomography, like DCE-MRI, can be preferred to identify numerous features associated with tissue type (normal or diseased tissue) based on a range of parameters, including tumour heterogeneity [9]. Currently, the size of the tumour determines the response to NAC in breast cancer [10]. The tumour size, on the other hand, could not be modified right once; it would take long durations [11,12].

In our previous work [13,14], we mainly concentrated on forecasting breast tumour reply to NAC therapy by means of image processing and deep learning (DL) methods employed for the volume of DCE-MRI. A multi-input convolutional neural network (CNN) has recently been created to allow for a baseline and post-treatment assessment to categorise patients with full pathological response [15]. We intend to train a large set of deep CNNs using DCE-MRI data to generate predictions regarding the pre-treatment, treatment, and response for breast tumours.

A non-invasive method of visualising cancer morphology and function by sensing the enhancement velocity of the contrast material is proposed. Several research projects have aimed to quantify DCE-MRI traits and connect them with genetic features, breast cancer subtypes, treatment response, and relapse-free survival (RFS). An innovative technique called DCE-MRI used by researchers Yamamoto *et al.* revealed gene- and disease pathway-related lncRNA biomarkers linked to early metastasis, which led to the discovery of genetic and molecular illness pathways. Despite advances, there are still challenges that prevent this technology from being used in therapeutic settings.

Tumour heterogeneity also exists across tumours, and in addition, it exists inside each tumour. More often than not, the disease can be found both within an individual tumour as well as throughout the tumour population. According to a study on spatially isolated regions, different gene expression signatures can be seen inside a single tumour of favourable and unfavourable prognoses [12]. Regarding varied locations within a tumour may have different imaging results. On MRI, there are a variety of dynamic enhancement patterns. Clustering trials have attempted to detect dynamical signals from tumour subregions of pixels and to further study them in order to recognise a certain dynamic pattern of prognostic or NAC response image features [9,13,14] in persons suffering from breast cancer. DCE-MRI resolution, however, means that each of the measured pixels may reflect a spatially mixed partial pixel by pixel, due to its restricted imaging capacities. The pressure–volume effect, which consists of numerous components, dynamic patterns of RMI occur in those areas of breast tumours [15].

An appropriate portrayal of this effect on DCE-MRI is required to emphasise tumour heterogeneity. An uncontrolled strategy to break down a series of tumours utilising a class-activation map (CAM) method for the detection of tumours with a different cinematic pattern was employed in the earlier work. The procedure was not monitored. There is also no indication that imagery in a tumour can enhance breast cancer survival in patients.

The purpose of this work is to estimate breast cancer prognosis by assessing intra-tumoural heterogeneity utilising deconstructed DCE-MR images, as well as imaging properties within these heterogeneous regions. There are other gene signatures connected to prognostic imaging characteristics found. These gene expression signatures are then compared with a separate dataset to see if they are associated with RFS or overall survival (OS).

A more prevalent way to assess breast cancer is DCE-MRI. Signal improvement with DCE-MRI-weighted T1 can be assessed using semi-quantitative approaches or pharmacokinetic models for quantification of changes in tissue contrast agent concentrations for estimating signal intensity changes. By virtue of containing pharmacokinetic parameters, it provides metrics pertaining to the changes in the microcirculation perfusion and permeability as well as allowing cross-site comparison. However, higher spatial resolution is required for the purpose of pharmacokinetic research, which necessitates trade-offs with spatial resolution. Perfusion and vessel permeability estimation involve both T1 relaxation time (T10) and vascular input function (VIF). Providing exact measurement methods for T10 and including local VIF in kinetic models of model-based DCE-MRI is difficult.

NAC has come a long way in the two decades preceding this one as a standard treatment for breast cancer patients. The treatment of people with breast cancer was initially done with NAC. Patients in the hope of healing are not regarded as suitable for surgery with respect to which the stage of the disease can be treated. NAC, however, has been used in patients in recent years. Mastectomy initially indicated the hope of a successful recovery. After NAC, the patient would become a best supportive care candidate [1].

The OS rate was reported to be 82% in a recent randomised trial [2] with a progression free-for-all survival rate at 4 years of age of 65% and a locoregional recurrence at 10%. These statistics show that NAC led to a considerable period of disease-free survival (DFS) for the majority of the patients. Over a period of time, though, a significant proportion of people recovered. Researchers have examined the impact of DCE-MRI features in the prediction of which patients would be less likely to have a DFS or total survival following NAC-OS.

They set out to investigate whether previous DCE-MRI data obtained might serve as a predictive guide for DFS and OS in patients with breast cancer and compare the results with other conventional predictors of survival such as grade, tumour necrosis metastasis stage, tumour type, and the presence of oestrogen (ER) and oestrogen receptor status.

The multiresolution fractal analysis in this case was used to investigate whether or not the following multiresolution fractal analyses were accurate for predicting the difference in the early response to neoadjuvant chemotherapy treatment (NACT) in volumetric DCE-MRI pharmacokinetic parametric maps: run length matrix, grey-level co-occurrence matrix, and single-resolution fractal analysis.

19.2 Materials and methods

19.2.1 Dataset

A total of 763 axial slices from the volume of DCE-MRI were utilised in this retrospective investigation to train and evaluate the DL architecture proposed by 42 patients with advanced breast tums. Before contrast was added, DCE-MRI was the only available method for post-contrast imaging. The clinical cohort used for this study was obtained from our radiology associate institution and followed by an ethical protocol authorised by the Committee on Ethics. While 14 of the test subjects responded with a full pathological complete response (PCR), 28 showed a complete response (CCR) (non-PCR). A cohort of 14 case studies served as the validation for the models (6 responders and 8 no responders). Before surgery, all patients had an MRI done, and after that, they went through many therapies. Only DCE-MRI images taken during the first treatment have been used in this study.

19.2.2 Neoadjuvant chemotherapy

NAC treatment in this trial included three rounds of chemotherapies followed by docetaxel (100 mg/m^2) given every 3 weeks for 3–4 cycles or paclitaxel (80 mg/m^2) given weekly for 3 months. Trastuzumab was given concurrently with chemotherapy treatment to individuals with HER2-positive malignancies [18].

19.2.3 MRI acquisition and parameters

The Siemens 1.5-T scanner with customised four-channel circularly polarised (CP) breast coile (Siemens, Erlangen, Germany) was used to carry out multi-parameter MRI scans at a number of time points (TPs) [16,17]. Six 3D FLASH pictures were taken from the dynamically enhanced MRI with a flip-angle of 25°, TR/TE = 0/4, 9, 39 ms, acquisition matrix of 512 × 154 × 64 and resolution of 0, 6.1, and

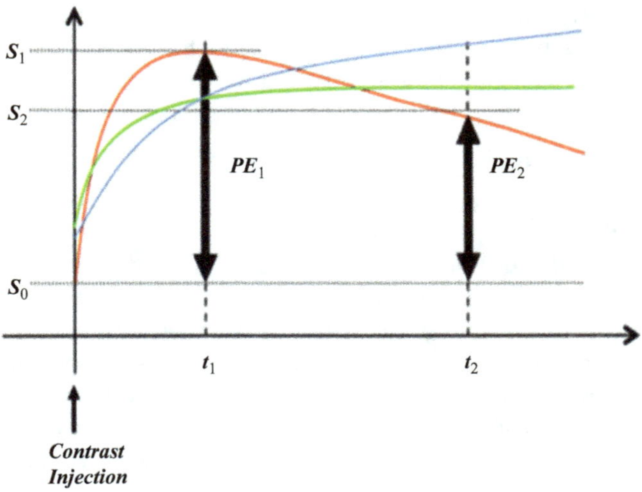

Figure 19.1 Signal intensity–time curve for DCE-MRI for early (PE1) and late (PE2) per cent enhancement

0.2 mm in DCE-MRI, before and after managing, 0.3 mmol/kg of gadodiamide (Ominiscan) [19].

19.2.4 Image processing

All volumes of DCE-MR were treated with bias correction [20] before the CNN design and training phase. The volume of each tumour was cut through a box around the region of interest using an experts radiologist (VOI). The cut volumes obtained both before and after the first treatment have been aligned based on a refined registration developed before [12], to offer similar parallel inputs. Due to the impact of the area around breast tumours, the volumes were divided into two sub-datasets: 1 – tumours with no segmentation and 2 – tumours with segmentation. The tumours in the second method were divided into a recent work with a U-Net profound learning architecture [21]. DCE-MRI subtraction sequences were employed to acquire the best segmentation precision, and the resulting mask was used for the first postwar volumes. All the input volumes cut down and recorded are 120 × 120 because only the photos of the same size are accepted by CNN. The data pre-processing that was conducted on the input photographs is shown in Figure 19.1.

19.2.5 Data augmentation

To eliminate overfitting and to carry out a well-coordinated training programme, on the original data, data augmentation techniques were used [22]. This entailed using a random linear algorithm. Rotations, translations, flips, zooming and other transformations for each training are preferred to produce new training samples epoch. As a result, over 100 training epochs, or around 110 training epochs have been produced. The system created and used 76,300 distinct samples from CNN. Nonlinear tumours were created with the goal of maintaining realistic tumour forms. In this investigation, no transformations were used.

19.3 CNN architectures

19.3.1 Single-input architecture

A backend was used with TensorFlow 1.8 and a DL network was used with Python 3.5 based on Keras 2.15. The original architecture as illustrated in Figure 19.3 consists of six blocks with three kernel convolutionaries each having three 3 × 3 layers, increasing the number of functional channels in subsequent levels. A linear rectified activation function (RELU) follows each convolution [23]. After each RELU feature, a max-pool layer [24] of 2 × 2 was employed to minimise the feature matrix size. A final classification result was acquired by stacking an

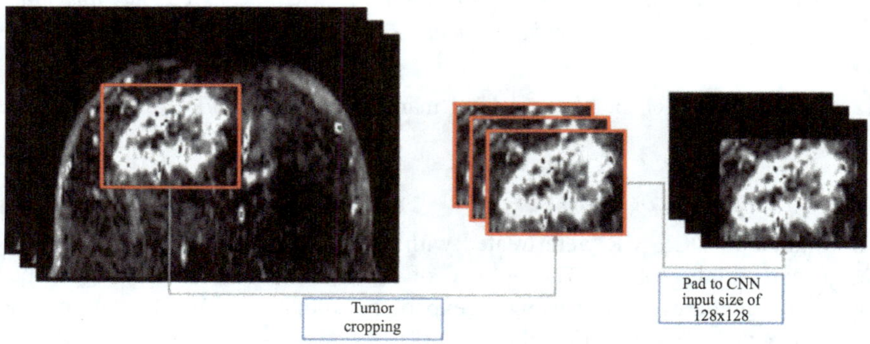

Figure 19.2 Data pre-processing

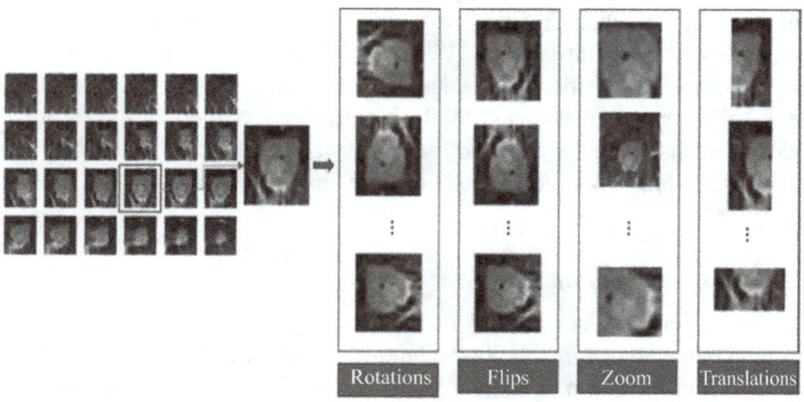

Figure 19.3 Data augmentation process based on random linear transformations

additional fully connected layer on top of the deep neural network (PCR or non-PCR). A 50% drop-over was employed before this fully connected layer to limit the number of redundant features [25]. This structure was built using photos collected before chemotherapy and after the first chemical therapy (post-treatment). Therefore, two unique models were developed to categorise respondents and non-respondents patients on the basis of independent data.

19.3.2 Multiple inputs architecture

A multi-input architecture was designed to improve the link between the gathered examinations and train the model with the most data possible. This architecture is divided into two sub-architectures, each with comparable layers to the prior single architecture. A flattened layer was applied to the retrieved topographies from the baseline and post-treatment DCE-MR images before applying a fully connected layer, as shown in Figure 19.4. After that, a concatenation was used between the two characteristics that have been flattened. To the best of our knowledge, this is the first time this DL architecture has been used in the literature. This DL parallel CNN might look at baselines at the same time and post-treatment evaluations to determine if the patient is PCR or non-PCR classification. To categorise a responder as either a responder or a non-responder. All of the slices of an autonomous subject were provided to the patient. The source of all categories is then fed into the DL model provided as an outcome.

The identical networks were trained using various inputs:

- Parallel inputs without the region of the segmented tumour.
- Parallel inputs within the region of the segmented tumour.

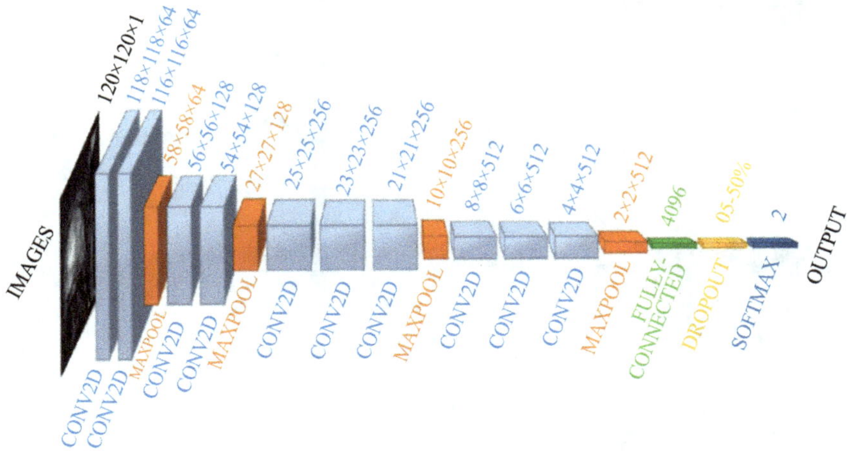

Figure 19.4 Single-input architecture

19.4 Method evaluations

Based on biopsy analysis [27], the founding truth of the investigation is the complete pathologic reaction (PCR). The accuracy of the model built was evaluated using the region under the AUC (ROC) curve, and the sensitivity and the specificity of each model are generated. The data were analysed to identify the optimal model for DCE-MR in the comparison between PCR and NAC in breast cancer. The 95% confidence Mann–Whitney *U*-test [28] was performed on the AUC values acquired for each model to give statistically validated data and assess if a difference is statistically significant between achieved AUC values. These quantitative measures are complemented by a visual tracking system to pick out the tumour regions that have the most relevant qualities for PCR and non-PCR selection. The methodology employed for this study was the use of a gradient-based approach for location [29]. After each convolutionary layer, the images that were overlaid over the retrieved feature heatmap were shown using the best constructed DL model.

19.5 Results and discussion

Mapping the therapeutic system in relation to how breast cancer responds to neoadjuvant therapy may affect treatment outcomes. For therapy initiation, PCR may be predicted and alternative methods should be employed. Various methods,

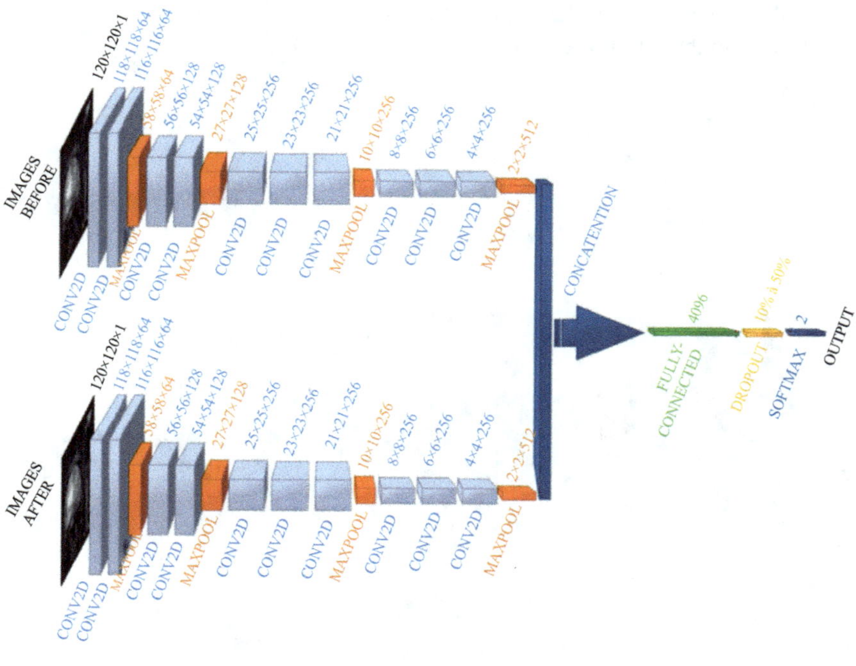

Figure 19.5 Multiple inputs architecture

including radiomics [30] or DL for baseline testing alone, were proposed (pre-NAC). Only a few of those studies included assessments in both pre- and post-NAC models. They also failed to include a visual map showing where the most critical elements that affect the classification of PCR and non-PCR are located. This study conceptualised and constructed a unique profound CNN architecture with one or more inputs. A single input from basic or post-NAC examinations is used for the initial iteration of this model. The model's AUC without segmented pictures at the beginning was larger than the post-NAC model AUC. Consequently, the breast tumour response to NAC may be predictable using only baseline DCE-MR scans.

The second model, created from the merged model, performed better in terms of accuracy and AUC. On the other hand, the AUC was considerably smaller for segmented photos with two versions than when no segmented images were being used. The local surroundings of the tumour may therefore play a critical role in determining NAC therapy response. The tumour location can be especially effective for providing information on lymphocytic infiltration according to our collaborative radiologist and current medical studies [31–36]. Indeed, lymphocytic infiltration and immunological response are frequently associated with the peripheral region [37–43]. Furthermore, a greater immune response is linked to improved post-chemotherapy outcomes [44–49]. As a result, cancers in non-responders' peripheral regions may signal a weak immune response [50–56]. Our data showed that the AUC value improved when images from only a baseline examination were included, while images from only a post-treatment assessment were not [57–61]. Since NAC response may be predicted using a baseline evaluation, baseline assessments are

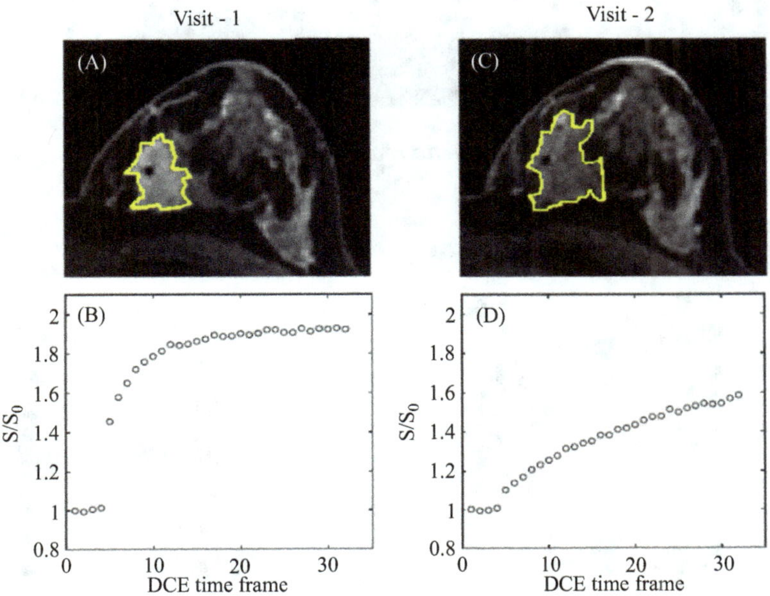

Figure 19.6 Post-contrast DCE-MRI image slice

enough to predict NAC response in breast cancer [62–66]. In contrast, integrating the pre- and post-treatment assessments to measure the first chemotherapy effect could be a more creative approach [67–71].

We used gradient class activation maps (Grad-CAM) to demonstrate the possibilities of our CNN input images, a technique which helps most people to predict NAC response [72–77]. We found that in DCE-MRI, CNN has refined all the areas of the tumour and emphasised its surrounding area for patients without PCR, whereas CNN is focussed on all the regions of the tumour and refines its extra area for patients with PCR [78–81]. The results show the Grad-CAM in PCR and non-PCR patients [82–85].

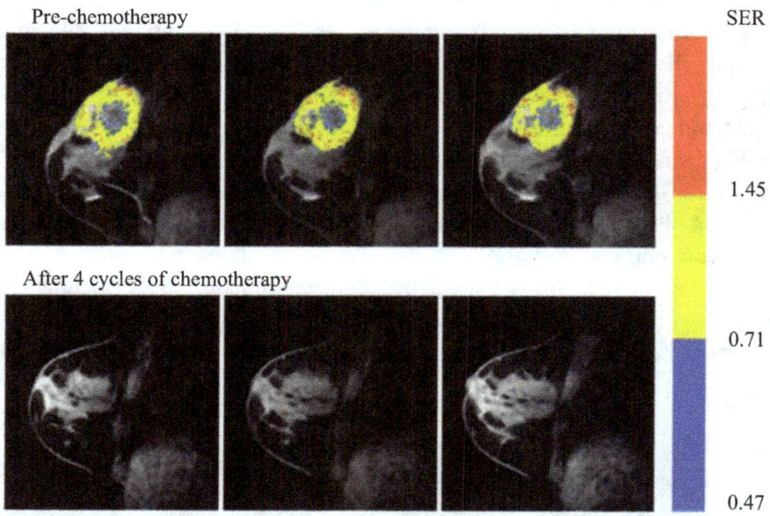

Figure 19.7 Magnetic reasoning imaging samples

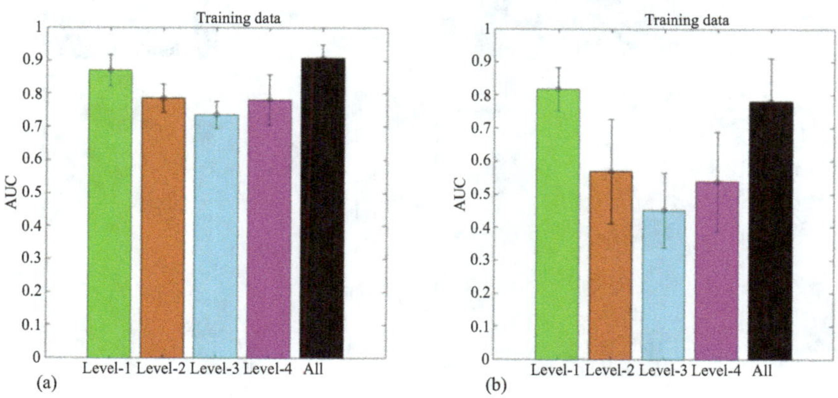

Figure 19.8 Multi-inputs fractal analysis of DCE-MRI data

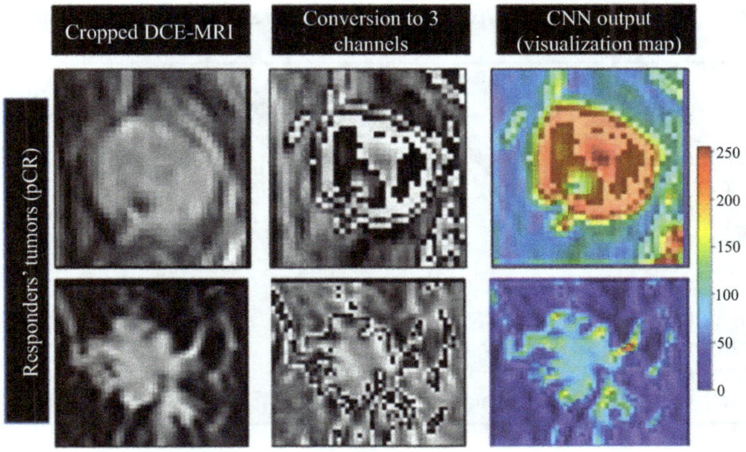

Figure 19.9 Randomly selected visual features Grad-Cam's results of PCR data

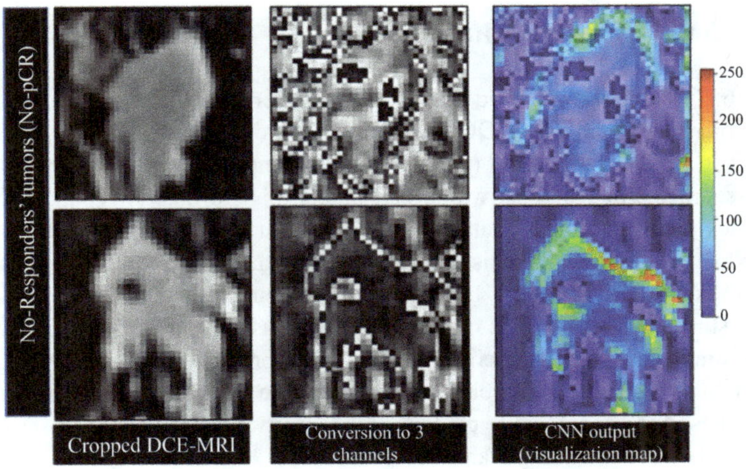

Figure 19.10 Randomly selected visual features Grad-Cam's results of non-PCR Data

Despite the positive findings produced by the DL model used in this study, it contains a number of limitations [86]. To begin with, the dataset used is quite little. To forecast the response from patients who had undergone NAC, as well as to gather new information, our research will focus on collecting annotated data from patients who had such treatment [87]. Table 1 compares previous methodologies to the proposed DL model. The amount of imaging data accessible is restricted by the number of pre- and post-treatment MRI tests [88]. Second, the information offered by this method necessitates a series of manual data pre-treatment steps. Although

Table 19.1 Quantitative results for using six different developed DL models

	AUC	Accuracy (%)	Sensitivity (%)	Specificity (%)
Single-input CNN Pre-treatment with segmentation	0.73	70	82.3	67.1
Single-input CNN Post-treatment with segmentation	0.70	71	81.4	67.3
Single-input CNN Pre-treatment without segmentation	0.79	82	82.1	67.9
Single-input CNN Post-treatment without segmentation	0.80	83	83.2	68.3
Multi-inputs CNN with segmentation	0.75	71	83.5	69.5
Multi-inputs CNN without segmentation	0.92	89	92.7	80.1

cropping the tumour volume and applying to align the corresponding slices, a 3D affine registration is used [85–88].

19.6 Conclusion and future scope

The deep CNN model was used to evaluate whether there were any non-reactants and recipients to NAC using DCE-MR pre- and post-treatment photos (PCR). A comparison was made between models which had inputs of pre-NAC or post-NAC tumour segmentation, models which had inputs of both post-NAC and pre-NAC tumour segmentation, and models that had no tumour segmentation. Although other models exhibit improved precision and AUC when compared to the segmented model, the combined model is, by far, the most accurate of these.

A distinctive characteristic of this study is the use of gradient class activations to show tumour areas of concern crucial to CNN predictions and provide greater interpretability than DL approaches without a visible strategic decision component. Radiologists and oncologists could employ activation maps to assess CNN recommendations for neoadjuvant therapy.

We plan to use this technology on more datasets that are still receiving annotation and ethical approval in the future. So those cancers that are needed will automatically crop and align. We wish to apply an area of interest detection technique based on DL to identify the tumour volume of interest. To allow radiologists and oncologists to have a comprehensive tool for their web applications, we want to add our DL approach to a cooperative web application.

References

[1] El Adoui M, Drisis S, Larhmam MA, Lemort M, and Benjelloun M (2017) Breast cancer heterogeneity analysis as index of response to treatment using MRI images: a review. *J Imaging Med* 9(4):109–119.

[2] Drisis S, El Adoui M, Flamen P, *et al.* (2020) Early prediction of neoadjuvant treatment outcome in locally advanced breast cancer using parametric response mapping and radial heterogeneity from breast MRI. *J Magn Reson Imaging* 51(5):1403–1411.

[3] Meisamy S, Bolan PJ, Baker EH, *et al.* (2004) Neoadjuvant chemotherapy of locally advanced breast cancer: predicting response with in vivo 1H MR spectroscopy-a pilot study at 4 T. *Radiology* 233(2):424–431.

[4] Liu SV, Melstrom L, Yao K, Russell CA, and Sener SF (2010) Neoadjuvant therapy for breast cancer. *J Surg Oncol* 101(4):283–291.

[5] Agrawal P, Sharma DK, Varun VL, *et al.* (2022) A survey on the scope of cloud computing. *Mater Today Proc*, 51(Part 1):861–864.

[6] Kul S, Cansu A, Alhan E, Dinc H, Gunes G, and Reis A (2011) Contribution of diffusion-weighted imaging to dynamic contrast enhanced MRI in the characterization of breast tumors. *Am J Roentgenol* 196(1):210–217.

[7] Adoui ME, Drisis S, and Benjelloun M (2017) Analyzing breast tumor heterogeneity to predict the response to chemotherapy using 3D MR images registration. In: *Proc Int Conf Smart Digit Environment (ICSDE)*. ACM, pp. 56–63.

[8] Theilmann RJ, Borders R, Trouard TP, *et al.* (2004) Changes in water mobility measured by diffusion MRI predict response of metastatic breast cancer to chemotherapy. *Neoplasia* 6(6):831.

[9] Rieber A, Brambs H-J, Gabelmann A, Heilmann V, Kreienberg R, and Kühn T (2002) Breast MRI for monitoring response of primary breast cancer to neo-adjuvant chemotherapy. *Eur Radiol* 12(7):1711–1719.

[10] Sannachi L, Gangeh M, Tadayyon H, *et al.* (2018) Response monitoring of breast cancer patients receiving neoadjuvant chemotherapy using quantitative ultrasound, texture, and molecular features. *PLoS ONE* 13(1):e0189634.

[11] Guenther LM, Rowe RG, Acharya PT, *et al.* (2018) Response evaluation criteria in solid tumors (RECIST) following neoadjuvant chemotherapy in osteosarcoma. *Pediatr Blood Cancer* 65(4):e26896.

[12] Cheung Y-C, Chen S-C, Su M-Y, *et al.* (2003) Monitoring the size and response of locally advanced breast cancers to neoadjuvant chemotherapy (weekly paclitaxel and epirubicin) with serial enhanced MRI. *Breast Cancer Res Treat* 78(1):51–58.

[13] Wassan S, Gulati K, Ghosh A, *et al.* (2021) Impact of digitalization on retailers and its future trends. *Mater Today Proc*, 51(Part 1):803–809.

[14] Cooper AB, Slack R, Fogelman D, *et al.* (2015) Characterization of anthropometric changes that occur during neoadjuvant therapy for potentially resectable pancreatic cancer. *Ann Surg Oncol* 22(7):2416–2423.

[15] El Adoui M, Drisis S, and Benjelloun M (2018) A PRM approach for early prediction of breast cancer response to chemotherapy based on registered MR images. *Int J Comput Assist Radiol Surg* 13:1233–1243.

[16] Mohiddin MK and Srilatha Indira Dutt VBS (2020) Mobility error prediction based LAB scheduling algorithm for optimizing system throughput in wireless sensor networks, *Int J Emerg Technol (IJET)* 11(2):1087–1092.

[17] El Adoui M, Larhmam MA, Drisis S, and Benjelloun M (2019) Deep learning approach predicting breast tumor response to neoadjuvant treatment using DCE-MRI volumes acquired before and after chemotherapy. In: *SPIE Medical Imaging, 2019: Computer-Aided Diagnosis*, vol. 10950, pp. 109502I.

[18] El Adoui M, Drisis S, and Benjelloun M (2019) Predict breast tumor response to chemotherapy using a 3D deep learning architecture applied to DCE-MRI data. In: *Seventh Int. Work-Conf. Bioinform Biomed Eng (IWBBIO), 2019*. Springer: Cham, pp. 33–40.

[19] Little RA, Barjat H, Hare JI, *et al.* (2018) Evaluation of dynamic contrast-enhanced MRI biomarkers for stratified cancer medicine: how do permeability and perfusion vary between human tumours? *Magn Reson Imaging* 46:98–105.

[20] Drisis S, Metens T, Ignatiadis M, Stathopoulos K, Chao S-L, and Lemort M (2016) Quantitative DCE-MRI for prediction of pathological complete response following neoadjuvant treatment for locally advanced breast cancer: the impact of breast cancer subtypes on the diagnostic accuracy. *Eur Radiol* 26(5):1474–1484.

[21] Drisis S, Flamen P, Ignatiadis M, *et al.* (2018) Total choline quantification measured by 1H MR spectroscopy as early predictor of response after neoadjuvant treatment for locally advanced breast cancer: the impact of immunohistochemical status. *J Magn Reson Imaging* 48(4):982–993.

[22] Kerwin WS, Zhao X, Yuan C, *et al.* (2009) Contrast-enhanced MRI of carotid atherosclerosis: dependence on contrast agent. *J Magn Reson Imaging* 30(1):35–40.

[23] Tustison NJ, Avants BB, Cook PA, *et al.* (2010) N4ITK: improved N3 bias correction. *IEEE Trans Med Imaging* 29(6):1310.

[24] Mohiddin MK and Srilatha Indira Dutt VBS (2019) Routing path estimation based on RWS method for competent energy dissipation employing X-layer network. *Int J Recent Technol Eng (IJRTE)* 8(2):6296–6303.

[25] El Adoui M, Mahmoudi SA, Larhmam MA, and Benjelloun M (2019) MRI breast tumor segmentation using different encoder and decoder CNN architectures. *Computers* 8(3):52.

[26] Schlüter J, and Grill T (2015) Exploring data augmentation for improved singing voice detection with neural networks. In: *ISMIR*, pp. 121–126.

[27] Giusti A, Cireşan DC, Masci J, Gambardella LM, and Schmidhuber J (2013) Fast image scanning with deep max-pooling convolutional neural networks. In: *2013 IEEE Int Conf Image Process*. IEEE, pp. 4034–4038.

[28] Kingma DP and Ba J (2014) Adam: a method for stochastic optimization. arXiv:1412.6980.

[29] Pham V, Bluche T, Kermorvant C, and Louradour J (2014) Dropout improves recurrent neural networks for handwriting recognition. In: *2014 14th Int Conf Frontiers Handwriting Recognit*. IEEE, pp. 285–290.

[30] Koutsoukas A, Monaghan KJ, Li X, and Huan J (2017) Deep-learning: investigating deep neural networks hyper-parameters and comparison of

performance to shallow methods for modeling bioactivity data. *J Cheminform* 9(1):42.

[31] Byrski T, Huzarski T, Dent R, *et al.* (2009) Response to neoadjuvant therapy with cisplatin in BRCA1-positive breast cancer patients. *Breast Cancer Res Treat* 115(2):359–363.

[32] Nachar N (2008) The Mann–Whitney U: a test for assessing whether two independent samples come from the same distribution. *Tutor Quant Methods Psychol* 4(1):13–20.

[33] Selvaraju RR, Cogswell M, Das A, Vedantam R, Parikh D, and Batra D (2017) Grad-cam: visual explanations from deep networks via gradient-based localization. In: *Proc IEEE Int Conf Comput Vis*, pp. 618–626.

[34] Fan M, Wu G, Cheng H, Zhang J, Shao G, and Li L (2017) Radiomic analysis of DCE-MRI for prediction of response to neoadjuvant chemotherapy in breast cancer patients. *Eur J Radiol* 94:140–147.

[35] Chaplin DD (2010) Overview of the immune response. *J Allergy Clin Immunol* 125(2):S3–S23.

[36] Abraham DC, Jones RC, Jones SE, *et al.* (1996) Evaluation of neoadjuvant chemotherapeutic response of locally advanced breast cancer by magnetic resonance imaging. *Cancer* 78(1):91–100.

[37] Mohiddin MK and Srilatha Indira Dutt VBS (2019) An optimum energy consumption hybrid algorithm for XLN strategic design in WSNs. *Int J Comput Netw Commun (IJCNC)* 11(4):61–80.

[38] Mahmoudi SA, El Adoui M, Belarbi MA, Larhmam MA, and Lecron F (2017) Cloud-based platform for computer vision applications. In: *Proc 2017 Int Conf Smart Digit environment (ICSDE)*. ACM, pp 195–200.

[39] Kelcz F and Santyr G (1995) Gadolinium-enhanced breast MRI. *Crit Rev Diagn Imaging* 36:287–338.

[40] Hulka CA, Smith BL, Sgroi DC, *et al.* (1995) Benign and malignant breast lesions: differentiation with echo-planar MR imaging. *Radiology* 197:33–38.

[41] Stomper PC, Winston JS, Herman S, *et al.* (1997) Angiogenesis and dynamic MR imaging gadolinium enhancement of malignant and benign breast lesions. *Breast Cancer Res Treat* 45:39–46.

[42] Mohiddin MK and Srilatha Indira Dutt VBS (2017) Minimization of energy consumption using X-layer network transformation model for IEEE 802.15.4-based MWSNs. In: *Proc 5th Int Conf Frontiers Intell Comput Theory Appl Adv Intell Syst Comput (AISC)* 515:741–751.

[43] Mussurakis S, Buckley DL, Drew PJ, *et al.* (1997) Dynamic MR imaging of the breast combined with analysis of contrast agent kinetics in the differentiation of primary breast tumours. *Clin Radiol* 52:516–526.

[44] Issa B, Buckley DL, and Turnbull LW (1999) Heterogeneity analysis of Gd-DTPA uptake: improvement in breast lesion differentiation. *J Comput Assist Tomogr* 23:615–621.

[45] Liney GP, Gibbs P, Hayes C, Leach MO, and Turnbull LW (1999) Dynamic contrast-enhanced MRI in the differentiation of breast tumors: user-defined

versus semi-automated region of- interest analysis. *J Magn Reson Imaging* 10:945–949.

[46] Kuhl CK and Schild HH (2000) Dynamic image interpretation of MRI of the breast. *J Magn Reson Imaging* 12:965–974.

[47] Wedegartner U, Bick U, Wortler K, Rummeny E, and Bongartz G (2001) Differentiation between benign and malignant findings on MR mammography: usefulness of morphological criteria. *Eur Radiol* 11:1645–1650.

[48] Yadu P, Mohiddin MK, Yadav PK, and Sahu AK (2017) A modified MPPT technique for drift reduction using neural networks. In: *Int Conf Innov Control Commun Inf Syst (ICICCI)*. IEEE Xplore, pp. 1–7.

[49] Furman-Haran E, Schechtman E, Kelcz F, Kirshenbaum K, and Degani H (2005) Magnetic resonance imaging reveals functional diversity of the vasculature in benign and malignant breast lesions. *Cancer* 104:708–718.

[50] Kuhl CK, Schild HH, and Morakkabati N (2005) Dynamic bilateral contrast-enhanced MR imaging of the breast: trade-off between spatial and temporal resolution. *Radiology* 236:789–800.

[51] Hulka CA, Edmister WB, Smith BL, *et al.* (1997) Dynamic echo-planar imaging of the breast: experience in diagnosing breast carcinoma and correlation with tumor angiogenesis. *Radiology* 205:837–842.

[52] Bone B, Aspelin P, Bronge L, and Veress B (1998) Contrast-enhanced MR imaging as a prognostic indicator of breast cancer. *Acta Radiol* 39:279–284.

[53] Mussurakis S, Buckley DL, and Horsman A (1997) Dynamic MR imaging of invasive breast cancer: correlation with tumour grade and other histological factors. *Br J Radiol* 70:446–451.

[54] Mukherjee I, Bansal A, Patra M, Pal R, and Mohiddin MK (2018) An Android application Studhelper for engineering students. *Int J Comput Sci Inf Security (IJCSIS)* 16(4):184–187.

[55] Stomper PC, Herman S, Klippenstein DL, *et al.* (1996) Invasive breast carcinoma: analysis of dynamic magnetic resonance imaging enhancement features and cell proliferative activity determined by DNA S-phase percentage. *Cancer* 77:1844–1849.

[56] Bone B, Szabo BK, Perbeck LG, Veress B, and Aspelin P (2003) Can contrast-enhanced MR imaging predict survival in breast cancer? *Acta Radiol* 44:373–378.

[57] Su MY, Cheung YC, Fruehauf JP, *et al.* (2003) Correlation of dynamic contrast enhancement MRI parameters with microvessel density and VEGF for assessment of angiogenesis in breast cancer. *J Magn Reson Imaging* 18:467–477.

[58] Tuncbilek N, Karakas HM, and Okten OO (2005) Dynamic magnetic resonance imaging in determining histopathological prognostic factors of invasive breast cancers. *Eur J Radiol* 53:199–205.

[59] Padhani AR, Hayes C, Assersohn L, *et al.* (2006) Prediction of clinicopathologic response of breast cancer to primary chemotherapy at contrast-enhanced MR imaging: initial clinical results. *Radiology* 239:361–374.

[60] Hylton NM (2001) Vascularity assessment of breast lesions with gadolinium-enhanced MR imaging. *Magn Reson Imaging Clin N Am* 9: 321–332.

[61] Furman-Haran E and Degani H (2002) Parametric analysis of breast MRI. *J Comput Assist Tomogr* 26:376–386.

[62] Esserman L, Hylton N, George T, and Weidner N (1999) Contrast-enhanced magnetic resonance imaging to assess tumor histopathology and angiogenesis in breast carcinoma. *Breast J* 5:13–21.

[63] Sinha H, Roy SK, and Mohiddin MK (2020) Optimal ordering equalization technique for MIMO wireless channel. *Int J Future Gener Commun Netw* 13 (4):54–60.

[64] Tofts PS (1997) Modeling tracer kinetics in dynamic Gd-DTPA MR imaging. *J Magn Reson Imaging* 7:91–101.

[65] Li KL, Henry RG, Wilmes LJ, *et al.* (2007) Kinetic assessment of breast tumors using high spatial resolution signal enhancement ratio (SER) imaging. *Magn Reson Med* 58:572–581.

[66] Knopp MV, Weiss E, Sinn HP, *et al.* (1999) Pathophysiologic basis of contrast enhancement in breast tumors. *J Magn Reson Imaging* 10:260–266.

[67] Kuhl CK, Mielcareck P, Klaschik S, *et al.* (1999) Dynamic breast MR imaging: are signal intensity time course data useful for differential diagnosis of enhancing lesions? *Radiology* 211:101–110.

[68] Liu PF, Debatin JF, Caduff RF, *et al.* (1998) Improved diagnostic accuracy in dynamic contrast enhanced MRI of the breast by combined quantitative and qualitative analysis. *Br J Radiol* 71:501–509.

[69] Li KL, Wilmes LJ, Henry RG, *et al.* (2005) Heterogeneity in the angiogenic response of a BT474 human breast cancer to a novel vascular endothelial growth factor-receptor tyrosine kinase inhibitor: assessment by voxel analysis of dynamic contrast-enhanced MRI. *J Magn Reson Imaging* 22:511–519.

[70] Tofts PS, Berkowitz B, and Schnall MD (1995) Quantitative analysis of dynamic Gd-DTPA enhancement in breast tumors using a permeability model. *Magn Reson Med* 33:564–568.

[71] Sujith AVLN, Khaja Mohiddin M, Varun VL, *et al.* (2021) A review on fast intra prediction mode decision algorithms in HEVC. *Design Eng (Toronto)* 2021(5):2176–2181.

[72] Eliat PA, Dedieu V, Bertino C, *et al.* (2004) Magnetic resonance imaging contrast-enhanced relaxometry of breast tumors: an MRI multicenter investigation concerning 100 patients. *Magn Reson Imaging* 22:475–481.

[73] Bottomley PA, Hardy CJ, Argersinger RE, and Allen-Moore G (1987) A review of 1H nuclear magnetic resonance relaxation in pathology: are T1 and T2 diagnostic? *Med Phys* 14:1–37.

[74] Chung M, Chang HR, Bland KI, and Wanebo HJ (1996) Younger women with breast carcinoma have a poorer prognosis than older women. *Cancer*; 77:97–103.

[75] Mayr NA, Taoka T, Yuh WT, *et al.* (2000) Magnetic resonance imaging in the assessment of radiation response in cervical cancer: regarding Hatano K *et al.* IJROBP 1999; 45:399–344. *Int J Radiat Oncol Biol Phys* 48:910–912.

[76] Tofts PS, Brix G, Buckley DL, *et al.* (1999) Estimating kinetic parameters from dynamic contrast- enhanced T(1)-weighted MRI of a diffusible tracer: standardized quantities and symbols. *J Magn Reson Imaging* 10:223–232.

[77] Partridge SC, Gibbs JE, Lu Y, *et al.* (2005) MRI measurements of breast tumor volume predict response to neoadjuvant chemotherapy and recurrence-free survival. *AJR Am J Roentgenol* 184:1774–1781.

[78] Maroor JP, Mohiddin MK, Nirmal A, *et al.* (2021) Empirical analysis of student performance using data mining approaches. *Design Eng (Toronto)* 2021(5):737–748.

[79] Giordano SH (2003) Update on locally advanced breast cancer. *The Oncologist* 8:521–530.

[80] Van de Hage JA, van de Velde CJH, Julien JP, *et al.* (2001) Preoperative chemotherapy in primary operable breast cancer: results from the European organization for research and treatment of cancer trial 10902. *J Clin Oncol* 19:4224–4237.

[81] Hayes C, Padhani AR, Leach MO (2002) Assessing changes in tumour vascular function using dynamic contrast-enhanced magnetic resonance imaging. *NMR Biomed* 15:154–163.

[82] Pickles MD, Lowry M, Manton DJ, *et al.* (2005) Role of dynamic contrast enhanced MRI in monitoring early response of locally advanced breast cancer to neoadjuvant chemotherapy. *Breast Cancer Res Treat* 91:1–10.

[83] Brix G, SemmlerW, Port R, *et al.* (1991) Pharmacokinetic parameters in CNS Gd-DTPA enhanced MR imaging. *J Comput Assist Tomogr* 15:621–628.

[84] Cocquyt VF, Villeirs GM, Blondeel PN, *et al.* (2002) Assessment of response to preoperative chemotherapy in patients with stage II and III breast cancer: the value of MRI. *Breast* 11:306–315.

[85] Gilles R, Guinebretiere JM, Toussaint C, *et al.* (1994) Local advanced breast-cancer—contrast-enhanced subtraction MR-imaging of response to preoperative chemotherapy. *Radiology* 191:633–688.

[86] Sasubilli SM, Kumar A, and Dutt V (2020) Machine learning implementation on medical domain to identify disease insights using TMS. In: *2020 Int Conf Adv Comput Commun Eng (ICACCE)*, pp. 1–4, doi:10.1109/ICACCE49060.2020.9154960.

[87] Sasubilli SM, Kumar A, and Dutt V (2020) Improving health care by help of internet of things and bigdata analytics and cloud computing. In: *2020 Int Conf Adv Comput Commun Eng (ICACCE)*, pp. 1–4, doi:10.1109/ICACCE49060.2020.9155042.

[88] Kotturu PK and Kumar A (2020) Comparative study on machine learning models for early diagnose of Alzheimer's disease: multi correlation method. In: *2020 5th International Conference on Communication and Electronics Systems (ICCES)*, pp. 778–783, doi:10.1109/ICCES48766.2020.9137872.

Chapter 20

Risk analysis and prediction of cancer associated with Type II diabetes: a review

K.N.Soumya[1] and P. Vigneshwaran[2]

Diabetes has become one of the most fatal disorders owing to the change in lifestyle, dietary habits, and lesser physical activity. According to a recent World Health Organization report, it is claimed that 422 million people are suffering from diabetes. Having said that, the Type II category of diabetes is more fatal in nature as it is characterized by the insulin resistance developed by the body. Moreover, Type II diabetes is known to severely impact the kidneys, eyes, and the heart. Also, a large number of researchers are conducting studies to find out if there is any correlation between diabetes and cancer. In this present work, we present a review of such studies and also present our endeavor toward cancer risk assessment and prediction in diabetes Type II patients.

20.1 Introduction

Change, as they say, is the only constant in the Universe. The socioeconomic development has brought a paradigm shift in the lifestyle of the citizens of developing countries like India. This has an impact not only on the urban population but also on semi-urban and rural populations as well. The change in lifestyle has brought changes in dietary habits, physical activity levels, etc. It is very trivial to state that this has led to what can be coined as lifestyle diseases though many of which are technically disorders and diabetes being more chronic and fatal. Diabetes is technically a disorder and not a disease, which is categorized by amplified levels of glucose in the blood leading to damage to the eyes, kidney, heart, and nervous system. Diabetes can be broadly categorized into two types: Type 1 and Type 2. In the former case, the pancreas produces little or no insulin by itself, whereas in the latter case, the body becomes resistant to insulin. It is imperative for the government to ensure access to affordable treatment for the patients for their survival. As per the latest reports by World Health Organization (WHO), about 422 million people, more specifically from low and middle-income countries, are suffering

[1]Department of Information Science & Engineering, Jain Deemed To Be University, India
[2]Department of Networking and Communications, SRM Institute of Science and Technology, India

from diabetes in the world, out of which 1.6 million deaths can be directly attributed to the aforementioned disorder. It is generally observed that patients with Type II diabetes tend to show cancer symptoms at a later point in time. Many researchers around the world have already carried out extensive research on both cancer and diabetes. However, both the efforts were put in their own silos which makes it challenging to study if diabetes can possibly cause cancer or lead to it. With the advent of high-performance computing (HPC), it has become possible to scale up the computations at a fraction of cost. It is our endeavor toward finding a plausible explanation for Type II diabetes patients to develop cancer symptoms that we present in this paper and also present a review of the efforts put in by fellow researchers around the globe toward the common cause.

20.2 Diabetes

Diabetes has become a major disorder in almost all the strata of the social fabric around the world, more especially so in India. In India, as per the National Diabetes and Diabetes Retinopathy survey, it is claimed that 11.8% of our population is affected by diabetes. By 2025, around 134 million people will be affected by diabetes. Currently, India has 77 million diabetes patients, i.e., every 1 in 6 Indians is diabetic, making India the second highest in the world. The major concern is that diabetes is a multidimensional causative factor that needs a multidimensional approach as opposed to the fragmented care that exists as of now. Diabetes can be normally categorized into two types, as discussed in the previous sections. Now let us understand these types.

20.2.1 Type I diabetes

It is a condition characterized by the immune system ending up destroying insulin-producing cells in the pancreas. These cells are termed beta cells, and the condition is usually diagnosed in young adults and children. Damage to the beta cells upsets the glucose metabolism where the glucose can no longer enter the cells as insulin is not present to facilitate the glucose metabolism and thus the glucose ends up in the bloodstream causing high blood sugar.

20.2.2 Type II diabetes

It is more severe and fatal than Type I diabetes. This condition persists for life. This condition is characterized by the resistance the body shows toward insulin. Type II diabetes has a severe impact on the kidneys, eyes, heart, and nervous system, and it is the most common type in many people.

20.3 Cancer

Cancer is a disease characterized by the uncontrolled replication/division of cells. While some types manifest in an exponential growth/division of cells, the others, on the contrary, might grow and divide slowly. Every cell in the body has a specific job and lifetime. Once the lifetime is over the cell gets signaled to die, only to get replaced by another by the body. Some of the major types of cancer are:

- **Carcinomas.** A carcinoma is characterized destruction of the healthy surrounding tissues which will eventually spread to other parts. This type of cancer generally begins in the tissues that are closer to the inner or outer surfaces of the body. Generally, this type of cancer occurs when the DNA of a cell is either altered or damaged. Some examples are prostate, breast cancer, lung, and colorectal cancer.
- **Sarcomas.** A sarcoma generally originates from the damaged cells of connective tissue. Broadly speaking, these include fat, cartilage, bone, etc. This can be found in any part of the body. A sarcoma has many associated subtypes characterized by the tissue and type of cell. Sarcoma can be broadly categorized into two types, viz., bone and soft tissue.
- **Leukemia.** Leukemia is also called blood cancer, wherein the ability of the body to fight infection is compromised. Leukemia is generally associated with problems in producing blood and usually affects the white blood cells (WBC). It is generally characterized by high number of abnormal blood cells.
- **Lymphomas.** Lymphoma is yet another type of cancer that originates in the lymphatic system. It is characterized by changes in lymphocytes and an abnormally high number of lymphocytes. Lymphoma can be broadly categorized into two types, i.e., Hodgkin's and non-Hodgkin's.

20.4 Related works

Zhu *et al.* [1] have worked on finding the relationship [2–5] between diabetes and colorectal cancer. This is based on a meta-analysis that reports the impact of diabetes on colorectal prediction and found in some cohort studies and the claims to have produced consistent results. They further claim to have carried out extensive searching on manifold records and performed a meta-analysis on the collected data of close to 2 million subjects. They have analyzed the existence rate and existence risk which are cancer-specific. The results of the study suggest that diabetic patients will have a reduced lifespan by at least 5 years, especially in subjects suffering from colorectal, colon, and rectal cancer by a significant factor of 18, 19, and 16%, respectively, as compared to that of non-diabetic subjects. They state that some of the studies [6–11] show that the overall survival of colorectal cancer patients with diabetes mellitus had shown a decreased risk of survival. However, they conclude that their meta-analysis [12–14] suggested that diabetes had a negative impact on colorectal cancer in overall survival aspects. A study on diabetes mellitus and breast cancer was performed by Peairs *et al.* [13] with the goal to perform a systemic review and meta-analysis to study the impact of diabetes on breast cancer. They have leveraged EMBASE and MEDLINE databases. They have used relative terms to diabetes mellitus and cancer for data mining. Their search included the terms like "glucose intolerance," "hyperglycemia," "cancer," "malignant neoplasm," etc. They further state that the two studies performed on cancer-specific mortality provided mixed results and further stated that Srokowski *et al.* [15] observed an increasing trend in breast-cancer-specific mortality with pre-existing diabetes mellitus. These subjects seem to have undergone chemotherapy. He further demonstrated that an increased number of women with diabetes mellitus showed indications of a more advanced stage of breast cancer than the non-diabetic subjects. The study also suggested that the subjects with pre-existing diabetes mellitus had more risk of repeated chemotherapy. Some of the studies [15–18] in similar lines that were conducted suggested the existence of a positive correlation between breast cancer and pre-existing diabetes mellitus, i.e., women already diagnosed with diabetes mellitus were more prone to having breast cancer, while the other researchers were carrying out studies to find out the plausible association of cancer with pre-existing diabetes mellitus condition. Tsilidis *et al.* [17] carried out a study with the goal to evaluate the validity of such a plausible association and also the assessment of the risk of fatality. They leveraged the data sources from PubMed, EMBASE, and Cochrane Database and then performed a meta-analysis across the data to find if the subjects with Type II diabetes are at risk of developing cancer or have a low survival score. During their study, they found that many meta-analysis reported that Type II diabetes has a positive correlation with an elevated risk of various types of cancer, viz., liver, breast, pancreas, etc. [19–21]. However, their study suffers from some misclassification as diabetes assessment was performed at baseline much before the cancer diagnosis and hence they are non-differential in nature. However, in the conclusion of their study, although they do not deny Type II diabetes being associated with various types of cancer based on many other literatures, they would have referenced and certainly expressed a caveat that the results of the studies showed a strong inclination to Type II diabetes being associated with risk of developing cancer or even worse death from cancer and demonstrated a light trace of bias. They further press the need for setting up consortiums and employ the better assessment techniques for Type II diabetes so as to facilitate more plausible conclusions.

Dániel Végh *et al.* [22] took up an interesting study on the association of Type II diabetes with oral tumors as Hungary has the largest number of oral cancer and also a large number of Type II diabetes incidences. They chose patients who had pre-existing malignant [23–25] form of cancer. The results of their study have shown that Type II diabetes was found in 25.9% of the subjects and impaired fasting glucose in 20.6% of the subjects who were diagnosed with oral cancer. The results further indicate that a total of 46.5% of the group of subjects diagnosed with oral cancer had a chronic metabolic disorder. The studies of most researchers indicate that type II diabetes is associated with an elevated risk of cancer. Also, some epidemiological data show that diabetes has an impact on cancers and more especially so in some hyper-local cancers like colorectal, pancreas, and liver. Hyperinsulinemia, characterized by a high degree of insulin resistance, is suspected to find the missing link. This is due to the hypothesis that insulin might have a possible mitogenic effect [26–29]. Also, some of the studies suggest that the hyperglycemic condition itself might be carcinogenic as it elevates oxidative stress

[30,31]. Hiroshi Noto *et al.* [32] have been studying these hypotheses to find a plausible link between cancer and Type II diabetes. A deep neural network (DNN) is an architecture of artificial neural networks (ANNs) which employs multiple layers between the input and output layers. The advantage of DNNs is that it always secures the correct mathematical aspect to convert the input to output irrespective of the data attributes being linear or nonlinear. With the HPC and GPUs becoming more accessible and economical it has become much easier to leverage such technology easily. Meng-Hsuen Hsieh *et al.* [33] have employed DNN to study the association of Type II diabetes with colorectal cancer [34–37]. They have leveraged k-fold cross-validation with the value of k being 10 as a metric in this case to determine the KPI-like loss function of predictor, high impact hyper-parameters, etc. Their DNN model consisted of a 37-dimensional input layer and three hidden layers of 30 dimensions, each of which resulted in one scalar output layer. They employed stochastic gradient descent [35–37] which they further claim to have optimized by leveraging accelerated gradient descent [38–40]. The high dimensionality of the data inherently means that the input layers will be densely connected and it becomes imperative for the activation function to be highly sensitive, and hence rectified linear unit [41] was chosen as the activation function for input and hidden layers. Since they were not interested in a binary classification, they use softmax activation [42] for the output layer. Accuracy was not a reliable metric of the predictor function owing to the unbalanced data distribution [43] and hence they used precision, sensitivity, and F1 score. In the concluding remarks of their study, they have stated that based on the DNN model which they had designed they could see the association between Type II diabetes and colorectal cancer in subjects with pre-existing Type II diabetes.

Given the fact that cancer has a large number of categories and impact, it becomes more feasible to perform the study to find a plausible association of Type II diabetes with a localized/site-specific cancer. Among many such studies, one study which holds our interest is the study of Type II diabetes being associated with breast cancer. Boyle *et al.* [44] have carried out studies to find out if there is a link between Type II diabetes and breast cancer. They initially studied the relative risk and their respective confidence intervals were derived statistically followed by calculating the corresponding variance. Their results suggest that their meta-analysis included as much as 40 risk estimates from their literature survey. The results further indicated that in the subjects with pre-existing Type II diabetes the summary relative risk (SRR) was found to be 1.27 with a confidence interval of 95%, and the sensitivity analysis produced statistically insignificant results. When the meta-analysis was restricted to the prospective group of subjects the difference in the SRR was negligible and the SRR being 1.23 with a difference of 0.04. On the contrary, when the same meta-analysis was done on the retrospective subjects, the SRR was significantly higher at 1.36. It is very important to note from their study that the SRR was non-tangential as far as the menopausal status of the subjects was concerned with an SRR value of 0.86, whereas the SRR in post-menopausal subjects was found to be 1.15 and the difference being statistically significant. To summarize their study, we can say that they found a much firmer association of diabetes with breast cancer [45–47], though, seemingly restricted to the post-menopausal [48] subjects. In some of the most recent studies, some researchers have endeavored toward leveraging ensemble learning methods to find a plausible association between diabetes and breast cancer. In this study, by Rudra *et al.* [49], researchers have used parameters like the gravida count, glucose [50,51] concentration, diastolic blood pressure, and results of an insulin test observed for 2 hours. They claim that their proposed model outperforms adaptive K-nearest neighbor, K-means + decision tree, and ensemble perceptron.

20.5 Performance analysis of existing methods

Early diagnosis of diabetes can be useful to improve the personal satisfaction of patients and upgrade their future. Administered calculations have been utilized to foster various models for diabetes location. Table 20.1 gives a perspective on the diverse profound learning and AI models prepared on Pima Indian Diabetes (PID) dataset and Indian Liver Patient Dataset (ILPD) with upgraded tuning boundaries. All procedures of grouping were tested in the python programming studio. The informational collection has been parceled into two sections (preparing and testing). The existing systems are modeled with 70% preparing information and tried with 30% excess information. For the performance comparison of existing methods, five unique models have been created to figure out how to distinguish whether the patient is diabetic or non-diabetic using the tuning parameters as shown in Table 20.1.

We have compared the performance of five existing methodologies proposed by various researchers such as support vector machine (SVM-straight), outspread premise work [radial basis function (RBF) bit SVM, logistic regression (LR), ANN, and multifactor dimensionality reduction (MDR) calculation are utilized. The tunning boundary and expectation precision is appeared in Table 20.2 and in Figure 20.1.

The datasets of PID and ILPD are assessed for every one of the five distinct models upon boundaries like exactness, review, area under the curve (AUC), and F1 score as shown in Table 20.2. To stay away from the issue of overfitting and underfitting, ten times cross-approval is finished. Exactness shows our classifier is the means by which regularly right in the finding of if the patient is diabetic.

Table 20.1 Deep learning models

S. no.	Model name	Tuning parameters
1	SVM with linear kernel	$C = 1$
2	SVM–RBF kernel	$C = 1$ and Sigma = 0.107
3	LR	$C = 1$
4	ANN	Size = 10
5	MDR	Recode function to convert the value into 0, 1, and 2

Table 20.2 Prediction accuracy

S. no.	Predictive models	Accuracy	Recall	Precision	F1 score	AUC
1	SVM	0.91	0.89	0.88	0.89	0.90
2	RBF	0.89	0.83	0.85	0.83	0.85
3	LR	0.88	0.87	0.88	0.88	0.86
4	ANN	0.92	0.90	0.91	0.92	0.91
5	MDR	0.85	0.89	0.86	0.85	0.84

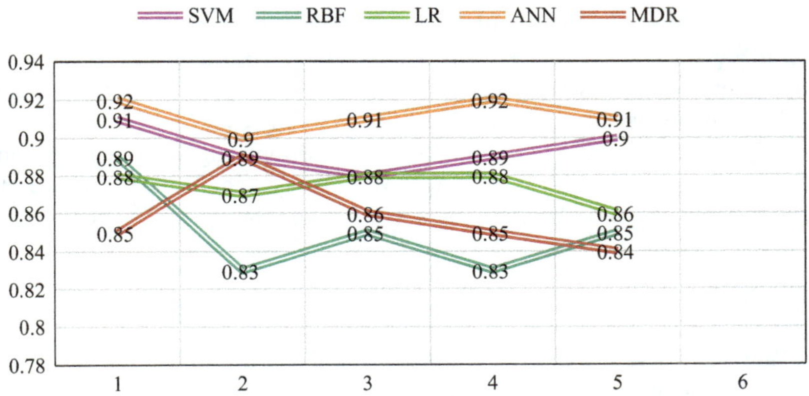

Figure 20.1 Performance comparison of existing methodologies

Accuracy has been utilized to decide the classifier's capacity to give the right of expectations of diabetes and disease. Review or affectability is utilized in our work to discover the extent of real sure instances of diabetes effectively distinguished by the classifier utilized. Particularity is being utilized to decide the classifier's capacity of deciding negative instances of diabetes. As the weighted normal of accuracy and review gives the F1 score, this score considers both. The classifiers of F1 score with almost 1 are named as the best ones. The beneficiary working trademark (receiver operating characteristics) bend is a notable device to imagine a paired classifier calculation's exhibition. It is a plot of genuine positive rate against bogus positive rate as the limit for appointing perceptions is shifted to a specific class.

20.6 Conclusion and future work

From the aforementioned review and analysis, we can say that a lot of commendable effort has been put in by a lot of fellow researchers around the globe. We deeply respect their work, and at the same time, we would like to share that we draw a lot of inspiration from their works. Most of the studies have not been able to substantiate firmly that a pre-existing Type II diabetes condition will lead to any form of cancer. We strongly believe that predicting the occurrence of any form of cancer in subjects with pre-existing Type II diabetes is critical in understanding the factors that lead to cancer, and also if we can predict between malignant and benign tumors, then it will improve the diagnosis and also facilitate a more effective line of treatment for such subjects. In the future, we endeavor toward finding a novel approach to establishing any plausible link between Type II diabetes and any form of cases by employing data-driven knowledge-based algorithms while leveraging deep learning techniques such as ANN which will facilitate us to find patterns even when there might be a non-linear relationship between the data attributes. We are endeavoring toward an approach where we can segregate the subjects based on the intensity of the cancer risk or the intensity of cancer itself. We are contemplating leveraging several specialized ML algorithms for specific tasks so as to improve the overall accuracy of the classification and prediction. We are also contemplating leveraging ensemble learning in our present work, where we will be primarily using SVM for classification and then a combination of AdaBoosting and other ML algorithms for prediction.

References

[1] Zhu B, Wu X, Wu B, Pei D, Zhang L, and Wei L (2017) "The relationship between diabetes and colorectal cancer prognosis: a meta-analysis based on the cohort studies," *PLoS ONE*, 12(4): e0176068. https://doi.org/10.1371/journal.pone.0176068.

[2] Chen L, Magliano DJ, and Zimmet PZ (2012) "The worldwide epidemiology of type 2 diabetes mellitus – present and future perspectives," *Nature Reviews Endocrinology*, 8(4): 228–236. https://doi.org/10.1038/ nrendo.2011.183.

[3] Klil-Drori A, Azoulay L, and Pollak M (2017) "Cancer, obesity, diabetes, and antidiabetic drugs: is the fog clearing?," *Nature Review Clinical Oncology*, 14: 85–99. https://doi.org/10.1038/nrclinonc.2016.120.

[4] Swarna SR, Kumar A, Dixit P, and Sairam TVM, "Parkinson's disease prediction using adaptive quantum computing," *2021 Third International Conference on Intelligent Communication Technologies and Virtual Mobile Networks (ICICV)*, 2021, pp. 1396–1401, doi:10.1109/ICICV50876.2021.9388628.

[5] Sasubilli G and Kumar A, "Machine learning and big data implementation on health care data," *Fourth International Conference on Intelligent Computing and Control Systems (ICICCS)*, 2020, pp. 859–864, doi:10.1109/ICICCS48265.2020.9120906.

[6] Sasubilli SM, Kumar A, and Dutt V, "Machine learning implementation on medical domain to identify disease insights using TMS," *2020 International Conference on Advances in Computing and Communication Engineering (ICACCE)*, 2020, pp. 1–4, doi:10.1109/ICACCE49060.2020.9154960.

[7] Sasubilli SM, Kumar A, and Dutt V, "Improving health care by help of internet of things and bigdata analytics and cloud computing," *2020 International*

Conference on Advances in Computing and Communication Engineering (ICACCE), 2020, pp. 1–4, doi:10.1109/ICACCE49060.2020.9155042.

[8] Tong L, Ahn C, Symanski E, Lai D, and Du XL (2014) "Temporal trends in the leading causes of death among a large national cohort of patients with colorectal cancer from 1975 to 2009 in the United States." *Annals of Epidemiology*, 24(6): 411–417.Epub 2014/02/18. https://doi.org/10.1016/j.annepidem.2014.01.005 PMID: 24529646.

[9] Fransgaard T, Thygesen LC, and Gogenur I (2016) "Increased 30-day mortality in patients with diabetes undergoing surgery for colorectal cancer," *Colorectal Disease: The Official Journal of the Association of Coloproctology of Great Britain and Ireland*, 18(1):O22–29. Epub 2015/10/16.

[10] Huang Y, Cai X, Mai W, Li M, and Hu Y (2016) "Association between prediabetes and risk of cardiovascular disease and all cause mortality: systematic review and meta-analysis," *BMJ (Clinical Research ed)*, 355: i5953. Epub 2016/11/25. PubMed Central PMCID: PMCPMC5121106.

[11] Tierney JF, Stewart LA, Ghersi D, Burdett S, and Sydes MR (2007) "Practical methods for incorporating summary time-to-event data into meta-analysis," *Trials*, 8: 16. Epub 2007/06/09. PubMed Central PMCID: PMCPMC1920534. https://doi.org/10.1186/1745-6215-8-16 PMID: 17555582.

[12] Zhu B, Wu X, Bi Y, and Yang Y (2017) "Effect of bilirubin concentration on the risk of diabetic complications: a meta-analysis of epidemiologic studies," *Scientific Reports*, 7: 41681. Epub 2017/01/31. PubMed Central PMCID: PMCPMC5278382. https://doi.org/10.1038/ srep41681 PMID: 28134328.

[13] Peairs KS, Barone BB, Snyder CF, *et al.* (2011) "Diabetes mellitus and breast cancer outcomes: A systematic review and meta-analysis," *Journal of Clinical Oncology*, 29(1): 40–46.

[14] Yancik R, Wesley MN, Ries LAG, Havlik RJ, Edwards BK, and Yates JW (2001) "Effect of age and comorbidity in postmenopausal breast cancer patients aged 55 years and older," *JAMA*, 285(7): 885–892. doi:10.1001/jama.285.7.885.

[15] Srokowski TP, Fang S, Hortobagyi GN, and Giordano SH (2009) "Impact of diabetes mellitus on complications and outcomes of adjuvant chemotherapy in older patients with breast cancer," *Journal of Clinical Oncology*, 27(13): 2170–2176.

[16] Fleming ST, Pursley HG, Newman B, Pavlov D, and Chen K (2005) "Comorbidity as a predictor of stage of illness for patients with breast cancer," *Medical Care*, 43: 132–140.

[17] Tsilidis KK, Kasimis JC, Lopez DS, Ntzani EE, and Ioannidis JPA (2015) "Type 2 diabetes and cancer: umbrella review of meta-analyses of observational studies," *BMJ*, 350: g7607. doi:10.1136/bmj.g7607.

[18] Riley RD, Higgins JPT, and Deeks JJ (2011) "Interpretation of random effects meta-analyses," *BMJ*, 342: d549. doi:10.1136/bmj. d549.

[19] Giovannucci E (2001) "Insulin, insulin-like growth factors and colon cancer: a review of the evidence," *Journal of Nutrition*, 131: 3109S–3120S.

[20] Monami M, Lamanna C, Balzi D, Marchionni N, and Mannucci E (2009) "Sulphonylureas and cancer: a case-control study," *Acta Diabetology*, 46: 279–284.

[21] Bowker SL, Yasui Y, Veugelers P, *et al.* (2010) "Glucose-lowering agents and cancer mortality rates in type 2 diabetes: assessing effects of time-varying exposure," *Diabetologia*, 53, 1631–1637. https://doi.org/10.1007/s00125-010-1750-8

[22] Végh D, Bányai D, Hermann P, Németh Z, and Ujpál M (2017) "Type-2 diabetes mellitus and oral tumors in Hungary: a long-term comparative epidemiological study," *Anticancer Research*, 37(4): 1853–1857.

[23] Wu CH, Wu TY, Li CC, Lui MT, Chang KW, and Kao SY (2010) "Impact of diabetes mellitus on the prognosis of patients with oral squamous cell carcinoma: a retrospective cohort study," *Annals of Surgical Oncology*, 17: 2175–2183.

[24] Faulds MH and Dahlman-Wright K (2012) "Metabolic diseases and cancer risk," *Current Opinion in Oncology*, 24: 58–61.

[25] Goutzanis L, Vairaktaris E, Yapijakis C, *et al.* (2007) "Diabetes may increase risk for oral cancer through the insulin receptor substrate-1 and focal adhesion kinase pathway," *Oral Oncology*, 43: 165–173.

[26] Werner H and Katz J (2004) "The emerging role of the insulin-like growth factors in oral biology," *Journal of Dental Research*, 83: 832–836.

[27] Carboni JM, Lee AV, Hadsell DL, *et al.* (2005) "Tumor development by transgenic expression of a constitutively active insulin-like growth factor I receptor," *Cancer Research*, 65: 3781–3787.

[28] Suba Z and Ujpal M (2006) "Correlations of insulin resistance and neoplasms," *Magyar Onkologia*, 50: 127–135.

[29] White MF (1997) "The insulin signalling system and the IRS proteins," *Diabetologia*, 40(Suppl 2): S2–S17.

[30] Stocks T, Rapp K, Bjorge T, *et al.* (2009) "Blood glucose and risk of incident and fatal cancer in the metabolic syndrome and cancer project (me-can): analysis of six prospective cohorts," *PLoS Medicine*, 6: e1000201.

[31] Salahudeen AK, Kanji V, Reckelhoff JF, and Schmidt AM (1997) "Pathogenesis of diabetic nephropathy: a radical approach," *Nephrology Dialysis Transplantation* 12: 664–668.

[32] Noto H, Goto A, Tsujimoto T, and Noda M (2012) "Cancer risk in diabetic patients treated with metformin: a systematic review and meta-analysis," *PLoS One*, 7(3): e33411.

[33] Hsieh MH, Sun LM, Lin CL, *et al.* (2018) "Development of a prediction model for colorectal cancer among patients with Type 2 diabetes mellitus using a deep neural network," *Journal of Clinical Medicine*, 7: 277; doi:10.3390/jcm7090277.

[34] Deng L, Gui Z, Zhao L, Wang J, and Shen L (2012) "Diabetes mellitus and the incidence of colorectal cancer: an updated systematic review and meta-analysis," *Digestive Diseases and Sciences*, 57: 1576–1585.

[35] Yuhara H, Steinmaus C, Cohen SE, Corley DA, Tei Y, and Buffler PA (2011) "Is diabetes mellitus an independent risk factor for colon cancer and rectal cancer?," *American Jouranl of Gastroenterology*, 106: 1911–1921.

[36] Sinagra E, Guarnotta V, Raimondo D, *et al.* (2017) "Colorectal cancer risk in patients with type 2 diabetes mellitus: a single-center experience," *Journal of Biological Regulators and Homeostatic Agents*, 31: 1101–1107.

[37] Jiang Y, Ben Q, Shen H, Lu W, Zhang Y, and Zhu J (2011) "Diabetes mellitus and incidence and mortality of colorectal cancer: a systematic review and meta-analysis of cohort studies," *European Journal of Epidemiology*, 26: 863–876.

[38] Sutskever I, Martens J, Dahl G, and Hinton G (2013) "On the importance of initialization and momentum in deep learning," *PMLR*, 28: 1139–1147.

[39] Martens J. "Deep learning via Hessian-free optimization," *Proceedings of the 27th International Conference on Machine Learning (ICML)*, 2010, pp. 735–742.

[40] Martens J and Sutskever I. "Learning recurrent neural networks with Hessian-free optimization," *Proceedings of the 28th International Conference on Machine Learning (ICML)*, 2011, pp. 1033–1040.

[41] Martens J and Sutskever I. "Training deep and recurrent networks with Hessian-free optimization," *Neural Networks: Tricks of the Trade*, 2012, pp. 479–535.

[42] Kingma DP and Ba J. *Adam: A Method for Stochastic Optimization*. Available online: https://arxiv.org/pdf/ 1412.6980.pdf (Accessed on 25 July 2018).

[43] Dozat T. "Incorporating Nesterov momentum into Adam," *Proceedings of the International Conference on Learning Representations, Workshop, San Juan, Puerto Rico*, May 2016, pp. 2–4.

[44] Boyle P, Boniol M, Koechlin A, and Robertson C (2012) "Diabetes and breast cancer risk: a meta-analysis," *British Journal of Cancer*, 107(9): 1608–1617.

[45] Coughlin SS, Calle EE, Teras LR, Petrelli J, and Thun MJ (2004) "Diabetes mellitus as a predictor of cancer mortality in a large cohort of US adults," *American Journal of Epidemiology*, 159: 1160–1167.

[46] de Waard F and Baanders-van Halewijn EA (1974) "A prospective study in general practice on breast-cancer risk in postmenopausal women," *International Journal of Cancer*, 14: 153–160.

[47] Bhargava N, Sharma S, Purohit R, and Rathore PS. "Prediction of recurrence cancer using J48 algorithm," *Second International Conference on Communication and Electrical System*, 2017, pp. 386–390.

[48] Verma D and Mishra N. "Analysis and prediction of breast cancer and diabetes disease datasets using data mining classification techniques," *International Conference on Intelligent Sustainable System*, 2017, pp. 533–538.

[49] Rudra S, Uddin M, and Alam MM (2019) "Forecasting of breast cancer and diabetes using ensemble learning," *International Journal of Computer Communication Informatics*, 1(1):1–5.

[50] Srivastava N, Hinton G, Krizhevsky A, Sutskever I, and Salakhutdinov R (2014) "Dropout: a simple way to prevent neural networks from overfitting," *Journal of Machine Learning Research*, 15: 1929–1958.

[51] He H and Garcia EA (2009) "Learning from imbalanced data," *IEEE Transactions on Knowledge and Data Engineering*, 21: 1263–1284.

Index

www.ingramcontent.com/pod-product-compliance
Lightning Source LLC
LaVergne TN
LVHW011831200926
848484LV00002B/344